Radar Origins Worldwide

History of Its Evolution

in 13 Nations

Through World War II

by
Raymond C. Watson, Jr.; Ph.D., P.E.

Order this book online at www.trafford.com
or email orders@trafford.com

Most Trafford titles are also available at major online book retailers.

© Copyright 2009 Raymond C. Watson, Jr.
All rights reserved. No part of this publication may be reproduced, stored in a retrieval system, or transmitted, in any form or by any means, electronic, mechanical, photocopying, recording, or otherwise, without the written prior permission of the author.

Note for Librarians: A cataloguing record for this book is available from Library and Archives Canada at www.collectionscanada.ca/amicus/index-e.html

Printed in Victoria, BC, Canada.

ISBN: 978-1-4269-2110-0 (Soft)
ISBN: 978-1-4269-2111-7 (Hard)

Library of Congress Control Number: 2009941127

Our mission is to efficiently provide the world's finest, most comprehensive book publishing service, enabling every author to experience success. To find out how to publish your book, your way, and have it available worldwide, visit us online at www.trafford.com

Trafford rev. 11/11/2009

 www.trafford.com

North America & international
toll-free: 1 888 232 4444 (USA & Canada)
phone: 250 383 6864 ♦ fax: 812 355 4082

CONTENTS

Preface	ix
Acknowledgements	xi
Dedication	xiii

Ch. 1 – Introduction and Background   1
- The Mystery Rays   2
- Early Radio Evolution   3
  - Initial Equipment   6
  - Initial Applications   7
- Maturing of Radio   10
  - Entry of Electronics   11
  - High-Power Transmitters   13
  - Radio During World War I   15
  - Radio Between the Wars   16
  - Vacuum-Tube Improvements   18
  - Research Laboratories   20
- Development of Television   23
  - Electromechanical Television   24
  - Electronic Television   26
- Precursors of Radar   29
- Bibliography and References for Chapter 1   34

Ch. 2 – Early RDF in Great Britain   37
- Radio Research Station, Slough   37
- From Death Ray to RDF Origins   41
  - The Initial Proposal   43
  - The Daventry Experiment   46
- History Making at Orfordness   48
- Bawdsey Manor and Initial Systems   52
  - Initial Chain Home RDF System   54
  - Airborne RDF Systems   62
  - The Start of Army RDF   68
  - Academic Connections   71
- The Royal Navy Goes It Alone   73
- The War Through Early 1941   77
- Dundee and the AMRE   81
- St. Athan and Airborne RDF   83
- Christchurch and the ARDD/ADRDE   88
- Swanage and the TRE   94

The Ground-Based Intercept	98
The PPI – A Map on the CRT Display	100
Improved Identification	102
Ground-Based Navigation Systems	102
Countermeasures and Intelligence	104
Bibliography and References for Chapter 2	108

Ch. 3 – Early Radio-Echoing in America — 111

Military Research Laboratories	111
Naval Research Laboratory	112
Signal Corps Laboratories	118
Emergence of Radar	120
Navy Radar	121
Detection by Interference Beat	122
Pulsed Systems	123
First Fielded Naval Systems	127
Army Radar	130
Microwave and Doppler Techniques	130
Pulsed Detection Technique	131
First Fielded Army Systems	132
Evans Laboratory	135
Bibliography and References for Chapter 3	138

Ch. 4 – Microwaves Bridge the Atlantic — 141

The Multi-Cavity Magnetron	143
Continued War and Its Conclusion	147
Early Microwaves in the U.S.	153
The Tizard Mission	155
The High-Power Magnetron	157
Other Radar-Related Secrets	159
The Radiation Laboratory	160
Initial Microwave Radar in Great Britain	164
Microwave Radar for the Royal Navy	165
Microwave Radar for the Army	167
Microwave Radar for the RAF	170
The Start of Microwave Radar Mapping	173
Initial Radiation Laboratory Projects	174
Project One – Airborne Microwave Radar	175
Project Two – Microwave Gun-Laying Radar	177
Project Three – Long-Range Navigation	180

Contents

Continued Developments in Great Britain	182
The Admiralty Signal Establishment	183
The TRE at Great Malvern	188
The ADRDE/RRDE at Great Malvern	195
Continued Developments in America	198
Meter-Wave Systems	199
Combined Research Group – IFF	201
Radio Research Laboratory – Countermeasures	203
VT Proximity Fuse	205
More American Magnetron-Based Radars	207
P-Band Systems	208
L-Band Systems	209
S-Band Systems	210
X-Band Systems	214
Radar Maintenance	217
Closure	222
Bibliography and References for Chapter 4	224
Ch. 5 – Funkmessgerät Development in Germany	**229**
Technologies Development	230
GEMA	233
Seetakt Radar	236
Freya Radar	237
Telefunken	239
Würzburg Radar	240
Lichtenstein Night-Fighter Radar	242
Lorenz	243
Land-Based Radars	244
Airborne Reconnaissance Radar	245
Guidance Systems	246
Siemens & Halske	248
Other Organizations and Contributors	249
Luftwaffe Radar Laboratories	249
Reich Postal Ministry	250
Private Firms	251
Special Development Leaders	252
Other Significant Activities	255
Air Defense System	255
Identification Systems – IFF	257
Microwave Radar	258
Radar Countermeasures	260
Radar Maintenance	268

Closure 269
Bibliography and References for Chapter 5 270

Ch. 6 – Radio-Location Development in the Soviet Union ... 273
Historical Background 273
    The Revolution and the USSR 274
    Higher Education 276
    Emergence of Radio 277
    Research Centers 279
Early Radio-Location Development 281
    Radio-Location Work in Leningrad 282
    Initial Radio-Location Systems 285
    Radio-Location Work in Kharkov 292
Wartime Radar in the USSR 296
    Ground-Based Radar Systems 297
    Airborne Radar Systems 305
    Naval Radar Systems 306
Closure 307
Bibliography and References for Chapter 6 309

Ch. 7 – RRF Development in Japan 311
Historical Background 311
Magnetrons and Other Early Developments 312
The Military and Radar Evolution 316
    Developments by the Imperial Army 319
    Developments by the Imperial Navy 324
Closure 333
Bibliography and References for Chapter 7 335

Ch. 8 – Early Radar Development in Other Countries 337
The Netherlands 337
Italy 341
France 345
Australia 352
New Zealand 355
Canada 362
South Africa 370
Hungary 373
Bibliography and References for Chapter 8 376

Index 379
The Author 403

PREFACE

Who invented radar? What were the evolutionary developments? When and where did these events occur, and why did they come about? This book contains my attempt to answer the *who, what, when, where,* and *why* of radar origins worldwide.

Of necessity, the book also includes a bit of *how*. Technical terms are included (hopefully they are suitably explained), but the extensive mathematics and physics of this technology has been left out – there is but a single equation.

Worldwide? Yes! Contrary to parochial views sometimes expressed, there were actually eight countries with independently developed radar (or what would later be given this name) before World War II. These were the United States, Great Britain, Germany, the USSR, Japan, the Netherlands, France, and Italy.

In addition, the basic development in Great Britain was conveyed to Australia, Canada, New Zealand, and South Africa (all technically advanced Commonwealth countries); here pre-war indigenous systems were also pursued. The rudiments came about independently in Hungary, with equipment emerging early in the war years. In all, 13 nations made this development in utmost secrecy, most believing that they were alone in having this capability.

The name "radar" came from the acronym for Radio Detection and Ranging, derived as a cover for the highly secret technology in America. Radio was used for detection in many earlier devices, but the records show that the first equipment for detection plus ranging (distance and direction measurement) was developed at the U.S. Naval Research Laboratory in Washington, D.C. A short time later, similar developments were made in Great Britain, followed by Germany.

Although it initiated the technology, the United States was slow to recognize its importance and fund the full development. In contrast, the new technology was immediately embraced in Great Britain and quickly matured into ground-, air-, and sea-based systems. Called RDF for cover, this largely allowed the much smaller Royal Air Force to defeat the *Luftwaffe* (German Air Force) during the vital Battle for Britain.

Excellent *Funkmessgerät* (radio measuring device; radar) systems were also developed in pre-war Germany. Hitler, however, considered this to be a defensive technology, and directed that offensive equipment be given production priority; World War II might well have ended differently without this restriction.

Both Japan and the USSR had outstanding scientists and engineers engaged in developments that could have led to good radar systems, but it was only after gaining access to British and American systems that this technology was matured. (The Japanese captured a few systems early in the war, and the USSR was given the equipment for strengthening its capability for holding Germany in check on the Eastern Front.)

This book is an outgrowth of my prior book, *Solving the Naval Radar Crisis* (Trafford Publishing, 2007), concerning the problem of radar implementation in the U.S. Navy at the start of World War II. That book includes a brief historical survey of the development of this technology in several countries. As information was collected for this earlier effort, I realized that, although there were many books and extended papers on the subject, there was no comprehensive treatment of worldwide radar development suitable for the general reader.

A book by Louis Brown, *A Radar History of World War II* (Institute of Physics Publishing, 1999), comes close to meeting this need, but it is lacking in certain time-line aspects, and "military imperatives" often dominate. Nevertheless, his is an excellent treatise, and I have drawn much from it for the present effort.

Before his untimely death, Dr. Brown reviewed my previous writing and encouraged me in its enlargement. He noted that, "It was free of the great radar myths that still fill many accounts: 'Before Rad Lab there was nothing.' 'We invented it in Britain and everyone copied it from us.' 'German radar was second rate and the Japanese did not have any.' "

This present book is a distillation of important information derived from hundreds of sources. The most significant of these sources are cited in the Bibliography and References section at the end of each chapter. Many paragraphs, and even sentences, contain material from multiple sources. Citing these with specific references would have greatly decreased the readability; thus, this usual mark of historical research has been regretfully omitted.

Great care has been given to establishing timelines for the major accomplishments, as well as in crediting the performing organizations. With the large number of sources, however, there were many conflicts in attributions and dates, and much effort was devoted to their resolution.

In essentially every activity, the names of key participants are given; about 450 individuals are cited in the Index. Almost all of these persons are recognized for their technical contributions, not the conduct of war. Many brief biographical sketches are included; these are indicated in the Index by underlining the applicable page number.

Raymond C. Watson, Jr.

ACKNOWLEDGEMENTS

As noted in the Preface, information for this book came from a large number of sources. The most significant of these are cited in the Bibliography and References section at the end of each chapter. These are mainly printed sources, but a few are documents that are also, or only, available on the Internet. Here it is noted that many other Internet sites with applicable information were also found; care, however, was exercised in using such material since its authenticity could seldom be determined.

Information from refereed publications of professional societies was considered to be the most creditable. Books, articles, and similar materials written by the original participants were usually believed to be accurate, but were sometimes parochial and thus incomplete by not broadly acknowledging contributions by others.

Particularly Recognized Sources

Special credit must be given to William H. Penley, a senior participant in the British developments almost from their start. Dr. Penley collected a broad range of papers and photographs concerning these activities and made them available to the public on the Web site: www.penleyradararchives.org.uk . He also provided the author with a partially completed autobiography that contained much valuable information.

A book, *Radar Development to 1945* (Peter Peregrinus Ltd, 1986), edited by Russell Burns, contains 40 brief papers that primarily give first-person accounts of many of the important activities. These papers were very useful, but sometimes reflected memory problems incurred by the four or more decades between the original events and the time of writing.

The Pulse of Radar (The Dial Press, 1959) is the autobiography of Sir Robert Watson-Watt, the initiator of this technology development in Great Britain. This book was a good source of first-person information, and gave an insight into the environment of this wartime endeavor.

Captain Linwood S. Howeth was commissioned to write a detailed book, *History of Communications-Electronics in the United States Navy* (Government Printing Office, 1963). This provided well-referenced information on early developments in this branch of the U.S. Military, and was especially valuable in establishing the time-line for radar. The full book may be found at: http://earlyradiohistory.us/1963hw.htm

Radar Origins Worldwide

A Radar History of World War II (Institute of Physics Publishing, 1999) by Louis Brown was noted in the Foreword. Two other books that were of considerable value were *Technical History of the Beginnings of Radar* (Peter Peregrinus Ltd, 1986), by Seán S. Swords, and Henry E. Guerlac's *Radar in World War II*, almost 1,200 pages in two volumes (American Institute of Physics, 1987). All three of these books have excellent, detailed information, but were not written for the general reader.

Gregory J. Goebel is a prolific writer of Internet documents. His "Wizard War: WW2 and The Origins of Radar" was very helpful in identifying sources of material for this effort. The URL for this book-length work is www.vectorsite.net/ttiz.html .

Over the past years, the IEEE History Center, operated by Rutgers State University, has made an outstanding collection of oral histories. A number of these were accessed in the writing of this book. Their web site includes the following statement:

> The IEEE History Center is determined to preserve as source material for the future historians of technology the personal memories of pioneering electrical and computer engineers, the technologists who transformed the world in the 20th century. . . . With the advent of the Web . . . it is suddenly possible for us to make these wonderful documents available to anyone, any time.

It is noted that these oral histories are not formal interviews. They are essentially unedited, and contain the bias and lapse of memory that would be expected in such materials. The URL for these documents is www.ieee.org/web/aboutus/history_center/oral_history/oral_history.html .

Picture Sources

There are many pictures contained in this book; I consider them very important in communicating the information. It is customary to identify the sources of pictures; however, with the large number of pictures, the size of their captions, and ambiguity in many electronic documents, the source identification for each was not practical.

Many of the pictures came from public-domain Web sites; these are often of low resolution, but suitable for the small sizes contained herein. A number of such pictures, particularly those from the early era of Navy electronics, are from the above-noted book by Howeth. Several of the pictures of specific equipment came from Goebel's document, and essentially all other pictures of hardware came from web sites devoted to particular units.

DEDICATION

To the Pioneers
Who Made Radar a Reality

William R. Blair (1874-1962) United States

Mikhail A. Bonch-Bruyevich (1888-1940) U.S.S.R

Edward G. Bowen (1911-1991) Great Britain

William A. S. Butement (1904-1990) Great Britain

Henri Gutton (1905-1984) France

Hans E. Hollmann (1899-1960) Germany

Yoji Ito (1901-1950) Japan

Rudolph Kühnhold (1903-1992) Germany

Kinjiro Okabe (1896-1984) Japan

Pavel K. Oshchepkov (1908-1992) U.S.S.R.

Robert M. Page (1903-1992) United States

Johannes E. Plendl (1900-1992) Germany

Wilheim T. Runge (1895-1987) Germany

Abram A. Slutskin (1881-1950) U.S.S.R.

Albert H. Taylor (1874-1961) United States

Ugo Tiberio (1904-1980) Italy

Robert A. Watson Watt (1892-1973) Great Britain

J. L. W. C. von Weiler (1902-1988) The Netherlands

Arnold F. Wilkins (1907-1985) Great Britain

Leo C. Young (1891-1981) United States

Radar Origins Worldwide

Antennas on U.S. Aircraft Carrier

British Army Mobile Radar

British Chain Home Antennas

German Antiaircraft Radar

U.S. Shipboard Radar Room

U.S.S.R. Army Radar

Early U.S. Army Radar

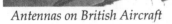
Antennas on British Aircraft

Japanese Naval Radar

Chapter 1

INTRODUCTION and BACKGROUND

World War II began on September 1, 1939, when, without warning, German tanks and planes swept into Poland. It expanded on December 7, 1941, with Japan's surprise attack on the United States at Pearl Harbor. The conflict in Europe, Northern Africa, and the Middle East continued until May 8, 1945 (V-E Day) when Germany surrendered, and finally ended in the Far East with the surrender of Japan on the following August 15 (V-J Day). In these six long and bloody years, 60 million people lost their lives, and hundreds of cities were reduced to ruins.

Two very different technologies, radar and the atomic bomb, were matured in utmost secrecy during this war, and each had a profound effect on the conflict. All of the five major powers – the United States, Great Britain, and the USSR leading the Allies, and Germany and Japan at the head of the Axis forces – had initiated development of both of these technologies, each believing that they were the sole originators.

Research toward an atomic bomb started when the scientific community learned in 1938 that physicists in Germany had "split the atom." Thankfully, God allowed the United States to be first in completing such a bomb. This horrible weapon was then used for a good purpose in bringing the war to a conclusion, saving millions of lives – military and civilian – that would have been lost in the planned invading the Japanese homeland.

The July 11, 1934, issue of *The New York Times* quoted Nikola Tesla as saying that he had invented a "death ray" that could "drop an army in its tracks and bring down squadrons of airplanes 250 miles away." In Great Britain, radar development was a direct result of examining the reality of such a ray. It is very interesting that in Japan this sequence was reversed; during the war, they had a major effort to convert radar to a death-ray weapon, and, except for America's atomic bomb, they might well have succeeded.

These two technological developments followed different paths in each country. In the U.S., they eventually came together in the final military act of the war. The two bombs dropped on Japan detonated at specific altitudes, triggered by small radar sets attached to their sides. Ironically, their Yagi radar antennas were a Japanese invention.

The origins of radar make fascinating history. Many magazine and journal articles, academic papers and presentations, popular and scholarly books, and even movie and television scripts on this subject have been written since this technology became public in 1945. Entering "radar" plus "history" into Google will result in many millions of hits! Nevertheless, little is available that puts the "who, what, when, where, and why" of radar origins together on a world-wide basis.

Add just a bit of "how," and you have this book.

THE MYSTERY RAYS

By the early 1930s, the basic technologies that would lead to radar were known to radio researchers throughout much of the world. Radio for information transfer and entertainment had become ubiquitous, and many people understood that transmitting stations emitted invisible signals that were picked up by receivers and converted to sound, Morse code, or even pictures. In 1935, however, when the news media reported that the military in several countries were developing "mystery rays" capable of "seeing" enemy aircraft and boats through fog, clouds, or dark, few recognized this as simply another application of radio.

Essentially as soon as radio came into being, investigators had found that the emitted signals were reflected from objects and, if detected, could be used to indicate the presence of the object. The first patent for detection by radio was filed in 1904, and many others followed over the next 30 years.

These radio-based detection devices, however, were not really radar. The name radar came from the acronym RADAR, coined by the U.S. Navy in 1940 for the then-secret "radio detection and ranging" technique. Ranging, or measurement of distance and direction to the object, is equally important to detection in a radar device, and this capability was first achieved in the mid-1930s.

Radar was a technology whose time had come. It was developed independently and essentially at the same time by researchers in the United States, Great Britain, Germany, Japan, Russia, Italy, France, and The Netherlands. Shortly, the developments in Great Britain spread in the United Kingdom to Australia, Canada, New Zealand, and South Africa. During the war, there was an independent development in Hungary. The tremendous potential value of these systems to the military was quickly recognized; therefore, essentially all of the early developments were closely held as national secrets

The reader will find that I have shown my own bias in this writing endeavor. Developments in the United States and Great Britain have been emphasized. This is partially due to my personal involvement with radar during World War II as a U.S. Navy instructor, but mainly because

I believe that the combined developments in these two nations – and after late 1940 they were basically combined – far exceeded the efforts in other countries.

The material on early origins in Great Britain is covered in more detail than that in America. The developments in Great Britain were under the urgency of an impending, then actual, war. Range and Direction Finding (RDF) – a cover name for their technology – came into being to "save the nation." It was their "most secret" weapon of the war, and the expenditure for it was more than for any other development. In addition, it was more "colorful" than the comparable activity in America.

Was, indeed, radar that important in World War II? There is little question but that it was the deciding factor in winning the Battle of Britain – stopping the massive attacks by the German *Luftwaffe* on English cities. Correct use of the newly installed radar could have made a great difference in the disaster at Pearl Harbor, but six months later the benefit was clearly shown in America's successful Battle of Midway.

Throughout the war, radar played a central role in all air, land, and sea activities. The Manhattan Project for developing the atomic bomb cost the United States about two-billion then-year dollars, but radar development and its implementation was an even larger endeavor, with a cost of some *three*-billion dollars.

It has often been said that radar won the war and the atomic bomb brought the peace.

EARLY RADIO EVOLUTION

As the name indicates, radar is an application of radio; thus to understand and appreciate radar origins, radio evolution itself should be examined.

The basic foundations of radio originated in Europe. The early contributors included Michael Faraday (1791-1867), James Clerk Maxwell (1831-1879), and Heinrich Hertz (1857-1894). Totally self-educated and with almost no ability in mathematics, Englishman Faraday experimentally discovered electromagnetic induction and postulated that electromagnetic fields extend into empty space around a conductor.

Maxwell, a Scotsman educated at the University of Edinburgh and Cambridge University, presented a mathematical model for electromagnetic waves in 1864, showing that a few equations could fully express electric and magnetic radiations.

German physicist Hertz, who earned a doctorate from the University of Berlin, first demonstrated the existence of electromagnetic waves in 1888. Concerning this, he commented: "I do not think that the wireless waves I have discovered will have any practical application."

Many others throughout the world thought differently; they were searching for means of improving the telegraph by the transmission of Morse signals without the use of wires – a wireless system. A number of inventors in America, Great Britain, and several other countries were successful to a degree in doing this, but it was an Italian who has received the most recognition.

Guglielmo Marconi

Guglielmo Marconi (1874-1937), a native of Italy and privately educated, built a wireless apparatus and performed experiments on his family's estate near Bologna, Italy. He had started his investigations after attending lectures by Augusto Righi, a University of Bologna physicist who was doing research on electromagnetic waves; Marconi's initial apparatus was patterned after Righi's laboratory equipment. In 1895, Marconi transmitted a wireless telegraphy signal a little more than a mile.

After an unsuccessful attempt to gain interest at the Italian Ministry of Posts and Telegraphs, Marconi took his apparatus to England, and in September 1896 demonstrated it to William H. Preece, Engineer-in-Charge of the British Post Office, and other witnesses. A short notice, "Telegraphy Without Wires," in the January 23, 1897 *Scientific American,* stated:

> A young Italian, a Mr. Marconi, has recently demonstrated to the London Post Office the ability to transmit radio signals across three-quarters of a mile, and if the invention was what he believed it to be, our mariners would have been given a new sense and a new friend which would make navigation infinitely easier and safer than it now was.

Preece arranged for Marconi to present a paper, "Signaling through Space without Wires," to the Royal Institute on June 4, 1897. With this, wireless communications, including its many extensions, took hold in Great Britain.

Marconi's apparatus included a spark-gap transmitter and a coherer receiver, neither of which he personally invented, but the combined system was granted British patent #7777 in 1900 for "improvements in transmitting electrical impulses and signals and an apparatus therefor."

This was by no means the only wireless patent in this period. Several methods could be used to "send" information through open space, including magnetic induction, conduction, electrostatic coupling, as well

Introduction and Background

as "Hertzian" waves. U.S. patents for elementary types of wireless communication apparatuses had been granted to a number of researchers, including Mahlon Loomis (1872), Amos E. Dolbear (1882 and 1886), Thomas A. Edison (1891), Isidor Kitsee (1895), and Nathan B. Stubblefield (1898). Lieutenant Bradley A. Fiske had experimented with wireless devices aboard American naval vessels as early as 1888. All of these, however, involved means other than the generation and reception of electromagnetic waves.

Many others had investigated practical applications of Hertzian waves. Dr. Oliver Joseph Lodge (1851-1940) of Great Britain demonstrated short-distance transmission of Morse-code signaling in August 1894, and received a U.S. patent in 1898 for "Electric Telegraphy" using wireless signals.

In a public demonstration at Calcutta in 1895, Jagdish Chandra Bose (1858-1937) ignited gunpowder and rang a bell at a distance of nearly a mile using electromagnetic waves. In India, Bose is called the "father of radio science."

At a meeting of the Russian Physical and Chemical Society in May 1895, Alexander Stepanovich Popov (1859-1906) demonstrated communications using radio waves. Since then, this day has since been celebrated in Russia as "Radio Day."

After being the only foreigner at Marconi's initial demonstration in England, Adolf K. H. Slaby (1849-1913), Professor at the Technical University of Berlin, developed the first experimental wireless link in Germany in August 1897.

Nikola Tesla (1856-1943), was born in Croatia, the son of priest in the Serbian Orthodox Churh. He attended lectures at several universities but never completed a degree. After employment as an engineer in Hungary and France, he immigrated to the United States in 1884 to work at Thomas Edison's power station, then started his own company, and still later joined Westinghouse. He became a citizen in 1891 and again formed a company where he did scientific research for the rest of his career. Here, his far-out ideas ultimately led to his being ostracized as a scientist

Tesla gave many papers on wireless technologies during the 1890s. He claimed to have transmitted electrical power over a 30-mile wireless link in about 1896, and he applied for a U.S. patent in May 1897 that included circuits tuned to the same frequency

Nikola Tesla

in both the transmitter and receiver, a technique predating that of Marconi's patent three years later. In 1943, the U.S. Supreme Court overturned Marconi's American patent and credited Tesla with the invention of radio.

Marconi's talents as an inventor were exceeded by his capability as a promoter and businessman. Notwithstanding all of the accomplishments of others, he did bring the world's attention to the wireless and is generally recognized as the originator of wireless communications. In 1897, he started what soon became the Marconi's Wireless Telegraph Company, with the stated purpose of attaining a worldwide monopoly in that field. In 1899, Marconi opened the first radio factory in England at Chelmsford in Essex.

Initial Equipment

All early wireless transmitters were of the spark-gap type. In this, a high voltage was injected into a circuit that contained an electrical gap (an open space). The circuit also contained an inductor (a coil) and a capacitor (then called a condenser) that would oscillate at a radio frequency (RF) as the spark jumped the gap. Since this was a direct-current (d-c) circuit, the RF oscillations would quickly damp out – trail out to zero amplitude. This type of circuit for generating electromagnetic impulses was first disclosed by Hertz in describing his experiment.

To generate a coded signal, such as in Marconi's apparatus, the voltage was pulsed on and off (normally at a few hundred times a second) by a vibrating or rotating breaker. The circuit was connected directly to an antenna, allowing the damped RF energy to be radiated. While the radiated signal was centered at the resonant frequency, its damped characteristic caused it to spread over a broad band, resulting in a "dirty" signal. Such a signal could not be modulated; thus, it was useful only for Morse-code communication.

Early Spark-Gap Transmitter

In 1898, Karl F. Braun (1850-1918), a physics professor at Strasbourg University in Germany, improved Marconi's apparatus by coupling the transmitter to the antenna through a transformer instead of having the antenna directly attached to the spark circuit. This allowed electrical matching of the transmitter and antenna circuits, significantly increasing the radiated power and hence the range. Marconi and Braun shared the 1909 Nobel Prize in Physics for their contributions to wireless development.

The first receivers used a coherer detector. This was a small tube filled with metal filings that rearranged orientation and decreased resistance upon being impressed with an electric signal. This, in turn, was coupled to a circuit containg a lamp, bell, headphone, or telegraph "inker" that displayed the signal on paper tape. To recover to the higher resistance, a vibrating taper was arranged to strike the coherer tube. The coherer, together with a resonating (tuning) circuit, was placed between the antenna and the ground (earth).

Early Coherer Receiver

The original detection device is attributed to Dr. Edouard Branly, a professor of physics and medicine at the Catholic University in Paris, who in 1880 published a paper on the conducting phenomena of iron filings; the basic principle, however, can be traced to publications as early as the 1830s. As wireless applications began, a number of researchers developed different types of this device. These included Marconi, who found mixtures of metals that were more sensitive to RF signals; Lodge, who also called the device a "coherer;" and Popov, who made the device self-recovering.

In his original experiment, Hertz used a very elementary spark-gap emitter and detected the radiation only a short distance away using a wire loop that contained a small gap. The transmitted electromagnetic wave generated a voltage in the loop, and this was indicated by a visible spark jumping the gap.

Initial Applications

After Marconi demonstrated his apparatus to the British Post Office, he briefly returned to Italy. There he showed his equipment to the Italian Royal Navy, successfully communicating from the naval dockyard in La Spezia to a cruiser that was 12 miles away at sea. Returning to England, he arranged to show it to the British Royal Navy. In mid-1899, he placed his equipment aboard two ships, and, in tests at sea, exchanged telegraph messages at a distance of 74 nautical miles.

Following Marconi's demonstration to the British Navy, he traveled to America for a similar demonstration in late 1899 to the U.S. Navy. In this, wireless telegraph tests were conducted between two warships some 35 miles apart.

The U.S. Navy then requested a quote from Marconi for supplying 20 wireless sets. He responded by establishing the Marconi Wireless Telegraph Company of America and offering to lease the equipment to the Navy for an unreasonably high price. This led to an enmity that remained until Marconi America was forced to close 20 years later.

A number of entrepreneurial inventors quickly set up shop in America, not only to build and sell wireless sets but also to offer communication services. The first was Harry E. Shoemaker, who in September 1899 established the first radio firm in the United States. Other notable pioneers included Dr. John S. Stone, William J. Clarke, Dr. Lee de Forest, Reginald A. Fessenden, and Walter W. Massie.

The first significant improvement in transmitters came with the rotary spark; in this, the spark was broken at a high rate, greatly decreasing the original "dirty" emission. Although still damped, it could be modulated (controlled in amplitude) to some extent.

On December 23, 1900, Reginald A. Fessenden at a U.S. Weather Department station in Washington, D.C., inserted a carbon microphone between a rotary-spark transmitter and the antenna, and spoke a message to his assistant a mile away. He made the following entry in his notebook: "This afternoon . . . intelligible speech by electromagnetic waves has for the first time in World's History been transmitted."

Reginald Fessenden

Reginald Aubrey Fessenden (1866-1932), a native of Canada, had only the equivalent of about two years of college when he came to the United States. However, after many inventions and publications while at several industries, he was appointed a professor at Purdue University and followed this as by instigating electrical engineering at what would later be Pittsburgh University. His elementary demonstration in 1900, followed by many successes in the next decade, certainly earned Fessenden credit for bringing amplitude-modulated (AM) radio into being.

His later accomplishments included the discovery in 1902 of the heterodyne principle – mixing of two signals of slightly different frequencies to produce a "beat" signal that was more easily processed. He was also the first to send a telephonic message across the Atlantic, and to broadcast the first radio "program" (both in 1906).

The first use of the wireless in warfare was during the Boar War in South Africa, just after the turn of the century. The British Royal Navy had sets from Marconi, and the local Boar Forces ordered sets from the German firms, Siemens & Halske. Communication distances were only a few miles, and neither side was very successful in their use. Since eavesdropping was easy, the British experimented with sending their Morse-code messages in arcane languages such as Hindustani, a ploy repeated by the American Navajo "wind-talkers" during World War II.

The best-known early demonstration of long-distance wireless occurred on December 12, 1901. At Poldhu, Cornwall, Marconi used a transmitter 100 times more powerful than his earlier unit. He set up his receiver in the trans-Atlantic cable facility at St. John's, Newfoundland. With the antenna held aloft by a kite, Marconi repeatedly received the letter "S" (three dots in Morse code) sent across the Atlantic, a distance of about 2,100 miles.

With his trans-Atlantic demonstration, Marconi showed that electromagnetic waves could "bend" around the Earth, thus making this technology useful for long-distance communications. Prior to this, it was believed that these waves, like light rays, always traveled in straight lines and could not be detected beyond the horizon.

To explain this wave bending, in 1902 American Arthur E. Kennelly and Englishman Oliver W. Heaviside independently proposed that a layer of ions (charged atoms and molecules) was above the atmosphere and reflected back to earth certain radio waves below some critical frequency. Successive reflections between this layer and the Earth's surface could bounce a signal across the Atlantic. Initially called the Kennelly-Heaviside layer, this is now known as the ionosphere and is composed of several layers with heights depending on a variety of conditions.

The term "frequency" was not commonly used at that time – the radiated signal was usually designated by "wavelength." Wavelength (λ) and frequency (f) are related in that their product equals the speed of wave propagation. For electromagnetic waves in the atmosphere, the useful equation is then:

$$\lambda \text{ (meters)} = 300{,}000 / f \text{ (kHz)} \text{ or } 300 / f \text{ (MHz)}$$

It should be noted that the earlier unit of measure for frequency was cycles per second (cps), leading to the terms kilocycles and megacycles, commonly abbreviated as kc and Mc, respectively. Hereafter, for consistency, frequency units in hertz (Hz) will be used (1 Hz ≡1 cps).

The RF spectrum is considered to be primarily between 3 kHz and 3 GHz, normally designated in decade bands as follows:

3 kHz – 30 kHz:	Very Low Frequency (VLF)
30 kHz – 300 kHz:	Low Frequency (LF)
300 kHz – 3 MHz:	Medium Frequency (MF)
3 MHz – 30 MHz:	High Frequency (HF)
30 MHz – 300 MHz:	Very High Frequency (VHF)
300 MHz – 3 GHz:	Ultra High Frequency (UHF)

These ranges and designations have varied through the years and in different countries.

Starting in 1902, the U.S. Navy bought equipment from the American start-up firms, as well as from Slaby-Arco, Braun-Siemens-Halske, DuCretet-Roger, and Rochefort; the first two being German companies and the last two French. The German companies merged in 1903 to form Telefunken, a firm that contributed much to the technology. A Radio Test Shop was set up at the Washington Navy Yard for comparing equipment performance; this shop also started in-house developments of new equipment.

In Great Britain and America, Marconi's companies rapidly set up shore stations and supplied equipment for ships. With backing from Lloyd's of London, they dominated this market. Initially, they only leased the equipment and supplied their own trained operators. As in the U.S., the British Royal Navy would not lease the equipment, but nevertheless obtained essentially all of their sets from Marconi.

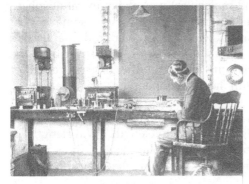

Early Wireless Station

Dating from early days of the telegraph, all non-military wireless operations in Great Britain were totally controlled by the Post Office, and, with their close ties with Marconi, there were few other suppliers.

MATURING OF RADIO

Two new technologies for generating continuous (undamped) waves began to emerge in 1902. Valdemar Poulsen at the Copenhagen Telephone Company in Denmark developed the arc transmitter, the first high-frequency generator without moving parts. At about the same time, Ernst F. W. Alexanderson at General Electric in America developed a high-speed alternator that was could generate continuous signals at low radio frequencies. It would be several years, however, before either of these technologies gained a prominent position in communications.

By the end of 1903, in the U.S. 75 commercial stations were constructed or in planning by Marconi America and a number of domestic firms. Many shipping companies had installed sets for communicating with their vessels. Following the 1900 hurricane in Galveston – the greatest natural disaster in American history with over 8,000 persons killed – the U.S. Weather Bureau started an extensive wireless network for reporting. The U.S. Navy Department had 20 shore stations in operation and planned more than 5 times this number for the near future. The Army Signal Corps was establishing a network connecting the major posts throughout the country.

Entry of Electronics

In 1904, Dr. John A. Fleming (1849-1945) in Great Britain invented the two-element vacuum tube that he called a thermonic valve because it used heat to control the flow of electrical current. Somewhat later, Dr. Lee de Forest in the U.S. filed a patent application for a wireless detector that was also a two-electrode vacuum tube. This led to a controversy concerning the invention of the diode. The courts eventually ruled that both devices were natural extensions of the 1883 discovery by Thomas Edison (for whom Fleming once worked) in which he observed that an electrical charge would collect on a metal plate adjacent to a heated filament in a vacuum (the "Edison effect").

John Fleming

The diode could rectify the radio signal coming from the antenna, thus allowing the output to be heard on a headphone. While this was sometimes used to replace the coherer, it would only serve where the transmitter signal was very strong and thus did not gain much early acceptance.

In Germany, Dr. Arthur R. B. Wehnelt invented the oxide-coated cathode and used this in a thermonic device for rectifying alternating currents. His 1904 patent application, however, did not mention its use for radio-wave detection, and he was unable to apply his device in that application.

Another device for signal rectification was the silicon crystal, patented in 1906 by Dr. Greenleaf W. Pickard in America. When contacted with a sharp point (called a "cat's whisker") at a critical region, it performed as a detector. Galena, the natural form of lead sulfide, worked almost as well and became very popular with experimenters. In Germany, Professor Karl F. Braun patented the crystal receiver in 1906.

De Forest added a control grid inside his diode, and in January 1907 filed a patent application on a three-electrode Audion or triode vacuum tube. This device, without a doubt the most far-reaching invention in radio history, found limited application when initially developed, apparently because de Forest did not fully understand its principle.

Lee de Forest (1873-1961) was raised in Alabama and earned his doctorate from Yale in 1999, writing a dissertation concerning Hertzian waves on open-ended transmission lines. During his career, de Forest

Lee de Forest

worked at many locations, received over 300 patents, was almost continuously involved in lawsuits, and was defrauded by a number of business partners. His invention of the triode vacuum tube certainly justifies designating him as the "father of the electronic age."

The first application of triodes was in 1910 when de Forest used a telephone transformer to couple two of them in series to produce the first amplifier. This was of great interest to American Telephone & Telegraph (AT&T) for its long-distance telephone networks, and triode patent rights for applications other than wireless were bought by AT&T's manufacturing firm, Western Electric.

In experimenting with an amplifier, de Forest accidentally fed the output back into the input producing a "squeal," and the vacuum-tube audio oscillator was born. Although he did not immediately apply for a patent on what became known as the regenerative oscillator, by October 1912 de Forest had increased the frequency to bring the first triode radio-frequency oscillator into being.

The presence of residual gases – thought by de Forest to be necessary – made early triodes unreliable and short-lived, but in the period 1912-1914, Dr. Harold D. Arnold at Western Electric and Dr. Irving Langmuir at General Electric greatly improved the triode and found how to achieve a near-perfect vacuum. Manufacturing of "hard" vacuum triode tubes, and their incorporation into circuits, soon spread into many countries.

The initial commercial triodes were poor amplifiers of RF signals; the transit time of electrons between the elements was too large. By making the elements closer together, this limitation was partially overcome, and their first applications were in the front end of receivers.

The regenerative receiver was invented by Edwin H. Armstrong in 1914 when he was in college; in this, the output of the detector was back into the incoming circuit, allowing the signal to be amplified many times by the same tube. In 1916, Ernst Alexanderson at General Electric patented the tuned radio-frequency (TRF) receiver; this had one or more RF stages before the detector that could be individually tuned to the desired station.

1916 Triode

About the same time, Fessenden's heterodyne principle was applied by Frenchman Lucien Levi to reduce the incoming signal to a lower frequency where it could be better amplified. In England, Henry J. Round, Marconi's leading assistant, developed a receiver in which a single tube served as both an oscillator and heterodyne mixer. Although relatively unrecognized, Round was responsible for many of Marconi's developments and was awarded 117 patents.

In Germany, a number of small companies, as well as units of old-line firms such as Siemens & Haiske, were involved in wireless hardware by the turn of the century. In 1903, Emperor Wilhelm II, infuriated that a wireless message from him had been refused by a Marconi station, encouraged the merging of several of the German firms to form Gesellschaft für drahtlose Telegraphie mbH, better known as Telefunken. From this time on, this company was the leader in wireless equipment in Germany.

Dr. Georg Graf von Arco was the first technical director and managing director of Telefunken; he eventually was granted more than 100 patents in wireless technology. The German Army and the Imperial Navy were the first customers of the new firm. In 1905, Telefunken delivered rotary-spark transmitters for coastal stations in Norddeich and Nauen. Continued development eventually led to the breaking of Marconi's monopoly in maritime communications.

High-Power Transmitters

During the early 1910s, Poulsen arc transmitters gained acceptance in America. Although typically with only about 30 kW power, they nevertheless exceeded in communication efficiency rotary-spark devices having much higher power. The output frequency of the Alexanderson alternator was gradually increased, allowing the generation of usable RF signals with hundreds of kilowatts in power. This unit, however, was massive and was restricted to large communication stations.

Poulsen 30-kW Arc Transmitter

Alexanderson 100-kW Alternator

Radar Origins Worldwide

The adjective "radio" apparently originated in France to describe something that emits (e.g., radioactive materials). It was picked up and used interchangeably with "wireless," especially in America. Following the 1906 International Wireless Telegraph Convention, it was intended that radio be solely adopted. In 1911, it was used in naming the Institute of Radio Engineers, and shortly thereafter standardized by the U.S. Government. At about this same time, it began to also be used as a noun. The name wireless, however, continued to be used, particularly in Great Britain.

Following the sinking of the RMS *Titanic* on April 14-15, 1912, with the loss of over 1,500 lives, the importance of reliable radio communications was widely recognized. By the middle of the decade, maritime systems were required by all major countries, and internal radio communications were commonplace. The September 1912 issue of *Popular Mechanics* had an article describing Marconi's "round-the-world" wireless network then under construction.

Spark transmitters were still used exclusively in all Marconi stations, but elsewhere they were rapidly being replaced by arcs or alternators, some with hundreds of kW power. High-power vacuum tubes were not yet available, so electronic transmitters with voice modulation were a rarity. Vacuum tubes for receivers and amplifiers, however, had become standard.

As a transmitter, the arc was qualitatively inferior to the alternator. The arc emitted many harmonics and had a slight frequency variation. In contrst, the alternator had no harmonics and transmitted but one sharply defined, stable frequency. The arc, however, was much simpler in construction, could be readily enlarged to almost any power, and was lower in cost. These two types of transmitters, each used for the purposes best adapted, would have remained supreme for years except for the eventual introduction of the high-power triode vacuum tube.

Other alternators were dominant in continental Europe. In Germany, Dr. von Arco at Telefunken developed the "mechanical transmitter," equivalent to the alternator. Dr. Rudolph Goldschmidt made an improvement by using multiple windings tuned to successively higher frequencies, allowing an increase in frequency without increasing the rotational speed of the machine.

Another approach was to pass an alternator's output through one or more stationary frequency converters, producing radio waves at double or quadruple the frequency from the alternator. A Telefunken alternator generated near 10 kHz; then two cascading frequency converters successively doubled this to 40 kHz. Alternators designed by Maurice Latour in France had two or three sections ganged in a row, allowing higher power levels with moving parts of practical size.

Introduction and Background

The possibility of war drove communication progress in Europe and America. Using the best available technologies, the U.S., British, and German navies had good radio communication systems; those of France and Russia were satisfactory. Except for some short-range land sets, communication was still by Morse code. A few aircraft had radios, but these were almost exclusively used for observational purposes.

Radio During World War I

War began in Europe in August 1914, with Great Britain, France, and Russia allied against Germany and Austria-Hungary. The "Great War" was one of the defining events in modern Great Britain. From a technological standpoint, it marked the debut of many new weapons such as tanks, airplanes, and the zeppelin, as well as the maturing of submarines, land and sea mines, poison gas, machine guns, and long-range canons. In 1915, British Admiral John "Jacky" Fisher stated, "The war is going to be won by inventions."

It was the first war to involve new electrical technologies, such as field telephones, searchlights for aircraft detection, and shipboard gun-directing motors. It was also the first to use radio communications. Instant messaging was possible between the headquarters and fighting forces on land or sea.

British 1918 Field Wireless Set

The most important advancement, however, was the transmission of voice rather than code, made possible by electron tubes in oscillators and amplifiers; this received limited use near the end of the war.

In Great Britain, Marconi remained the primary supplier to the Royal Navy for shipboard and shore-station wireless equipment, but there was also research and development at the Experimental Department of the Signal School in Portsmouth. The War Department had established the Signals Experimental Establishment (SEE) at Woolwich Common, London, and the Royal Aircraft Establishment (RAE) at Farnborough was responsible for developing sets for the Royal Flying Corps. In addition to Marconi, a number of other large commercial firms now existed.

In the U.S., the Navy Radio Laboratory, an adjunct to the National Bureau of Standards, was responsible for both adapting commercial radio equipment for Government use and developing new items. As the war loomed closer, this laboratory and the Radio Test Shop in Washington were heavily engaged in preparatory projects. The Navy Aircraft Radio Laboratory was set up shortly before the war, then moved to the Anacostia Naval Air Station in Washington, D.C. The Army established the Signal Corps Radio Laboratories at Camp Vail, New Jersey. A broad range of industrial firms provided manufacturing in Navy and Army projects.

After German U-boats sank the British liner RMS *Lusitania* with a loss of 128 Americans and then later torpedoed three American merchant vessels, the U.S. entered the war on April 6, 1917. Upon declaration of war, all communication stations within American jurisdiction were placed under the control of the U.S. Navy. During the war, the Navy's well-established network of stations was augmented by a 500-kW arc transmitter in Annapolis, Maryland, and, later, a 1,000-kW arc transmitter in Bordeaux, France.

The U.S. Naval Communication Reserve was formed in early 1917, and its members were then called to active duty at the start of the war. Many of these reservists, all experienced commercial radio operators or amateurs (Hams), were assigned to oversee the commercial stations taken over by the Navy. Lieutenant Commander (Dr.) H. Hoyt Taylor was in charge of the Marconi America trans-Atlantic station at Belmar, New Jersey. In future years, Taylor and two others at the station, Leo C. Young and Louis A. Gebhard, would make outstanding contributions. The facility at Belmar would later become a research center for the Army.

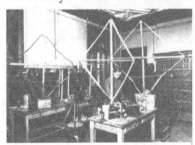

Signal Corps Laboratory, Paris

During the war, radio was used by all of the combatants for communicating with their warships, but on the ground, telephone and telegraph lines still dominated. Both Great Britain and the U.S., however, had considerable work on equipment for eaves-dropping and direction-finding. Henry Round of Marconi developed a highly sensitive receiver, and set up a network of tracking stations around Britain. The U.S. Signal Corps had a field laboratory near Paris for such work, and it was there that then-Captain Edwin Armstrong invented the super-heterodyne circuit, the fundamental circuit of receivers still used today.

Radio Between the Wars

The war was officially over with the signing of an Armistice on November 11, 1918. During the war, it had been decreed that no foreign company would be allowed to hold more than 20 percent of an American firm; this sealed the fate of Marconi America. Also, all German radio patents issued in the U.S. had been seized. As a result, in October 1919, General Electric, with guidance and support from the government, led in forming Radio Corporation of America (RCA) to handle these holdings.

This new entity – jointly owned by General Electric, AT&T/Western Electric, Westinghouse, and several smaller corporations – acquired Marconi America. This brought together all of these stockholders' radio patents, as well as those seized from Germany; RCA then held more than 2,000 patents. Although it was formed to be a marketing outlet for the stockholding firms, almost immediately RCA initiated its own research and manufacturing operations, soon becoming America's largest source of consumer, commercial, and governmental radio products.

The National Bureau of Standards was assigned the call letters of WWV for testing of broadcasting equipment. In May 1920, WWV began broadcasts of Friday evening music concerts. The station's 50-W transmitter operated at 600 kHz and could be heard about 25 miles. An article in the August 1920 *Radio News* reviewed the Bureau's "Portaphone," a portable radio receiver designed to allow people to "keep in touch with the news, weather reports, radiophone conversations, radiophone music, and any other information transmitted by radio."

The broadcasts of WWV were initially intended to provide a reference signal at a precision frequency, stated as having an accuracy of "better than three-tenths of one percent." (Today, the station's many frequencies are controlled to within 1 part in 10^{13}. The well-known time announcement in telegraphic code was added in October 1945, and voice announcements began on January 1, 1950.)

During the mid- and late-1910s, a number of experimental AM stations in the U.S. operated by amateurs, colleges, government units, and commercial firms occasionally sent out "entertainment." One of these, 8ZZ owned by Frank Conrad, was taken up by Westinghouse and on October 27, 1920, became KDKA Pittsburgh, the Nation's first commercial station with regularly scheduled broadcasts. (It should be noted that KDKA was preceded in the Americas by similar stations in Montreal, Canada, and in Buenos Aires, Argentina.)

KDKA initiated an avalanche in commercial broadcasting and subsequent demand for receivers. By the end of 1922, there were 569 broadcasting stations and an estimated 2.5 million receivers. The U.S. Department of Commerce initially allocated 833 kHz (360 m) as the single frequency for commercial stations, resulting in tremendous interference. By 1924, this allocation had expanded to become the band from 550 to 1500 kHz in 10-kHz steps.

In addition to the broadcast market, there was a similar demand for parts and equipment from the burgeoning ranks of Ham radio amateurs. With thriving commercial sales saturating industry and the loss of wartime funding in the military, the needed upgrading of now-obsolete radio equipment had to be done throughout the 1920s by the struggling Army and Navy development centers. Even under these limitations, however, by the end of the 1920s the military communication systems had been fully brought up to date.

There was a similar situation in Great Britain. During the decade following the war, developments in AM radio as an entertainment medium dominated the commercial field. In 1920, the Marconi Company began daily half-hour broadcasts of music and news on experimental station 2MT at Writtle, Essex. Great pressure from the public for "entertainment" broadcasting followed. In May 1922, the Post Office (with total control of radio in Great Britain) allowed the establishment of the British Broadcasting Company (BBC), an umbrella organization of receiver manufacturers and others interested in setting up broadcasting stations.

Marconi 2MT Transmitter

In November 1922, the first BBC station, 2LO operating at 815 kHz (375 m), opened in London. Seven other stations followed, including a centrally located transmitter at Daventry, Northamptonshire. By the following May, there were 564 ownership firms. All of these operated on the same wavelength and with a power of 1.5 kW. The early broadcasting stations were funded from fees for "receiving licenses"; by the end of 1923, about 500,000 such licenses had been bought, mainly for the highly popular crystal sets. The Greenwich Time Signal ("The Pips") began in February 1924. The BBC was nationalized in 1927, becoming the British Broadcasting Corporation.

By the mid-1920s, essentially every developed nation in the world had stations providing entertainment broadcasting. This was so dominant that the period has often been called "the decade of the radio."

Vacuum-Tube Improvements

With the advent of radio broadcasting and subsequent increased commercial value in transmitters, both General Electric and Western Electric started intensive research in higher-power vacuum tubes. Rapid improvements quickly followed. By early 1921, General Electric had developed a 1-kW tube with a heat-conducting metal base, shortly followed by a 3.5-kW water-cooled tube from Western Electric. In 1922, General Electric released a 20-kW water-cooled tube, with its first use in a new communication station on Long Island.

A major improvement in receiver tubes came in 1929 with the introduction of the tetrode – a four-element device. The extra element, a screen grid, isolated the input grid circuit from the output plate, resulting in much more stable operation and delivering considerably greater amplification for each stage. Although invented by Dr. Walter Schottky at Siemens in Germany, the tetrode reached production status through development at General Electric. This was the first real advancement in vacuum tubes since the triode and quickly became the standard in both TRF and super-heterodyne receivers.

In standard vacuum tubes, the electron transit time between electrodes limits the frequency that they can handle. Primarily for use in television, special tubes with very small inter-electrode spacing were developed to operate at higher frequencies. For receivers, RCA supplied Acorn tubes starting in 1934. Dr. Arthur L. Samuel at Bell Telephone Labs (BTL) developed similar receiver tubes, as well as the Doorknob transmitter tube. General Electric developed the Lighthouse

Lighthouse Tube

Acorn Tube

tube, used in both receivers and transmitters. These and similar tubes could function up to about 1000 MHz (1 GHz), but the output power of transmitter tubes decreased significantly beyond 500 MHz.

Although at this time there were no immediate applications, there was a drive to develop tubes capable of operating in the UHF portion of the spectrum – in the centimeter-wavelength region. To overcome the inter-electrode limitations, special tubes were developed to handle the electrons, using a technology called velocity modulation.

In 1916 at General Electric, Dr. Albert W. Hull invented what he called the magnetron, a velocity-modulation device using a resonant cavity in a magnetic field. In theory, it could produce oscillations in the UHF region (determined by the size of its cavity), but it delivered low power at these frequencies. For the next quarter century, a high-power magnetron was the "holy grail" of UHF tube researchers

A different type of velocity-modulation tube was developed in 1920 by Drs. Heinrich G. Barkhausen and Karl Kurz in Germany. Although the Barkhausen-Kurz tube was also low-power, it operated well at near the UHF region and was widely used in both experimental transmitters and receivers.

Research Laboratories

Research laboratories played a central role in evolving radio to radar. This section briefly reviews those in the United States and Great Britain involved in this evolution. Research laboratories in other countries are described in the related chapters.

In Great Britain, governmental laboratories in that period with involvement in radio were those operated by the Department of Scientific and Industrial Research (DSIR), and those under the United Kingdom War Office.

Relevant research laboratories under the DSIR were the National Physics Laboratory (NPL) and the Radio Research Station (RRS). Under the War Office were the British Army, Royal Navy, and Royal Air Force. Central research laboratories for the Royal Army were at the Signals Experimental Establishment (SEE), for the Royal Navy at the Experimental Department of His Majesty's Signal School (HMSS), and for the Royal Air Force at the Royal Aircraft Establishment (RAE).

As the war approached, three highly specialized research activities devoted to radar (then called range and direction finding) were established in Great Britain, one each for the Army, Navy, and Air Force. These went through a variety of names.

In the United States, government research laboratories involved with radio in the 1930s were the National Bureau of Standards (NBS), under the Department of Commerce, and two military laboratories under the Department of War: the Naval Research Laboratory (NRL) operated by the Navy and the Signal Corps Laboratories (SCL) of the Army. While the two military laboratories functioned independently of each other, both had a close relationship with the NBS, having, in fact, functional laboratories located within the NBS facilities.

Just before the start of WWII, the National Defense Research Committee (sometimes called Council), reporting directly to the U.S. President, formed the Radiation Laboratory at MIT, serving both the Army and Navy with advanced research and development in radar.

All of the major manufacturing firms sponsored in-house research in electronics. The largest suppliers of electrical and electronic products in Great Britain were Metropolitan-Vickers, British Thompson-Houston

Introduction and Background

(BTH), General Electric Company (GEC), and Western Electric. Others included Marconi, MEI, Ferranti, A. C. Cossor, and Pye Radio. All of these had significant research and development operations, often in close alliance with university laboratories.

In America, the electrical giants during the 1930s were RCA, General Electric, Westinghouse, and Western Electric. Some of the medium and smaller firms that specialized in electronics included Raytheon, Philco, General Radio, National Radio, Collins Radio, Hammarlund, Hallicrafter, and Aircraft Radio. While there were connections with universities for research, the commercial laboratories tended to operate in isolation.

American Telephone and Telegraph (AT&T) was another American giant that was in a class by itself. It was established in 1885 to run the first long-distance telephone network in the Unites States. By 1891 it had grown to such a size that it was able to purchase a controlling interests in Western Electric – one of the largest hardware firms in the world – and made it the exclusive developer and manufacturer of AT&T equipment.

In 1925, AT&T established a research organization that would profoundly affect electronics and communications worldwide – the Bell Telephone Laboratories, Inc. (BTL). The BTL took over the work previously performed by the research division of Western Electric and research activities of the AT&T engineering department. It wasowned equally by AT&T and Western Electric.

An initial major mission of the BTL was development of technologies for long-distance cable and radio networks that would be built by Western Electric, but, to do this, it also conducted basic research on a par with the best university laboratories. The BTL soon became internationally recognized as the world's preeminent industrial research facility and as a center of scientific and engineering excellence, with seminal papers published in *Bell System Technical Journal*. A number of BTL scientists were Nobel Laureates.

In the early 1930s, research laboratories at universities were far less extensive than during the war and the following years. Funding was scarce, and relatively few students were in university graduate programs. Physics departments and their associated laboratories were considerably more prevalent than in electrical engineering – the two disciplines where advancements in electronics would be expected. Although the university laboratories in America and Great Britain did not, in general, play a significant role in advancing the applications of radio, they were the primary source for senior personnel in the industrial and governmental activities, and quickly turned to military work as the war approached.

Research at the MIT must be noted. As early as 1926, Dr. Edward L. Bowles of the Electrical Engineering Department started experimental studies of radio antenna patterns and infrared radiation with specific application to the military. In 1932, Dr. Wilmer L. Barrow initiated major research in millimeter and microwave radio transmission and hardware, providing the foundation of what would eventually become MIT's Radiation Laboratory. There was also outstanding work in MIT's Instrumentation Laboratory, started by Dr. Charles S. Draper in 1932. This laboratory pioneered the development of gyroscopically based gun-pointing and firing systems for shipboard and fighter-aircraft use.

By the 1930s, very few private, non-commercial laboratories still existed in America. One that must be mentioned was financed and operated by Alfred Lee Loomis at Tuxedo Park, an exclusive residential area about 60 miles north of New York City. Educated as an attorney and very wealthy from years as a Wall Street tycoon, Loomis also had an insatiable appetite for scientific research. He formed the Loomis Laboratories at Tuxedo Park in the 1920s, and personally paid for a cadre of highly qualified researchers to pursue studies in a wide variety of areas, including the foundations of sonar and microwave radar.

Tuxedo Park Laboratories

As the U.S. mobilized for war, Loomis was anxious for his researchers to make significant contributions to these efforts. Dr. Karl T. Compton, president of MIT, suggested that Loomis Laboratories join in a project on distance finding by radio. Loomis suggested using microwaves in a plane-detector for anti-aircraft weapons that would "include a fairly simple computer which would control the gun directly." A team at Loomis Laboratories was assembled and began research in microwave technology. Within a short time, a simple Doppler microwave object-detection system was demonstrated.

Another private organization that should be mentioned is the Hammond Radio Research Laboratory, established by John Hays Hammond, Jr. The son of a wealthy family and a protégé of Thomas Edison and Alexander Graham Bell, Hammond graduated from Yale and started his privately funded research activities in 1911.

Introduction and Background

In early research, Hammond concentrated on developing an "automatic course stabilizer" for the remote guidance of naval vessels. He satisfactorily demonstrated the system to the Army Coast Artillery Board by piloting a small craft from an airplane at five-miles distance, earning him the title, "The Father of Remote Control."

During the 1930s, Hammond continued with military projects including a multi-channel radio, radio communications secrecy, altitude determining systems, mine protection devices, submarine sound transmission, and incendiary shells. Producing 437 patents and ideas for over 800 inventions, he was one of America's premier inventors.

There were a number of other private laboratories scattered throughout the world. Typical of these was the *Forschungslaboratorium für Elektronenphysik* (Research Laboratory for Electron Physics) in Berlin, mainly funded by Baron Manfred von Ardenne. Founded in 1928 by the then-21 year old von Ardenne and operated by him until it was bombed in 1945, it took out about 600 patents in radio, television, electron microscopy, medical electronics, nuclear technolog, and other areas.

The Carnegie Institution of Washington (CIW) should also be noted. Andrew Carnegie, then often called the richest man in the world, founded the CIW in 1902 as an organization for scientific discovery. The Department of Terrestrial Magnetism (DTM) was formed to map the geomagnetic field of the Earth, but became involved with a variety of research activities.

After the formation of the NRL, there was considerable interaction between this and the CIW. In 1925, Dr. Gregory Breit and Merle A. Tuve, researchers at the DTM, were supported by Louis Gebhard and Leo Young of the NRL in developing a pulsed transmitter for successfully measuring the height of the ionosphere. This was an important step in the radio-to-radar evolution. Tuve would later lead DTM's radar-based proximity fuse program.

DEVELOPMENT OF TELEVISION

The development of television was a major stepping-stone on the path to radar. As television evolved, it required amplifiers of wide bandwidth (video circuits), precision pulse-forming and timing electronics, and high-frequency transmitters. To display the picture, high-resolution cathode-ray tubes were needed. All of these were also necessary for radar. If television had not existed or been under development, these would have added to the burden of radar creation.

Like radar, television is an application of radio; thus, radio had to come first. Television, however, was not the first technology for electrically transmitting pictures, both still and moving. It all started with the telegraph, the first device for using electricity for transferring information. The basic form, that still exists today, was conceived by American Samuel F. B. Morse in 1831. Englishman Charles Wheatstone came up with another device and in 1838 gave a public demonstration, but the Morse telegraph eventually dominated.

The telegraph was extended by Scotsman Alexander Bain in 1843, adding a capability of sending two-dimensional materials – drawings and written information, the fax forerunner. It was 1902 before Dr. Arthur Korn, a German-born American, perfected the first practical fax machine, and this was refined by Frenchman Edouard Belin for sending news photographs. In 1926, RCA made available a radio system for transmitting photographs.

Electromechanical Television

The basic concept of transmitting moving pictures by electrical means also preceded the wireless. Paul J. G. Nipkow, a university student in Germany, received the first patent for an electrical imaging machine in 1884. His system, called an "electric telescope," used a rotating disk with a spiral of holes that scanned a viewed object onto a photocell, then sent the electrical signal to a display using a neon bulb behind a similar disc directly connected to the scan driver.

Nipkow Scanner

Constantin D. Perskyi, a Russian scientist, introduced the name "television" in 1900 at a meeting in Paris to describe devices that make pictures using electricity.

The separation of the Nipkow disc scanner and a similar display device required a means of precisely synchronizing the two rotation units, and took some time to evolve. The use of radio for transmitting pictures using the Nipkow disc, more often called mechanical television, was used by essentially all developers in this field for three decades. The speed of disc rotation established the imaging rate, and the number of holes on the spiral determined the number of scanning lines (usually 20 or less). The image quality was limited by the number of lines and the responses of photocells and neon bulbs.

In Great Britain, Scotsman John Logie Baird developed a Nipkow-disc television system in the early 1920s. He first showed it to the public in 1926, when it produced 30 lines of scan at 5 frames per second. The next year, he gave a demonstration over 438 miles of telephone lines, and followed this in 1928 with broadcasts over BBC, as well as a wireless channel from London to New York.

The most prominent mechanical television system in America came from Charles Francis Jenkins. He developed a system using a Nipkow disc scanner and a rotating drum in the separate receiver. In 1928, he formed an experimental television station in Wheaton, Maryland, and sold sets that produced 48 lines and 15 frames per second. The U.S. Navy became a user of Jenkins's scanners, sending weather maps from Washington to ships.

Both General Electric and the Bell Telephone Laboratories also developed mechanical television systems similar to that of Jenkins. Ernst Alexanderson (of alternator fame) led the work at General Electric. In 1928, General Electric began regular broadcasts on an experimental station with a 48-line, 18-frame system, and sold home receivers with a 3-by-4-inch projection screen.

The first television system using a cathode-ray tube (CRT) for displaying the received image was developed by the Russian scientist, Dr. Boris Lvovich Rosing. He filed a German patent application in 1907, and an article on his system was published in 1911 by the *Scientific American*. Television in Russia, however, remained Rosing's laboratory development for 20 years. In 1931, others began test broadcasts using a 30-line electro-mechanical system. Rosing, accused of not applying television in the best interest of the then-USSR, was exiled to the far north city of Archangel, where he died two years later.

Television systems using electro-mechanical scanners and displays were developed in several additional countries. Experimental broadcasts began in Germany in 1929, but were without sound for several years. Germany was to host the 1936 Olympic Games, and the Nazi leaders, anxious to have them broadcast using television, pressured the various radio industries to develop a system. A 180-line apparatus using a Nipkow-disk camera and all-electronic receivers was built by Manfred von Ardenne in 1933, and was field tested in the Berlin area. Network electronic service started in 1935, with demonstration rooms being set up around the country for the public to watch the games the next year

A laboratory devoted to television development was established in 1931 at Montrouge, France. Rene Barthelemy was named the technical head, and in a short time had a Nipkow-disc system producing a 60-line picture. By the end of 1933, this had increased to 60 lines and Paris Television started a daily hour-long broadcast. The definition increased to 180 lines by the end of 1935. The operation continued on a very part-time basis until September 1940 when France was occupied by the Germans.

Kenjiro Takayanagi, an engineer teaching at the Hamamatsu Technical High School in Japan, began television experiments using a Nipkow-disc apparatus in 1925. Like Rosing in Russia, he used a CRT for the display, and in 1928 gave a demonstration at a television conference of the Tokyo Electrical Academy. In the same time period, developments were underway by Tadaoki Yamamoto and others at Waseda University. In 1930, Takayamgi joined the NHK (Radio Japan) Science & Technology Research Laboratories and, in cooperation with Yamamota successfully developed a system having 100 lines and 20 frames per second.

Electronic Television

By 1935, mechanical television had essentially reached a dead end. Image resolution remained low, and transmission and reception standards were lacking. Fortunately, all-electronic television – from camera tube to receiver display – was rapidly coming into being.

The concept of television scanned, synchronized, and displayed by electronic means was originated in 1908 by Alan A. Campbell-Swinton, a radiologist in Scotland. In a letter published in the June 1908 issue of *Nature*, he described the "Distant Electric Vision." This was eventually popularized as "Campbell-Swinton's Electronic Scanning System" in the August 1915 issue of Hugo Gernsback's popular magazine, *Electrical Experimenter*. (Here it should be noted that Campbell-Swinton's concept of television – and it was no more than a concept – did not apply well to electronic television as it eventually evolved.)

Vladimir Zworykin

Vladimir K. Zworykin (1889-1982), who had worked as an assistant to Rosing in Russia, came to America in 1919, and joined Westinghouse a year later. Working in off-hours, he conceived the Iconscope – a CRT-based camera tube – and in 1923 submitted a patent application that was granted five years later. He proposed a formal television project at Westinghouse, but it was rejected by his supervisor, who suggested that he devote his time to more practical endeavors.

Still working off-hours, Zworykin returned to experiments that he had done with Professor Rosing in Russia on a CRT-based display unit. At an IRE convention in 1929, Zworykin demonstrated a picture tube called the Kinescope. That same year, he completed his Ph.D. degree and shortly thereafter joined RCA to direct their new electronics research laboratory in Camden, New Jersey.

Philo T. Farnsworth (1906-1971), a highly intelligent Utah farm boy, became extremely interested in electronics, reading all that he could find on the subject. In 1920, at age 14, Farnsworth conceived, and disclosed in writing to his high-school chemistry teacher, a television camera and display tube, both based on the CRT, forming a full electronic television system.

Philo Farnsworth

It was 1926 before Farnsworth could obtain financial support and set up a small laboratory in San Francisco. At the heart of his system was his camera tube called the Image Dissector; his receiver was called an Oscillite. In January 1927, patents were filed for a "Television Receiving System" and a "Television System." Farnsworth tested the full system in September 1927. This event sounded the death knell of the mechanical, rotating-disc scanner systems and initiated a rivalry between Farnsworth and Zworykin for the eventual title of "father of modern television."

To further develop the Farnsworth system, Television Laboratories, Inc., was established in 1929. Zworykin visited the laboratories and was given full information on Farnsworth's equipment. When Zworykin reported back to Westinghouse, little interest was displayed, and he was again admonished to "work on something more useful."

Television Laboratories moved to Philadelphia in 1931, first working under an arrangement with Philco, and then setting up as a new firm, Farnsworth Television. In this same period, Zworykin left Westinghouse and joined RCA to head its Camden Research Laboratory, with a major mission of developing all-electronic television.

In 1932, RCA demonstrated an improved camera tube, the Iconoscope that provided 120 lines of resolution. The stage was thus set for a long, intense battle between Farnsworth and RCA over improvements and television patent rights. Zworykin improved the Iconoscope to 343 lines, and Farnsworth upped his Image Dissector to 441 lines, both at 30 frames per second. By 1937, there were a number of experimental stations operating in the U.S., setting the 441-line, 30-frame standard. A number of firms were selling television receivers.

The U.S. Patent Office ruled in February 1935 in the case of *Farnsworth vs. Zworykin* that "Priority of invention is awarded to Philo T. Farnsworth." RCA appealed, and it was 11 years before the final settlement in full favor of Farnsworth took place. Although Farnsworth was eventually granted the basic patent for electronic television, Zworykin and RCA must be given credit for bringing this technology to the practical stage.

By 1934, British electronic firms EMI and Marconi had jointly created an all-electronic television system. It produced 405 scanning lines at 25 frames per second and used the Iconoscope camera tube through a license from RCA. This system was The BBC officially adopted this system in 1936, and a regular schedule of broadcasts began.

Desperate to have the 1936 Olympics shown using all-electronic television, the Germans acquired rights from RCA to use an Iconoscope camera tube. Although the initial system still used the relatively crude 180-line picture, they gained the attention of the general public massed at 20 new television parlors installed for the big event. The next year they followed the American standard and had a 441-line system on the air.

A few other countries also adopted all-electronic television about this same time. In 1933, Tokyo Denki Company of Japan conducted an experiment using a 120-line system employing a Farnsworth Image Dissector tube. This was followed in 1935 by Takayanagi's team at HNK that internally built an iconoscope and soon demonstrated an all-electronic system with 245 interlaced lines at 20 frames per second. By 1937, this was improved to match the American standard of 441 lines and 30 frames per second, and NEC and Toshiba developed the first commercial television receivers.

In France, three firms started experimental all-electronic stations in 1935, each using a different standard: 441 lines for Gramont, 450 lines for the Compagnie des Compteurs and 455 for Thomson. In 1938, the government established a common standard of 455 lines. By 1939, there were about 300 individual television receivers, some of which were available in a few public places.

In the USSR, equipment manufactured and installed by RCA began regular television broadcasts at 343 lines in late 1938. Although a domestic receiver was produced, it was not available to the general public; only Soviet officials could obtain these sets.

Cathode-ray tubes were used for the picture display in television receivers, as well as in monitors on the cameras and their controls. CRTs of the early 1930s were imported to the U.S. from Telefunken in Germany at high costs. They normally had screens only a few inches in size and burned out after 30 or so hours. Dr. Allen B. Du Mont devised a low-cost, long-lasting cathode-ray tube for television and other applications. In the middle of the decade, his DuMont (different spelling) Laboratories developed CRTs into a relatively inexpensive product with a lifetime in thousands of hours, and by the end of the decade they were producing a "standard" 12-inch tube.

PRECURSORS OF RADAR

The fundamental concept of reflecting electromagnetic waves predates the invention of radio. As far back as 1886, Heinrich Hertz gave laboratory demonstrations to show that electric inductors could reflect his waves. This led to many activities in which radar was "almost" invented. A few of these precursors that are frequently mentioned in the literature are briefly recounted.

At age 22, Christian Hülsmeyer formed a company in Dusseldorf, Germany, to apply Hertz's reflections to detecting the presence of ships. In 1904, Hülsmeyer registered German and foreign patents for an apparatus, the *Telemobilskop* (Telemobiloscope). An article, "The Telemobiloscope," was published in a 1904 issue of *The Electrical Magazine*. This was described as an anti-collision system using a 50-cm wavelength spark-gap transmitter and a coherer detector. The radiated signal was beamed by a funnel-shaped reflector that could be aimed. The receiver used a separate vertical antenna with a semi-cylindrical movable reflecting screen.

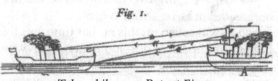
Telemobiloscope Patent Figure

Hülsmeyer gave a public demonstration of his system, receiving reflections from a ship nearby on the Rhine River and causing an electric bell to ring as long as the ship was in the transmitted beam. Later improvements increased the detection distance to 3,000 meters. Although a number of patents were granted, Hülsmeyer was never able to gain financial backing or to sell his patents to industry, so the effort was dropped.

During the First World War, another German, Richard Scherl, apparently without knowledge of Hülsmeyer's previous work, designed the *Strahlenzieler* (Raypointer), a wireless device using echoes for detection. Assisted by a well-known science-fiction writer and engineer, Hans Dominik, they successfully produced an experimental set working at about 10-cm wavelength. Scherl sent details of his apparatus to the Imperial German Navy in February 1916, but his offer was rejected as "not being of importance to the war effort."

Radar Origins Worldwide

In 1922, Dr. A. Hoyt Taylor and Leo C. Young with the U.S. Naval Aircraft Radio Laboratory at the Anacostia Naval Air Station were testing a 60-MHz transmitter and receiver that they had developed. In driving the receiver around in a car, they noticed that the buildings would cause the signal from the transmitter to fade in and out. To get away from this, they placed the transmitter and receiver at fixed sites on opposite sides of the nearby Potomac River. At first, the signal was strong and steady; then, as Young later stated,

> We began to get quite a characteristic fading in and out – a slow fading in and out of the signal. It didn't take long to determine that this was due to a ship [the wooden SS *Dorchester*] coming up and around Alexandria.

Taylor reported this interference-beat detection to the Bureau of Engineering with the following suggestion:

> Destroyers located on a line a number of miles apart could be immediately aware of the passage of an enemy vessel between any two destroyers of the line, irrespective of fog, darkness, or smoke screen.

No action was taken on this at that time, but this would later lead to one of the most important developments in Navy history.

While delivering an address in June 1922 to a joint meeting of the American Institute of Electrical Engineers and the Institute of Radio Engineers, Guglielmo Marconi made the following suggestion:

> I have noticed the effects of reflection and deflection of these [radio] waves by metallic objects many miles away. . . . It seems to me that it should be possible to design apparatus by means of which a ship could radiate or project a divergent beam of these rays in any desired direction, which rays, if coming across . . . another steamer or ship, would be reflected back to a receiver . . . and thereby immediately reveal the presence of the other ship in fog or thick weather.

In Germany, Dr. Heinrich Löwy conducted extensive research on distance-measuring devices. In July 1923, he filed a U.S. patent application for connecting an antenna alternately between a transmitter and a receiver for "measuring of the distance of electric conductive masses and for ascertaining of the height of flying vehicles." His proposed apparatus was described as radiating very short trains of waves and using the transit-time between transmitted and returned (reflected) signals to measure the distance. Although there is no indication that such a device was ever built, the description is close to that for the first practical radar system.

Dr. Gregory Breit and Merle A. Tuve, researchers in the Department of Terrestrial Magnetism (DTM) at the Carnegie Institution of Washington, were studying characteristics of the ionosphere (then called Kennelly-Heaviside layer). In using a crystal-stabilized, high-frequency transmitter developed by Louis Gebhard at the NRL, they had problems in extracting height information from the continuous returned signal. In 1925, NRL's Leo Young suggested using pulsed transmission, and built a square-wave control for this purpose. With the transmitter at the NRL and the receiver several miles away at the DTM, the experiments were successful, demonstrating distance measurement using pulsed transmission.

Gregory Breit & Merle Tuve

During the late 1920s, several aircraft-based altitude-sensing systems using FM radio were developed. At General Electric in 1928, Jetson O. Bentley filed a patent application for an apparatus involving a transmitter with a linearly varying frequency aimed toward the ground The reflected signal was mixed with a portion of the initial signal, giving a beat signal proportional to the two-way radiation time, from which the altitude could be determined. Russell C. Newhouse wrote a thesis at Ohio State University on this subject in 1929, then joined BTL where patents were obtained on his concepts. Neither of these systems, however, were fully developed for almost a decade and contributed little as precursors to FM radar.

While conducting outdoor tests on a receiving antenna at the Naval Research Laboratory in June 1930, Lawrence A. Hyland observed interference beat between the direct signal from a high-frequency transmitter and the signal reflected from a passing airplane. Leo Young noted the similarity to the observations made in 1922 by himself and Hoyt Taylor. This was reported to the Bureau of Engineering, and the Bureau responded by directing the NRL to continue with the investigations. However, no funding was provided for the continued effort, so it was given a low priority.

In Great Britain, William Alan S. Butement, a scientific officer at the Woolwich Research Station of the Signals Experimental Establishment, considered the potential of using pulsed 50-cm signals for detection of ships. In January 1931, Butement and his associate, P. E. Pollard, prepared a memorandum on this subject, and it was entered as a Coastal Defense Apparatus in the Inventions Book maintained by the Royal Engineers. They also built and tested a system that functioned over a short distance. For reasons that are undefined, the War Office did not give it consideration. (Butement later applied their concepts in developing fielded systems, as well as the radar fuse for projectiles.)

In 1932, RCA was testing an experimental television transmitter operating at 44 MHz (6.8 m) on the 85th floor of Empire State Building in New York City. When measuring the signal strength at the RCA Building some distance away, it was observed that the "stop-and-go" motion of traffic in streets between the transmitter and receiver could be discerned by variations in the signal level. Measurements of the speed of isolated vehicles were sometimes made from the Doppler shift of the reflected continuous-wave signal. While these observations were reported in conferences, they were not applied in developing detection systems.

Dr. Carl L. Englund, Arthur C. Crawford, and William W. Mumford at the BTL conducted studies of propagation phenomena, including the effect of single trees, woods, wired houses, fluctuations from moving bodies, and aircraft-to-ground links. In 1933, they published a paper, "Some Results of a Study of Ultra-short-wave Transmission Phenomena," in the *Proceedings of the IRE*. In discussing the field fluctuations from moving bodies, re-radiation from aircraft was highlighted and the following was noted:

> For ordinary airplane heights a high-energy transformation loss in the re-radiation process can occur and still give marked indications in the receiver meter. The airplane re-radiation was noticed at various subsequent times, sometimes when the airplane itself was invisible.

In the summer of 1934, a small group of engineers from the U.S. Army Signal Corps gathered at a promontory on New York's lower bay to watch an experiment by Drs. Irving W. Wolff and Ernest C. Linder of the RCA Camden Laboratory research staff. The two scientists had brought with them a microwave transmitter using a low-power magnetron built by Linder, a receiver, two four-foot parabolic antennas, and an audio amplifier with a loudspeaker.

As the assemblage watched and listened, the antennas were aimed toward a small boat passing some 2,000 feet away. An audio tone emerged from the speaker as the boat went by; this was generated by beating the Doppler-shifted reflection against some of the transmitter signal. As the antennas were turned to follow the boat, the tone continued until the boat passed out of range.

Using low-power microwave equipment, researchers under Dr. William R. Blair at the Signal Corps Laboratories (SCL) at Fort Monmouth, New Jersey, also observed beat-interference phenomena, detecting a moving truck at a distance of 250 feet. In subsequent experiments using an SCL-built copy of a Hollmann tube from Germany with a 5-watt output at 50 cm (600 MHz) and a receiver placed some 12 miles away, reflected Doppler signals were generated from a passing boat. In his 1934 annual report, Blair commented on this method, which was only effective when there was motion between target and detector:

> It appears that a new approach to the problem is essential.
> Consideration is now being given to the scheme of projecting an interrupted sequence of trains of oscillations against the target and attempting to detect the echoes during the interstices between the projections.

Thus, the principle of pulse detection was stated. The Army researchers, however, did not reduce this to demonstrative equipment for two more years.

Dr. Chester W. Rice, the son of a previous General Electric president, was a very prolific inventor at the GE Advanced Technology Laboratory in Schenectady, New York. He had earlier been the co-inventor of the moving-coil loudspeaker and types of antennas used for long-distance radio communications. Prone to working all night, he actually did much of his work in a special laboratory built at his home.

In 1936, the *General Electric Review*, a respected journal, included a paper by Rice on applications of microwaves that described in detail his 4.8-cm (6.3-GHz) apparatus for measuring distance, bearing, and velocity of target aircraft and ships. A 4-watt magnetron was used in the transmitter and a Barkhausen-Kurz tube in the receiver, both made at GE. The unit was pulsed for distance measurements, and operated CW for velocity determination (via Doppler). Perhaps the most significant part was a rotating display, servo

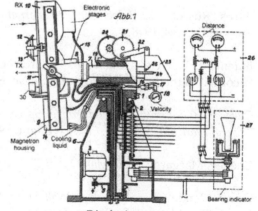

Rice's Apparatus

linked to the antenna, and very similar to the future Plan Position Indicator (PPI).

Multiple patents were filed in the U.S., Great Britain, France, Belgium, and Germany on various parts of the apparatus. A full system was built, and successful tests were made by Rice on nearby targets. Little else, however, seems to have been done at GE with Rice's invention. Although GE was highly involved with future radar programs, there are no indications that they ever sought to apply their patents in these developments.

Bibliography and References for Chapter 1

Aitken, Hugh G. J.; *Syntony and Spark: the Origins of Radio*, Princeton University Press, 1979

Aitken, Hugh G. J.; *The Continuous Wave: Technology and American Radio, 1900-1932*, Princeton University Press, 1985

Baker, William John; *A History of the Marconi Company*, Methuen, 1970

Breit, G, and M. A. Tuve; "A Test of the Existence of the Conducting Layer," *Phys. Rev.*, vol. 28, p. 554, 1926

Brown, Louis; *A Radar History of World War II – Technical and Military Imperatives*, Inst. of Physics Pub., 1999

Burns, Russell W.; *Television: an International History of the Formative Years*, IEE Press, 1998

Butement, W. A. S., and P. E. Pollard; "Coastal Defense Apparatus," Recorded in the *Inventions Book of the Royal Engineers Board*, Jan. 1931

Carnearl, Georgetta; *A Conqueror of Space; An authorized Biography of the Life and Work of Lee de Forest*, H. Liveright, 1930

Cheney, Margaret; *Tesla: Man Out of Time*, Prentice-Hall, 1981

Conant, Jennet; *Tuxedo Park: A Wall Street Tycoon and the Secret Palace of Science That Changed the Course of World War II*, Simon & Schuster, 2002

Introduction and Background

Drury, Alfred T.; "War History of The Naval Research Laboratory," in the series, *U.S. Naval Administrative Histories of World War II*, Department of the Navy, 1946

Englund, C. L., A. C. Crawford, and W. W. Mumford; "Some Results of a Study of Ultra-short-wave Transmission Phenomena," *Proc. IRE*, vol. 21, no. 3, p. 464, 1933

Ewell, Cyril Frank; *The Poulsen Arc Generator*, Van Nostrand, 1923

Fagen, M. D. (editor); *A History of Engineering and Science in the Bell System; National Service in War and Peace (1925-1975)*, Bell Telephone Laboratories, 1978

Farnsworth, Philo T.; "Television by Electron Image Scanning," *J. Franklin Inst.*, vol. 218, p. 411, 1934

Fessenden, Helen M.; *Fessenden, Builder of Tomorrow*, Coward-McCann, 1940

Fessenden, R. A.; "Wireless Telegraphy," *The Electrical Review*, vol. 60, p. 252, 1907

Fessenden, Reginald Aubrey; "Autobiography," *Radio News*, May through November, 1925

Fort Monmouth Historical Office; *A Concise History of the U. S. Army Communications-Electronics Life Cycle Management Command and Fort Monmouth, New Jersey*, CE LCMC Historical Office, 2005;
http://www.monmouth.army.mil/historian/pub.php

Howeth, Linwood S.; *History of Communications-Electronics in the United States Navy*, Government Printing Office, 1963;
http://earlyradiohistory.us/1963hw.htm

Hull, Albert W.; "The Effect of a Uniform Magnetic Field on the Motion of Electrons Between Concentric Cylinder," *Phys. Rev.*, vol. 18, p. 31, 1921

Hülsmeyer, Christian; "Hertzian-Wave projecting and receiving apparatus," British patent No. 13,170 granted Sept. 22, 1904

Jabbari, B.; "Introduction to the Classic Paper [June 1922] by Marconi," Proc. IEEE, vol. 85, no. 10, p.1523, 1997

Jenkins, Charles F.; "Radio Vision," Proc. IRE, vol. 15, no. 11, p. 958, 1927

Kilbon, Kenyon; *A Short History of the Origins and Growth of RCA Laboratories, 1919 to 1964,* David Sarnoff Research Laboratory, RCA, 1964; http://www.davidsaranoff.org/kil.htm

Killer, Peter A.; *The Cathode-Ray Tube: Technology, History, and Applications,* Palisades Press, 1992

Lehmann, G. L., and A. R. Vallarino; "Study of Ultra-High-Frequency Tubes by Dimensional Analysis," Proc. IRE, vol. 33, no. 10, p. 663, 1945

"Marconi Calling," an extensive Web site dedicated to the life, science, and achievements of Guglielmo Marconi, Marconi Corporation, 2001; http://www.marconicalling.com

"Mystery Ray Locates 'Enemy': U.S. Army Tests Detector for Hostile Ships and Planes," *Popular Science,* Oct. 1935

"Mystery Ray: Telefunken firm in Berlin reveals details of a 'mystery ray' system capable of locating position of aircraft through fog, smoke and clouds," *Electronics,* September 1935

"Mystery Rays 'See' Enemy Aircraft," *Modern Mechanix,* Oct. 1935

Rice, Chester W.; "Transmission and reception of centimeter radio waves," *General Electric Review,* vol. 39, p. 363, 1936

Schatzkin, Paul; *The Boy Who Invented Television,* TeamCom Books, 2002

Swords, S. S.; *Technical History of the Beginnings of Radar,* Peter Peregrinus Ltd, 1986

Tyne, Gerald F.; *Saga of the Vacuum Tube,* Howard W. Sams, 1977

Zworykin, Vadlimir K.; "Description of an Experimental Television System and the Kinescope," Proc. IRE, vol. 21, p. 1655, 1933

Chapter 2
EARLY RDF IN GREAT BRITAIN

The Department of Scientific and Industrial Research (DSIR) of the United Kingdom was established in 1917 with the aim of enhancing research by means of government funding. It encompassed the National Physics Laboratory (NPL), which had been set up in 1900 at Teddington on the outskirts of London to serve as the measurement standards laboratory. The activities of the NPL, however, had expanded considerably and, in association with universities, involved research in many areas.

Following the First World War, there was a great surge in radio technology, and the military users were anxious to upgrade their systems. Funds for the needed research and development, however, were scarce. In 1920, the DSIR formed the Radio Research Board (RRB) to prevent duplication of effort in radio research between the different armed services. The Board was given the responsibility to "direct any research of a fundamental nature that may be required, and any investigation having a civilian as well as a military interest." The RRB Chairman was Admiral of the Fleet Sir Henry Jackson, himself a radio pioneer.

Much of the development of military radio equipment centered at the Experimental Department of the Admiralty's Signal School and the War Department's Signals Experimental Establishment. The General Post Office, responsible for domestic communication control since taking over the telegraph lines in 1868, also had research in radio equipment.

Major commercial suppliers of receivers, transmitters, and other electronic equipment in Great Britain during this period included General Electric Company (GEC), Gramophone (later becoming EMI), Marconi, Metropolitan-Vickers, W. G. Pye, Radio Communications, Western Electric, British Westinghouse, British Thompson-Houston (BTH), A. C. Cossor, and Ferranti.

RADIO RESEARCH STATION, SLOUGH

During the First World War, the Air Ministry had established a Meteorological Office, situated in the Royal Aircraft Establishment at Farnborough. When the RRB was formed, it took authority over the work of the Meteorological Office, as well as the nearby Aldershot Wireless Station, and set up the station for the study of atmospherics. The NPL agreed to share in the research, as well as continue certain work of an allied nature that it already sponsored at Aldershot. Robert A. Watson Watt, who had been with the Meteorological Office, was named as the Supervisor of research in atmospherics.

The NPL also agreed to form a Wireless Division for research in the cause of errors in direction finding of emission sources (lightning and radio transmitters) and related problems. An "electromagnetically quiet" site was needed for this work, and such a site was found in Ditton Park, where the Admiralty already operated an observatory. With the influence of Admiral Jackson, access to the West Park area of the observatory was obtained.

Ditton Park is located on the outskirts of Slough in Berkshire, some 22 miles west of the heart of London. Moat-surrounded Ditton House or Castle, dating back to the early 1500s, is central to the Park; North and West Parks extend considerable distances from the center. In 1915, Ditton Park had been taken over and used for the Admiralty Compass Observatory.

The NPL Wireless Division initiated research in facilities constructed in West Park in 1920. Robert H. Barfield supervised the work in direction finding; there was also work in field-strength measurements conducted by J. Hollingsworth. Over the next decade, Barfield and a small number of coworkers, with the encouragement and personal review of Admiral Jackson, produced several important papers.

In 1924, the Air Ministry decided to re-occupy the Aldershot site. The RRB then directed that the research equipment and personnel at the Wireless Station be moved to Ditton Park, locating in the North Park. Although adjacent to the NPL research facilities in the West Park, there was essentially no working relationship between the two activities. They both had their own small buildings but did share a large common workshop.

Ditton Park Laboratory

To improve the relationship and increase research efficiency at Ditton Park, in December 1927, the direction-finding, field-strength, and upper-atmosphere research were merged, forming the Radio Research Station (RRS), Slough, an entity under the DSIR but under the general supervision of the NPL. Watson Watt was appointed as the first Superintendent, and James F. Herd, who had been with the operation for many years, was the Deputy. Barfield and Hollingsworth continued to lead their specialized areas, while Watson Watt and his team made significant contributions to instrumentation for detecting and tracking lightning.

Robert Alexander Watson Watt (1892-1973) was born and raised in Brechin, Scotland. He was a descendant of James Watt, developer of the steam engine. His father was a carpenter and an Elder of the Presbyterian Church; his mother, as described by Watson Watt, was "miraculous and a temperance reformer." His name is most often given as Watson-Watt, a hyphenated form he chose after being knighted in 1942.

Robert Watson Watt

Watson Watt studied at the University College, Dundee, receiving in 1912 the B.Sc. degree with special distinction in electrical engineering. Following graduation, he served three years at the University as a professor's assistant in natural philosophy. In 1915, he was employed by Air Ministry's Meteorological Office, located at the RAE at Farnborough, devising radio means for providing meteorological information to aircraft pilots.

When the newly formed DSIR absorbed the Meteorological Office and moved it to the Aldershot Wireless Station in 1917, Watson Watt became the supervisor of activities centering on radio techniques for locating thunderstorms. His first attempts at using a cathode-ray tube (CRT) for displaying the information was made at that time. He continued with this research when the Aldershot activities were moved to Ditton Park in 1924, then in 1927 was named Superintendent when the Radio Research Station (RRS), Slough, was formed.

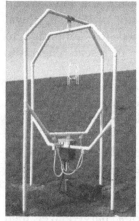

Square-Loop Antennas

Watson Watt and his team made significant contributions to instrumentation for detecting and tracking lightning. The equipment included two square-loop antennas set orthogonally for receiving, and a cathode-ray oscilloscope with a highly persistent phosphor screen for displaying the fleeting signals. In 1929, Watson Watt visited the laboratory of Manfred von Ardenne in Berlin, where the original Braun tube had been turned into the useful cathode-ray tube. Watson Watt bought several of these CRTs and used them to build special oscilloscopes.

Antenna outputs were connected to the horizontal and vertical inputs of an oscilloscope. The direction of the signal (with 180 degrees ambiguity) was determined by comparing the signal intensity from the two deflections. With two separated stations providing direction, the distance to the strikes could be determined by triangulation. Watson Watt reported that strikes as remote as 2,500 km (4,000 miles) had been detected, obviously involving Kennelly-Heaviside reflections.

Success in developing this instrumentation, particularly the display, is reflected in the monograph *Applications of the Cathode Ray Oscillograph in Radio Research*, by R. A. Watson Watt, J. F. Herd, and L. H. Bainbridge-Bell, published in 1935 by HMSO. The cathode-ray application is also described in "The Generation and Reception of Wireless Signals of Short Duration," a paper by Herd published in *Proc. Physical Society* in 1933.

Watson Watt also worked with Edward V. Appleton in developing instrumentation for examining the Kennelly-Heaviside ionized layers above the Earth, and in 1926 he had suggested that this region be called the ionosphere.

Edward Appleton

Edward Victor Appleton (1892-1965) was born in Bradford, England. He attended St. John's College at Cambridge University, receiving his B.A. degree in 1913, then, studied under the Sir J. J. Thompson (1906 Nobel Prize) and Lord Ernest Rutherford (1908 Nobel Prize). During the First World War, he served as an officer in Royal Engineers, then returned to Cambridge to begin a life-long career in researching radio waves.

In 1920, Appleton received an appointment in experimental physics at Cambridge's Cavendish Laboratory. Four years later, he was appointed Wheatstone Professor of Physics at King's College of London University and served there for 12 years. Most of his radio research, however, continued to be centered at the Cavendish Laboratory.

Starting in 1924, Appleton and his student, Miles A. F. Barnett, had conducted experiments that proved the existence of a charged layer, then called the Heaviside-Kennelly layer, about 63 miles (100 km) above the earth's surface. The experiments involved transmissions from a BBC station at Bournemouth, and a receiver located near Cambridge. The BBC cooperated by providing slight changes in the transmitter frequency; comparisons of the signals at different frequencies were used to determine the layer's height. Two years later, a second, electrically stronger layer was found at 150-225 miles (240-359 km), the layer allowing round-the-world communications.

Appleton's work at that time was supported by the RRB. In 1929, the experimental work was moved to the Radio Research Station, joining the upper-atmospheric activities in North Park. Robert Naismith was assigned by the RRS to assist with this research. L. H. (Labouchere Hillyer) Bainbridge-Bell, whom Watson Watt described as his best circuit designer, did much of the equipment development.

A transmitter with an upward-radiating antenna was set up a few miles away in Windsor Great Park (allowed under the condition that the antenna be "hidden among the trees"), and the signal reflected from the ionized layer was monitored at Ditton Park. By slowly increasing the signal frequency, a point was reached, thereafter called the critical frequency, for which the layer would no longer reflect. This led to calculations of the charge characteristics of the layer.

In 1930, pulse-modulation was incorporated in the transmitter, allowing a more precise height measurement of the returning layer. This was the technique first used in 1925 by Dr. Gregory Breit and Merle A. Tuve in America. Beginning the next year, a sequence of ionospheric soundings was started, continuing for many years and contributing greatly to improved world-wide radio communications. Appleton was knighted in 1941 and received the 1947 Nobel Prize in Physics for his "Investigations into the upper layer of the atmosphere."

FROM DEATH RAY TO RDF ORIGINS

The Great War of 1914-1917 was called "the war to end all wars." In less than a generation, however, it was realized that this would not be so. Germany was rearming, and preparations for the next conflict began. The general responsibilities for defense were divided as follows: the Air Ministry, defense of the air, through the Royal Air Force (RAF); the War Office, defense of the land and coastal regions through the Royal Army; and the Admiralty, defense of the sea through the Royal Navy.

The advent of bombers led Great Britain to establish an Air Fighting Zone along the south coast, with Army searchlight and anti-aircraft gun stations and RAF aircraft ready on airfields. To give warning, various technologies were tried without success, including radio detection of ignition sparks and infrared detection of engine heat.

The overall situation led to considerable pessimism over the defense of Great Britain. In a speech to Parliament in 1932, Stanley Baldwin, the future Prime Minister, stated the following: "I think it is well for the man in the street to realize that there is no power on earth that can protect him from being bombed."

The only detection technology with any promise, albeit small, was acoustic sensors for hearing engine noise. The Air Defense Experimental Establishment, under the Air Ministry, began work in 1933 on a network of large concrete collectors for concentrating and detecting the noise. Persons would monitor the noise and pass via the telephone anything heard to a central command post, giving at best a few minutes of warning. Blind operators were suggested, in the belief that their hearing would be more acute.

Acoustic Mirror

Parabolic Collector

As war clouds loomed over Great Britain, a large-scale air defense exercise was held during the summer of 1934, including mock raids on London and other potential targets. Although the routes of bombers were known in advance, over half of them reached their destinations without being detected. Both the Air Ministry and the Houses of Parliament were successfully "destroyed," while few "enemy" bombers were intercepted. Stanley Baldwin, now Prime Minister, summed up the feeling: "The bombers will always get through."

The likelihood of air raids and the threat of invasion by air and sea drove a major effort in applying science and technology to defense. In November 1934, the Air Ministry established the Committee for Scientific Survey of Air Defence (CSSAD) with the following official function:

> To consider how far recent advances in scientific and technical knowledge can be used to strengthen the present methods of defence against hostile aircraft.

Commonly called the "Tizard Committee" after its Chairman, Sir Henry Tizard, this group had a profound influence on technical developments in Great Britain.

Henry Thomas Tizard (1885-1955) was born at Gillingham, Kent, son of the Assistant Hydrographer to the Royal Navy. He studied at Magdalen College of Oxford University, graduating in 1908 with a degree in mathematics and chemistry. After a year of study at the

University of Berlin, he returned to Oxford to engage in research. With the outbreak of World War I, he joined the military, eventually serving as an experimental pilot in the Royal Flying Corps. He rose to the rank of Lieutenant Colonel and Controller for Aircraft Research at the Ministry of Munitions.

After the war he was made Reader in Chemical Thermodynamics at Oxford where he experimented in the combustion of aircraft fuels, devising what is now referred to as octane numbers. In 1920, he took a position with the Department of Scientific and Industrial Research (DSIR), responsible for coordinating research to meet needs of the three military services. He was appointed DSIR Secretary in 1927, but left in two years to become Rector of Imperial College, London, a position he held until 1942. He was knighted in 1937.

Sir Henry Tizard

In addition to Tizard, initial members of the CSSAD were Archibald Vivian (A. V.) Hill, a psychologist, mathematician, and Nobel Prize Laureate, then serving as a Research Professor at University College, London; Patrick M. S. Blackett, Professor of Physics at Birbeck College, London; Henry Egerton (H. E.) Wimperis, Director of Scientific Research at the Air Ministry; and, as the Committee's Secretary, Albert Percival (A. P.) Rowe, a physicist who was the Personal Assistant to Wimperis at the Air Ministry.

The Committee of Imperial Defense, established near the turn of the century, was the highest authority in Great Britain on defense matters. It was where the heads of the Government and the most powerful politicians in the country supervised and superintend actions and made sure that necessary funds were available. With the new importance of air defense, this Committee set up an Air Defense Research (ADR) Committee, through which the CSSAD was resourced. Tizard, himself, was made a member of the ADR Committee.

The Initial Proposal

The July 11, 1934, issue of *The New York Times* had an article concerning a "death ray" invented by the controversial scientist Nikola Tesla. It quoted Tesla as saying that his beam would "drop an army in its tracks and bring down squadrons of airplanes 250 miles away." While he was preparing for the first meeting of the Tizard Committee, this, or similar public articles, came to the attention of Wimperis. In mid-January, 1935, Wimperis asked Robert Watson Watt to meet with him and comment on the potential for directed radio beams to serve as such a weapon. Watson Watt, then Superintendent of the Radio Research Station, Slouth, said that he doubted that this would be possible, but he would consider it and give his answer in a few days.

For responding to Wimperis, Watson Watt asked his Scientific Assistant, Arnold F. Wilkins, to make simple calculations concerning the necessary energy level, and the corresponding generation and propagation of this energy. Wilkins' analysis showed that such a weapon was not feasible with foreseeable radio transmitting equipment. Remembering, however, a report from the General Post Office citing signal disturbances caused by an aircraft flying through a radio beam, Wilkins told Watson Watt that "this phenomenon might be useful in detecting enemy aircraft." He provided analytical backup to both parts of his reply.

Arnold Wilkins

Arnold Frederic ("Skip") Wilkins (1907-1985) was born and raised at Chester, the heritage city in Cheshire, England. He received his early education at the prestigious King's School in Chester, then attended Manchester University, where he earned the B.Sc. degree followed by study at St. John's College of Cambridge University, graduating with a B.A./M.A degree in physics. He joined Watson Watt's staff at the Radio Research Station, Slough, in 1931.

An early assignment for Wilkins was to select a new receiver for use in ionospheric measurements. For this, he compared several available sets with a receiver then in use by the General Post Office (GPO) at a communications station in Colney Heath. The engineers there showed him a report (GPO No. 232, June 1932) that described testing a very-high-frequency transmitting and receiving system. In this it was stated that aircraft flying near the test site caused variation in the intensity of the received radio signal.

Insofar as the GPO was concerned, this aircraft-induced interference was undesirable; no thought was given, at the time, to using it beneficially. Only a short time later, Wilkins used this in suggesting one of the most critical developments in the forthcoming defense of Great Britain.

Watson Watt sent Wimperis a memorandum saying that an analysis had shown that a radio beam weapon was not feasible, but concluded with the following:

> Attention is being turned to the still difficult, but less unpromising, problem of radio-detection as opposed to radio-destruction, and numerical considerations on the method of detection by reflected radio waves will be submitted.

This memorandum was discussed at the first meeting of the Tizard Committee on January 28, 1935.

A second memorandum, marked Secret and dated February 27, 1935, was titled "Detection and Location of Aircraft by Radio Means." Using calculations from Wilkins, this outlined how such a detection system could function, drawing on the techniques already in use at the Radio Research Station for examining ionospheric layers and in detecting thunderstorms.

Specific calculations were given concerning the potential energy of the transmitted beam, the reflection characteristics of a typical bomber, and the necessary receiver sensitivity. This is summarized as follows: If a transmitting aerial radiates at 6 MHz (50-meters wavelength), a bomber with a wing span of 25 meters (82 feet) would act as a half-wave aerial matched to the illuminating signal, reradiating (reflecting) the signal with good efficiency.

As to the possible range of detection, it was noted that the "common practice" of putting 15 amperes into the sending aerial would give a "large factor of safety . . . for ranges of the order of 10 miles at flying heights of about 10,000 feet." There is a reference to increasing the received signal "by at least tenfold by the provisions of a suitable beam array," as well as increasing the radiated energy by pulsing the transmitter, both leading to increased detection ranges.

It was noted that pulsing the transmitter could also provide a means of measuring the range:

> If now the sender emits its energy in very brief pulses, equally spaced in time, as in the present techniques of echo-sounding of the ionosphere, the distance between the craft and sender may be measured directly by observation on a cathode ray oscillograph directly calibrated with a linear distance scale.

This was based on the successful use of the cathode-ray apparatus at the RRS in direction-finding and measurements of the ionosphere.

Addressing the problem of secrecy, it was suggested that the work be conducted under the guise of an ionospheric research activity at a new location – Orfordness, a barren stretch of land on the North Sea coast.

The Daventry Experiment

The memorandum from Watson Watt was reviewed with great enthusiasm by the Tizard Committee; here appeared to be a solution to bomber detection. However, when presented to Sir Hugh Dowding, the Air Marshal, he said that he was not much impressed by calculations, but if a practical demonstration were given, he might allocate scarce funds for an initial development. In question was whether an aircraft would broadly reflect (not mirror-like) a signal sufficient for detection. It was agreed that if Wilkins could repeat the Post Office observations of reflections from a nearby aircraft, this would be satisfactory.

The BBC Empire short-wave transmitter (with call letters GSA) was located at Daventry, near the middle of England. It beamed a 10-kW signal at 49.8 meters, essentially the same wavelength as in the initial example. Wilkins outlined an experiment using a bomber flying along the beam and measuring the reflected signal at a ground station. To eliminate reception from the main beam, two spaced antennas would be phase-adjusted to neutralize the direct signal but allow reception of the reflected signal coming from a different direction.

Daventry Van

Using receiving equipment in a van from the RRS, Wilkins set up the antennas in a field near Weedon, about six miles from the Daventry transmitter. On February 26, 1935, a Hadley Page Heyford aircraft made several runs along the beam, with reflected signals distinctly received each time. Watson Watt and several others from the RRS Slough observed the successful demonstration, as did Rowe, representing the Tizard Committee. This is commonly referred to as the Daventry Experiment.

Watson Watt was so impressed with the success that he declared, "Britain has become an island again!"

Upon receiving a report from Rowe on the demonstration at Daventry, the Air Ministry classified the project as Highly Secret and, with the consent of the ADR Committee, £12,300 in initial research funds were allocated in March 1935. One of the most important technical developments in the history of Great Britain then got underway.

The initial equipment would be developed at the RRS, Slough, then moved to Orfordness for testing. The expected signal from an aircraft was calculated to be about 19 orders of magnitude less than that associated with ionospheric reflections; thus, the transmitter, receiver, and antenna, collectively, must be greatly improved. Since the most sensitive, super-regenerative receiver available was already in use by the RRC, the most effort would need to be associated with the transmitter and antenna. Wilkins, assisted by Bainbridge-Bell, would handle the receiver and antennas, and advertisements were placed for a person to develop the transmitter. Near the end of April, Dr. Edward G. Bowen was hired for this very important assignment.

Edward George ("Taffy") Bowen (1911-1991) was born and raised in the village of Cockett near Swansea, Wales. (Taffy was a common nickname for Welch people.) His father was a steelworker and also the organist in the local Congregational Chapel. Bowen developed an early interest in radio, and this assisted him in obtaining a scholarship to study physics at Swansea University College. Upon receiving a First Class Honors degree in 1930, he continued with graduate research on X-rays, earning the M.Sc. degree the next year.

E. G. "Taffy" Bowen

Bowen then received a scholarship to study under Professor E. V. Appleton at King's College, London University. Much of 1932 and 1933 was spent at the Radio Research Station, Slough, working with the cathode-ray direction finder and becoming known to Watson Watt. When he completed his Ph.D. degree from London University, Bowen responded to the advertisement from the RRS and, after being interviewed by Watson Watt and Herd, was selected for the position of Junior Scientific Officer.

When he joined the RRS, Bowen was unaware of the development project but immediately realized it was something very unusual.

> I was introduced to the provisions of the Official Secrets Act and he [Jimmy Herd] explained the penalty for the slightest deviations . . . literally to be 'hanged by the neck until life was extinct.'

HISTORY MAKING AT ORFORDNESS

A preliminary system was designed and built at the RRS by the team, based on pulsed transmission as used for probing the ionosphere. The transmitter would operate at 6 MHz (50 meters), have a pulse-repetition rate (PRF) of 25 Hz, a pulse width of 25 μs, and a target power of 100 kW. Their existing RRS transmitters for ionospheric research had a peak power of about 1 kW, so the output would need to be increased 100 times. For this, Bowen obtained high-voltage transformers and rectifiers designed for X-ray equipment and two NT46 silica triode valves manufactured by the Admiralty's Signal School.

In mid-May, 1935, the team moved the equipment to Orfordness. Involved were Wilkins, Bowen, Bainbridge-Bell, and his technical assistant of many years, George Willis. Before their arrival, power and telephone lines had been run, and six 75-foot wooden towers were erected for the transmitting and receiving antennas. The peninsula had been under the administration of the government since before WWI.

Orfordness Peninsula

Several buildings, including a small power facility, from an earlier ordnance test operation of the Air Ministry had been reconditioned, but there were no living quarters.

Orfordness, often written as Orford Ness, is a narrow, 19-mile peninsula in Suffolk along the coast of the North Sea, and almost the closest point in England to Germany. Connected to the mainland at the north end, it is otherwise separated from the land by the River Aldeit. Located about 90 miles northeast of London, the Orfordness peninsula was then mainly a wilderness of salt and mud flats, widening at the center with a crude

Abandoned Buildings

airstrip, a few abandoned buildings, and a lighthouse. Terminal equipment of an underwater telegraph cable installed between England and Holland in 1853 had originally been housed in the lighthouse base.

Lighthouse

Crown & Castle Hotel

Orford, located on the mainland opposite the lighthouse, was the nearest town. RRS people found lodging there or at private homes in the surrounding areas while visitors usually stayed at the Crown and Castle Hotel. Access to the usable part of the peninsula was from Orford by powered boat, operated under contract with the Air Ministry. Transportation on the peninsula was by an ancient fire truck (left over from earlier operations) and a Ford T-Model automobile.

Within a few days, the receiver and cathode-ray unit was assembled by Wilkins and Bainbridge-Bell while Bowen assembled the transmitter. Joseph E. Airey, leader of the RRS workshop at Slough, accompanied by Alex J. Muir from his shop, came up to string the antennas. The transmission antenna was a center-fed dipole. The receiving antenna was of the type used in the RRS lightning-monitoring facility, except with crossed dipoles instead of loops. This antenna was set between four receiving towers, set some distance from the two transmitting towers. Watson Watt remained in Slough but came to Orfordness on weekends.

By the end of May, testing of the equipment began. The transmitter, initially with 5,000 volts on two NT46 output valves, produced about 25 kW power. Although Bowen had been warned not to exceed this power, he reasoned that pulsed operation would allow the valves to recover between pulses, so higher voltage and subsequent power was possible. The voltage was then increased to 10,000 volts, resulting in the planned power of near 100 kW. This generated arcing on the antenna insulators, and the arrangement had to be enlarged.

Still, no echoes from aircraft were observed. There were, however, impressive returns from the ionosphere, including some that had triple-bounced as much as 1,000 miles from continental Europe. Although disappointed with the lack of aircraft echoes, the team was generally pleased with the ionospheric returns – after all, this was the guise under which the Orfordness facility was operating. Bainbridge-Bell, however, became generally pessimistic concerning the project; Watson Watt then sent him back to Slough (only to later have him rejoin the team). Willis remained, and was joined by another technical assistant, Robert H. A. Carter.

Watson Watt invited the Tizard Committee to visit Orfordness on the weekend of June 15 and examine the equipment. Unknown to the team, he had also arranged for an aircraft to fly nearby, hoping to show the equipment in full operation. During the demonstration, Watson Watt was certain that he saw an echo signal, but this was not seen by anyone else. The Committee members were shown spectacular ionospheric displays, however, and left satisfied with the progress.

Unlike his usual routine, Watson Watt had not returned to Slough after the weekend but stayed at Orfordness for further tests. According to Bowen's notes, on Monday, June 17,

> No equipment changes had been made from the day before, but when the gear was run up a clear echo immediately appeared at 17 miles.... Good echoes were received as it moved up and down the coast. We were all wildly excited.

Watson Watt telephoned the Felixstowe Air Station and verified that it was one of their aircraft. He then asked that the plane be returned to the Orfordness area, and this was done. In Bowen's words, "We had a further hour of unadulterated pleasure as we followed [on the cathode-ray tube] his passage up and down the coast."

In another record of the event, this one dated June 20, Bowen stated the following:

> It is perhaps worthy of note in an historical survey of this kind that the first echoes were seen on that date by Watson Watt, Wilkins, and Bowen, the aircraft concerned being a Scapa flying boat flying at 17 miles.

This date was evidently entered in error; Watson Watt also recorded it as the earlier date.

It is historically correct that on June 17, 1935, radio-based detection and ranging was first demonstrated in Great Britain. Watson Watt, Wilkins, and Bowen are generally credited with initiating radar in this nation.

Development of the system continued over the next several months. Daily flights of an aircraft from the nearby RAF Martlesham Heath Airfield were arranged. The transmitter voltage was slowly increased still further without undue stress being shown on the NT46 valves, and a power of 200 kW was attained. On July 24, echoes were correctly interpreted as a formation of three aircraft at a distance of 40 miles.

In August 1935, Rowe coined the acronym RDF as a cover for the work, meaning Range and Direction Finding but suggesting the already well-known Radio Directing Finding.

Bowen, then 24 years old and single, has described working at Orfordness as follows:

They were idyllic times. In spite of the crude conditions under which we lived and the lack of amenities, for me it was one of the happiest periods of my life. It was a time of high achievement and very obvious progress, and we did not care a jot about the absence of refinement in our private lives. The work itself was enormously satisfying.

In weekend meetings at the Crown and Castle Hotel, Watson Watt, Wilkins, Bowen, and a series of official visitors discussed future plans, including distinguishing enemy and friendly aircraft, systems for searchlight control and gun-laying, directing fighter aircraft against night-time bombers, airborne systems for sea searching, and the important need for detecting low-flying aircraft. Also discussed was a chain of stations to protect the Thames Estuary.

A memorandum, dated September 9, 1935, was prepared by Watson Watt and submitted to the Tizard Committee. While describing progress on the work at Orfordness, the main portion outlined a proposal for a full coastal-defense system:

> A chain of stations with transmitters every 20 miles along the coast to be defended, and with receiving installations at each alternate station.

He discussed the effects of target height, height of antennas, wavelength, transmitter power, and receiver sensitivity on the possible detection ranges. The report concluded with the following:

> The provision of a suitably situated central research and development station, of large size and with ground space for a considerable number of mast and aerial systems, is a first and highly urgent necessity.

On their next visit in October, the Tizard Committee witnessed tracking at near 60-mile range of an aircraft at 15,000 feet altitude. By the end of the year, tracking at 80 miles was achieved. Here it is noted that at that time there was strife within the Committee. Controversial Oxford Professor Frederick A. Lindemann, who had been added in July, saw no value in the RDF; he strongly advocated infrared detection of aircraft and the use of long wires hanging from balloons as deterrants. Eventually, he was replaced by Appleton, and any reservations toward RDF went away.

A height-finding (elevation-angle) capability was added, using horizontal, spaced dipoles, a technique developed previously by Wilkins at Slough for determining the downward direction of incoming transatlantic signals. For bearing (azimuth) measurements, the existing crossed-dipole technique continued to be used, but there was some consideration of incorporating two widely spaced receivers and a graphical technique for angle estimation.

The 50-meter wavelength interfered with commercial long-distance radio, so the transmitter was first changed to 26 meters, then to 13 meters, and then to four selectable wavelengths between 10 and 13 meters (23 and 30 MHz). Later, the frequency range was increased to between 20 and 55 MHz.

Bowen suggested that using a much higher frequency would allow focusing the transmitter beam. Then, rather than floodlighting, there might be a rotating "RDF searchlight" to pick up targets as well as the RAF aircraft, but this was beyond technical feasibility at that time.

The improvements completed all of the technical promises for an air-warning system as given in Watson Watt's original proposal. Meanwhile, the proposal for a chain of RDF stations made its way through the hierarchy.

Near the end of September, the ADR Committee approved the development of an RDF chain of five transmitting stations with receiving stations between them. The system was given the code name "Chain Home." Construction of the acoustical mirror system originally intended for this protection was stopped, and on December 19, 1935, the Treasury appropriated £60,000 for a five-station chain covering approaches to the Thames Estuary.

BAWDSEY MANOR AND INITIAL SYSTEMS

With the new mission, an extensive enlargement of the staff and experimental facilities was needed. On one of his weekend visits, Watson Watt had driven to the tiny village of Bawdsey, on the coast some 10 miles south of Orfordness. There he saw a manor house on an estate that might be for sale. On his next visit, Watson Watt drove Wilkins and Bowen to the site, and they immediately recognized it as ideal for the operations.

The house and out buildings would be sufficient for laboratory and office space as well as some quarters. The 180 acres of grounds, mainly 70 or more feet above the sea, would be perfect for an RDF station. In short order, it was purchased by the Air Ministry.

Bawdsey Manor House

In March 1936, an enlarged RDF development team began occupying the newly formed Bawdsey Research Station. Laboratories were set up in the White

Tower and the Stable Block of the Bawdsey Manor House. Watson Watt was named the Superintendent, and moved from Slough to a flat in the Manor House. It was decided that Wilkins would be responsible for building the Chain Home (CH) stations and Bowen would lead in developing other new RDF equipment.

Airey, who had originally come up from Slough to build antennas, became a permanent employee at Bawdsey, responsible for facilities and research equipment. He remained with the operations for many years as names and locations changed. Other organizations attached senior personnel to coordinate relationships; these included E. J. C. Dixon from the Post Office, W. G. Eastwood from the Admiralty, and G. H. Barker from the War Department

Food and general supplies came from the larger, nearby village of Woodbridge. Batchelor quarters were available in the Manor House, as was food service for all the personnel. Some married men lived in cottages on the grounds, and others lived in the Felixstowe Township across the Deben Estuary to the south, commuting by a ferry to the Station.

Felixstowe-Bawdsey Ferry

Recruits were mainly drawn from physics and engineering departments of universities across the country, but some transferred from other government laboratories. Initially, all civilian personnel continued as employees of the National Physics Laboratory, but effective August 1936, they became Air Ministry employees.

Watson Watt operated Bawdsey much like a university research laboratory. Usually oblivious of Civil Service red tape, the personnel worked hard but the hours were flexible. It was not unusual to find a cricket game on the lawn or swimmers at the sea shore during the day, but the same people might still be at work well after midnight. Bowen noted, "Some of the best technical discussions took place late at night in a timbered hall in front of a roaring fire."

It is surprising, considering the high security, but volunteers were recruited from London to work weekends at Bawdsey. This is described by Jack Nissen in his book, *Winning the Radar War* (St. Martin's Press, 1987). Then a 17-year old service technician at EMI, Nissen had been persuaded by the RAF to join a small group for part-time, unpaid work of "utmost importance" – EMI itself had not yet been brought into the secret work. He recounted his first visit there as follows:

He [Joe Airey] proceeded to give us a pep talk on security, speaking in the tones of a sergeant-major . . . making us promise on oath not to discuss with anyone anything we saw at Bawdsey. We didn't have the slightest idea what he was talking about.

Initial Chain Home RDF System

The prototype Chain Home system had units built by a number of contractors but finally assembled at Bawdsey. The transmitter was based on one developed earlier by the Post Office for short-wave radio communications, but modified for pulsed operation by Metropolitan-Vickers. Designated TF3, in its output it used two NT57 valves, a newer silica triode built by the Experimental Department of the Admiralty's Signal School. For the initial transmitting antennas, two 240-foot wooden towers were built, with dipoles strung between them.

The receiver, designated RF5, was based on those used at Slough and modified by A. C. Cossor Company. The receiving antennas assembly had dipole arrays at two levels, each with two sets of dipoles set at right angles. The different heights were used for estimating target elevation angle and thus, through using range, the height. The orthogonal arrangement was for estimating target azimuth angle or bearing, using the relative strength of signals from the two directions.

There were a few false starts, including failure in an attempted demonstration of the untested system at which Air Marshal Dowding was present. When made ready, however, the system performed as expected, measuring range to 80 miles with quarter-mile accuracy and elevation angle to within 15 percent. Going back to operations at the RRC and Orfordness, it was known that the method of determining azimuth resulted in considerable errors.

The azimuth angle, of course, was very important in directing fighter aircraft toward the target. With fixed receiving antennas, these could not directly indicate the target bearing with the needed accuracy; therefore, an indirect method was required. The CH System was initially planned to use two separate receiving stations, one between each transmitting station, and a method called "range cutting." This method involved drawing arcs on a map centered at each receiving station and with the radius being the respective range measurement. The intersection of the arcs would indicate the target bearing.

As the CH design progressed, it was realized that an instrument called a radio goniometer, invented early in the century, would allow the bearing determination using a single receiver and without the time-consuming, error-prone drawing of arcs. This accepted the inputs from a pair of fixed, single-turn, loop antennas, set perpendicular to each other. Alternatively, a pair of dipoles could be used in place of the loops. The output of each fed a small "field" coil matched to the antenna position. A third "search" coil was free to rotate within them. When this coil was turned so that its electrical output was minimized, its angular position indicated the target bearing.

Incorporation of the goniometer was very important to the ultimate successful operation of the CH System. The identification and position calculation of many targets would have been impractical using range-cutting methods.

Goniometer Knob

While actually working on RDF, the guise of performing research on the ionosphere was maintained. Watson Watt, Banbridge-Bell, Wilkins, and Bowen published a paper, "The return of radio-waves from the middle atmosphere" (*Nature*, 1937), establishing stratified layers of ionization and describing the need for moving the research from Slough:

> Since a wide range of work in progress at the Radio Research Station, Slough, made it very undesirable to introduce a new high-power emitter there, the Air Ministry very kindly offered facilities at their Orfordness Research Laboratory.

An RAF group under Squadron Leader Raymund George Hart was attached to Bawdsey in July 1936. Hart was well qualified for this assignment. He was a physics graduate, had studied electrical engineering in France, and first became a pilot during WWI. The group drew support from the Air Armament Establishment at Martlesham Heath Airfield, about 10 miles to the east. Hart's primary purpose was to establish a school to instruct personnel who would operate the first CH stations. He and his group were fully integrated into the operations, calling the scientists and engineers at the station "boffins." (This was possibly an alteration of puffin, a bird that is both serious and playful at the same time.) The name stuck for RDF participants throughout the war.

The RAF first exercised the Bawdsey prototype CH station in September 1936. About 100 aircraft were involved; half were bombers simulating raids on Baudsey and half were fighters directed toward intercepts. The exercise was a disaster; the equipment failed to detect the bombers at more than 15 miles, insufficient to allow interception. Air Vice-Marshall Hugh Dowding observed the test, and concluded that the system had not yet proved itself in practice. Dowding was then shown an experimental 6.7-meter system being developed by Bowen, and this detected the planes out to 50 miles. Dowing then changed his position, saying that the CH prototype could eventually function satisfactory.

In another exercise by the RAF the following April, the system tracked a flying boat at up to 80 miles for 80 minutes, and a detection at 112 miles was made. The Air Staff then said that the technology was now proven, and recommended to the Air Ministry that planning for the CH System be enlarged to provide 20 stations, covering the east and south coasts of Great Britain, but with priority given to the stations already under development.

The CH work accelerated and became a joint operation with the RAF. Sites for the initial five CH stations were selected: Canewdon, Dover, Dunkirk, and High Street in Suffolk, plus one and a coordination center at the Bawdsey Research Station. Specifications for the sites and equipment were drawn up and submitted to the Air Ministry for review and approval. Concerning the sites, the Ministry added the requirement that they "should not gravely interfere with the grouse shooting." Only in Great Britain could there be such priorities!

For security, final design and manufacturing of CH equipment was partitioned between different firms: Metropolitan Vickers (Met-Vick) for transmitters, Marconi for antennas, and A. C. Cossor for receivers and displays. All were held to the highest levels of secrecy, with work performed in closely controlled areas and by vetted engineers and mechanics.

The following is a very general technical description of the CH equipment. From the start of design until the end of the war, the various pieces of apparatus were continuously undergoing change. Thus, these descriptions are, at best, a snapshot in time, primarily based on the specifications provided to the manufacturing contractors in early 1938 for the first stations.

Each station would have a transmitter operating on four pre-selected frequencies between 20 and 55 MHz, adjustable within 15 seconds, and delivering a peak power of 200 kW. The pulse duration would be adjustable between 5 to 25 µs, with a repetition rate selectable as either 25 or 50 Hz.

For synchronization of all CH transmitters, the pulse generator was locked to the 50 Hz of the British power grid. The duty cycle of pulses would be staggered between different stations so that they would not interfere with each other when operating on the same frequency.

Dr. John M. Dodds, head of the Radio Laboratory, and John H. Ludlow led the efforts at Met-Vick, and a complete description of the transmitter circuits is given in their paper, "The C.H. radiolocation transmitters," (*Journal of the Institute of Electrical Engineers*, 1946). Jim Brown, then an apprentice at Met-Vick, has provided an excellent description of transmitter manufacturing in his book, *Radar: How it all Began* (Janus Publishing, 1996).

The transmitting arrays would be supported between 360-foot steel towers, with platforms at 50-, 200-, and 350-foot levels. Two separate arrays, initially requiring four, later three, towers, allowed operation at different bands of frequencies. The arrays were a curtain arrangement of horizontal dipole and reflector elements strung between hanging feed and end lines. Mesh (chicken-wire) reflectors were positioned behind the arrays to prevent wasted energy and to eliminate confusing signals to the rear. The array was designed to produce a "floodlight" radiation pattern about 60 degrees in width. Patterns from adjacent stations overlapped, given relatively continuous coverage.

360-ft Steel Tower

The station receivers were based on the pulse-polarization analyzers originally used at the RRS, consisting of a receiver and a cathode-ray tube display. The input would be from a radio goniometer (previously described). Four 240-foot wooden towers would support cross-dipole arrays at three different levels. Antennas at 95- and 215-foot levels would be used for low-level targets, and those at 45 and 95 feet for aircraft at high elevations. The same goniometer would be used for estimating both elevation and azimuth angles, with the operator selecting appropriate pairs of arrays. This technique would lead to considerable errors in both bearing and height, but data from adjacent stations would be consolidated at the Filter Center to greatly improve the accuracy.

The receiver would cover 20 to 50 MHz in two bands. There would be four selectable frequencies that could be changed by a single operator in no more than 10 seconds. The receiver would operate satisfactorily with the transmitter as close as 600 feet. To isolate the overall receiver from direct reception from the transmitted pulse, a suppressor device, activated by the leading edge of the pulse, would control the gain of the RF amplifier stage. The cathode-ray sweep was also triggered by the transmitter pulse, and the received pulse from the target was shown on the trace, graduated in distance units.

John W. Jenkins led the receiver and CRT display effort at Cossor and provided a detailed paper on this activity: "The development of

C.H. type receivers for fixed and mobile working," (*Journal of the Institute of Electrical Engineers*, 1946).

The data from dials, goniometers, and the cathode-ray tube had to be transformed into three-dimensional coordinates for use for directing the fighter aircraft. In addition, the equipment required continuous calibration, and this had to be entered with the data. Determining the coordinates required trigonometric calculations. This was a time-consuming, error-prone function for the operator.

To simplify the process, G. A. Roberts at Bawdsey developed an electro-mechanical analog computer. Called the Fruit Machine (a British expression for a slot machine), it allowed the operators to enter data by pressing keys. The computer then performed the calculations, and provided a readable output of target coordinates.

As construction of the CH stations proceeded, the Tizard Committee placed another requirement on the system: the information from the different stations must be communicated using secure telephone lines to a central location for analysis and coordination, similar to what had been planned for the acoustic-mirror system. Called the Filter Center or Room, this was initially set up at the Biggin Hill RAF Airfield, then later moved to Fighter Command Headquarters at Bentley Priory RAF Station at Stanmore. At the request of the Air Ministry, the Post Office developed special switching equipment and secure lines.

Target data from two or more CH stations would be plotted on a single chart, with the intersections assumed to be the most accurate. The process was called plan position filtering; hence the name Filter Room. Data from the Filter Room would be passed on to the various fighter operations, then passed on to the appropriate pilots using a VHF communications link. P. A. Merchant and K. M. Heron of the Post Office Electrical Engineering Department would later develop an electrical calculator to obtain the final answer from the sorted information.

As a part of the Hart's RAF school program, all aspects of the CH operation and its interface with air operations were examined. Hart became a personal participant in planning how the CH System would interact with the Fighter Command. Several analysts, originally at Biggin Hill, moved to Bawdsey, forming the Operational Research unit (so named to distinguish it from hardware research work). Led by Harold Larnder, this team's work became the foundation of analysis techniques called Operational Research (Operations Research in America).

In the summer of 1937, equipment delivery and assembly started, and by October all 20 of the initial CH stations, as well as the Filter Room, were in check-out operation. Before the end of the year, another RAF exercise was performed, this time involving all of the initial CH system.

It was a resounding success. On the basis this, £10,000,000 – an unheard sum at that time – was appropriated in secrecy by the Treasury for an eventual full chain of coastal stations.

As 1938 began, the RAF took over control of the CH stations and also developed a network integrating observer stations and the Filter Room. A Fighter Command post was established at Biggin Hill to use high-frequency radio links for directing pilots to the vicinity of targets. Their initial efforts – attempting to discretely intercept airliners arriving from Europe – were largely unsuccessful. In mid-1938, however, the full system of five stations and the Filter Room proved its value; during the Home Defense exercises, ground controllers directed interceptors to their targets in three-quarters of the attempts. In September, when British Prime Minister Neville Chamberlain made his infamous trip to Munich and attempted to appease Adolf Hitler, CH stations tracked his aircraft for 100, miles both into and out of Germany.

Typical Chain Home Transmitter Room

Typical Chain Home Receiver Room

Typical CH Tower Arrangement

Watson Watt became Director of Communications Development at the Air Ministry in May 1938, and A. P. Rowe was then appointed to take his place as Superintendent of the Bawdsey Research Station. Although holding a degree in physics, Rowe had little expertise in radio and electronics. He had inherited the RDF founders, including Wilkins and Bowen, but, as expressed by Bowen, "never, in subsequent years, did he reach any kind of rapport with the pioneers."

A. P. Rowe

Albert Percival (A. P.) Rowe (1898-1976) was born and raised at Launceston in Cornwall, England. His father was an agent for a sewing-machine company. Rowe attended the Portsmouth Dockyard School and studied physics at the Royal College of Science, University of London, earning the B.Sc.Honors degree in 1922.

When Dr. H. E. Wimperis became the first Director of Scientific Research at the Air Ministry in 1925, he selected Rowe to serve as his Personal Assistant. Rowe also lectured at the Imperial College of Science and Technology. In 1934, Rowe became concerned over the potential for war, and studied all of the files on air defense. Most of the previous work was related to acoustic mirrors and barrage balloons. He concluded that very little true scientific research had been carried out.

Based on his findings, Rowe had written a report to Wimperis stating, "Unless science could evolve some new method of aiding air defence, we were likely to lose the war if it starts within the next ten years." This led Wimperis to recommend that a high-level group be formed to examine the potential role of science in war preparation. The Air Ministry then formed the Committee for Scientific Survey of Air Defense with Sir Henry Tizard as the Chairman. Wimperis was named a member and brought Rowe with him as Committee Secretary.

Throughout the subsequent activities at Slough, Orfordness, and Bawdsey, Rowe served as the primary contact on the Tizard Committee. Thus, it was natural that he was eventually appointed to lead the RDF development.

Under Rowe, the overall operation at Bawdsey was significantly changed. While Watson Watt had led it in a "university" type environment, Rowe, being primarily an organizational administrator, quickly made changes. Firm lines of authority were established and strict procedures were imposed. However, Rowe's authoritarian manner was compensated by a strong belief in the free exchange of ideas. This he encouraged in regular meetings at Bawdsey, called the "Sunday Soviets," involving civilian and

Typical "Sunday Soviet"

military personnel of all levels and functions. Rowe has described the origin of this name, as well as some of the sessions, in his book, *One Story of Radar* (Cambridge U. Press, 1948).

One of the Sunday-Soviets topics concerned possible techniques for CH target identification. In Watson Watt's memorandum of February 27, 1935, he had noted the need to identify friendly targets. A number of target-identifying methods had been examined as the CH system progressed, but none adopted. Hart's RAF personnel and Lerner's OR group were very concerned that the CH system could not discriminate targets. Finally, in late 1938, Wilkins and Carter started the development of an aircraft-carried transponder for this purpose.

Called Identification Friend or Foe (IFF) and sometimes "Parrot," the transponder was a receiver that continuously swept through all frequencies used by the CH transmitters. When an interrogating pulse reached an IFF antenna – at one or more points on the aircraft – oscillations in the receiver would increase and serve as a small transmitter. This would show on the receiver at the interrogating CH station as a brief heightening of the returned signal, indicating the target as "friendly." The sets went into production by Ferranti as IFF Mk I, and installation on aircraft began in November 1939. One thousand of these sets were delivered in time for the Battle of Britain in mid-1940.

When Chain Home Low (CHL, described later) was identified to be added to the CH stations, this operated at higher frequencies that also needed to be covered. In August 1939, Dr. J. Rennie Whitehead initiated the development of another version of an IFF set, eventually designated as Mk II, covering both CH and CHL frequencies. Because of the importance of IFF, considerable ground and air testing was conducted, especially in selecting antenna locations on the many types of RAF aircraft.

Recognizing the inevitability of war, the British Government, in February 1939, sent messages to heads of state of several Commonwealth nations asking that they send representatives to England to be given information on a "most secret device" being developed for the detection of aircraft. This invitation was sent to Australia, Canada, New Zealand, and South Africa.

In response, Dr. David F. Martyn from Australia, Dr. John T. Henderson from Canada, and Dr. Ernest Marsden from New Zealand went to England for some time, being briefed on every aspect of RDF and examining the activities at Bawdsey. For some unknown reason, South Africa was not represented, but Marsden, as he was returning to New Zealand, stopped by South Africa and briefed Dr. Basil F. J. Schonland.

As described in Chapter 8, development of RDF systems was soon started in all four of these nations.

Airborne RDF Systems

The initial CH stations, with equipment eventually designated as AMES Type 1 (Air Ministry Experimental Station Type 1), had serious weaknesses: they had only outward-looking capability – bombers could not be tracked after crossing the coast – and. because their sensitivity greatly decreased at look-angles less than about two degrees, low-flying aircraft could escape detection. While Wilkins was leading the development of Chain Home, other groups at Bawdsey were pursuing possible corrections to these shortcomings, as well as addressing other areas of RDF.

Bowen was particularly concerned with the problem of directing fighter aircraft to the enemy aircraft when the final path was obscured by clouds or in night interception. For this, the pilot must come much closer to the target than could be directed by the ground-based CH equipment. An RDF system carried on the fighter aircraft was urgently needed.

The equipment being developed for CH was obviously too big and heavy for operating within an aircraft, and it also required antennas that were far too large. In mid-1936, Watson Watt assigned Bowen part-time to examine the prospects of an airborne RDF system. He first had discussions with engineers at the nearby Aeroplane and Air Armament Establishment; these people having been let into the secrecy for supporting Hart's group at Bawdsey. From them he obtained weight, size, power, and aerodynamic limitations. Based on this and other discussions with the Air Ministry, Bowen assumed that an airborne RDF system would be limited to 200 pounds and a volume of 8 cubic feet, must require no more than 500 watts of power, and have antennas about 18 inches or less in length.

These appeared to be impossible limitations, even for a system having an operating range of only 1,000 feet. Nevertheless, he began the design of such equipment. Working back from the antenna size, the operating wavelength would need to approach 1 meter (300 MHz), far shorter than in any ordinary equipment of that day. Hardware being designed for experimental television, then emerging in Great Britain, had the greatest promise.

EMI had an outstanding staff of scientists and engineers establishing a television base for the firm. They had just developed a receiver for 45 MHz (6.7 meters) and with a bandwidth of 1 MHz on a 3" x 18" chassis. EMI was approached about purchasing one of these units, but declined, fearing that their proprietary design would be stolen. Also, they were fully engaged in developing items for the coming television market, and did not want to get involved with government work.

Somehow, Bowen was able to obtain one of the EMI units, and this formed the heart of his airborne system. After adding a small cathode-ray tube with high-voltage supplied from an automobile ignition coil, the weight was only 20 pounds and the volume less than 2 cubic feet. This was a step toward solving the receiver problem, but the wavelength was still far too long.

Laboratory space was assigned in two towers (called Red and White) of the Bawdsey Manor House, and Bowen began assembling a team. Initially included were Dr. A. Gerald Touch, a physics researcher recruited from the Clarendon Laboratory of Oxford University, and Keith A. Wood, a talented technical assistant found in the local area. Both men continued with airborne RDF development throughout the war

It was decided to proceed by first developing a transmitter for operation on the ground, then testing the receiver in an aircraft. A 6.7-meter, pulsed transmitter was built, following the basic design of the unit at Orfordness. The power approached 40 kW, and the pulse width was near 3 µs. The receiver and transmitter were set up in the separate laboratory spaces, with dipole antennas on the roofs of the two facilities. This experimental system worked from the start, and by September 1936, it could detect aircraft at ranges up to 50 miles. It was this equipment that had "saved the day" during the first exercise by the RAF of the CH prototype system.

For security purposes, particularly in reports, Watson Watt called the CH the RDF-1 and the future airborne system RDF-2. The intermediate experimental system, with transmitter on the ground and an airborne receiver, was appropriately designated RDF-1.5. In late 1936, testing of the RDF-1.5 system began. The receiver was taken to Martlesham Heath and placed in a Hadley Page Heyford aircraft. Power was provided by dry batteries, and a half-wave (about 10 feet) dipole antenna was strung between the aircraft's wheels. The transmitter remained at Bawdsey. On the first test, flying over the transmitter at about 3,000 feet and with Bowen operating the receiver, echoes were obtain from aircraft at ranges up to 10 miles.

Hadley Page Heyford

Progress toward the airborne transmitter accelerated with the acquisition of Western Electric 316A valves, commonly called Doorknobs.

These had a potential of operating up to the desired 300 MHz. Percy Hibberd, a recently added engineer, built a small transmitter using a 315A, initially operating at 45 MHz to match the receiver. It had a pulse rate of 1,000 Hz, a pulse width of 2 to 3 µs, and an output power of a few hundred watts. The transmitter was added to the Heyford aircraft, with electrical power provided by a hodge-podge of different types of batteries, an alternator driven by the aircraft 6-volt system, and a Model-T spark coil for the cathode-ray high voltage. With Bowen and Touch aboard, there was success on the first flights. Even with the low transmitter power, there were clear signals from objects on the ground and from what was believed to be ships off the coast. The first British airborne RDF system had been flown in March 1937.

Bowen Testing First Airborne RDF Set

The next step was to reduce the operating wavelength, with the objective of 1 meter. Bowen and Hibbert worked on the transmitter while Touch addressed the receiver. By using two 315A valves, the output power increased to about a kilowatt. The pulse width was reduced to 1 µs and the operating wavelength to 1.25 meters (240 MHz), about the best that could be achieved with these valves. This gave the advantage of requiring half-wave dipole antennas to be only about 2 feet in length.

The receiver was also improved to 240 MHz by adding an Acorn tube from RCA in a frequency converter stage on the input. The 45 MHz original receiver then became the intermediate-frequency (IF) section of a super-heterodyne circuit. An IF of 45 MHz remained the standard of RDF receivers for many years.

For prime power, an engine-driven alternator was developed, giving a stabilized output of 250 volts at 50 Hz. A separate power converter supplied the various DC voltages needed by the transmitter and receiver. The adoption of the alternator also had the advantage that the RDF equipment could be run from the standard supply mains when on the ground. Addition of the alternator, however, required a change to an Avro Anson aircraft, which had more available engine power. Later, Met-Vick changed the alternator to supply 80 volts at 1,000 Hz.

Two Avro Anson aircraft with ground crews were assigned for further testing. While the Heyford was not certified to fly past the coast, the Ansons were Coastal Command aircraft and testing could be performed over either land or sea. The first test flight of the new 1.25-meter airborne RDF-2 system was made on August 17, 1937. Touch and

Wood operated the equipment and obtained clear echoes from ships at ranges up to 3 miles. Shortly thereafter, it was found that increasing the wavelength slightly to 1.5 meters (200 MHz) gave significant improvements. Thus, 200 MHz was adopted as the standard for airborne RDF systems.

Avro Anson

An RAF exercise was scheduled to start on September 5, 1937, in which aircraft of the Coastal Command would search for units of the British Fleet in the North Sea. It was an ideal opportunity to test the airborne equipment in detecting ships, and Watson Watt made arrangements for Bowen and Wood to participate as observers in one of their Ansons. On the first day, flying at 3,000 feet and under conditions of very low visibility, they easily found the cruiser HMS *Southhampton* together with the aircraft carrier HMS *Courageous* at a range of about six miles. Another first occurred at this time: aircraft took off from the carrier and, with the sea in the background, were observed at a range of about one mile by the airborne system.

The RDF-2 had, in general, proven itself for both intercept and sea-search applications. The equipment would need to be configured differently for the two uses, and different small groups were assigned for this effort. Starting at the end of 1937, the two projects were known by acronyms: AI (for Air Interception) and ASV (for Air-to-Surface Vessel). Touch and Wood continued with AI, and two new members, Robert Hanbury Brown and Brian White, were principals on the ASV team. Hanbury Brown was a brilliant young physicist from London University who had been recruited by Tizard for the original work at Orfordness. White was another engineering find from the local area.

Work continued on the basic RDF-2 equipment during 1938. The most significant change was the replacement of Doorknob valves in the transmitter with another Western Electric product – Type 3204. This increased the peak power to about 2 kW, resulting in a detection range for ships up to 15 miles. The transmitters had not yet been engineered; they were essentially a wooden board with two exposed power valves. There was still only a single receiver, and it was exchanged between systems as needed for experiments.

In early 1939, the AI system was configured to separately measure azimuth and elevation target directions. For this, four receiving antennas were mounted on the aircraft, providing four overlapping lobes, two for azimuth and two for elevation. This arrangement followed the basic design already incorporated by the German firm Lorenz in their widely used radio landing system. A rapidly rotating, four-way RF switch was built for selecting the antenna beam entering the receiver. Following discussions with the users, two CRTs were incorporated on the AI, one for azimuth and one for elevation, both as functions of range. These would be in an operator's position; the pilot would have a single CRT.

Expermental AI Antennas

In May 1939, the experimental AI system was designated Mk I. Six sets were needed for testing, and Metropolitan-Vickers and A. C. Cossor – both already involved with the CH system – were the logical choices for manufacturing. Met-Vick readily contracted for the transmitters, although they were entirely different from those that they were building, and A. C. Cossor agreed to build the receivers. Since there was only one EMI 45-MHz chassis, this could not be provided for copying, and Cossor worked from written performance specifications. When they delivered the prototype receiver, it was totally unsuitable – too large, too heavy, and with only a small fraction of the desired sensitivity.

In search of another source, Bowen turned to Appleton, his doctoral mentor and now a member of the Tizard Committee. Appleton recommended W. G. Pye Company of Cambridge. Upon visiting Pye, Bowen found that they, like EMI, had built a number of 45-MHz receivers in anticipation of the television market. Pye was already involved with Army RDF work (described later) and agreed to take on building the airborne receivers.

Pye 45-MHz TV Strip

The initial AI system had fundamental problems associated with both longest and shortest ranges. The 1.5-meter transmitter drove a dipole antenna with a reflector on the nose of the aircraft, radiating in a floodlight-type fashion. Reflected signals were then also received from the ground, masking the airborne target signals. Thus, the maximum range was limited to the same height of the aircraft – as little as a few thousand and seldom more than 15,000 feet. The adoption of multiple receiving antennas with narrow beams reduced, but did not eliminate, this maximum range problem.

The minimum range for the AI system came from characteristics of the transmitter pulse. The receiver was made inactive for the pulse duration; thus, the pulse shape also affected the off and on time. With the width and shape of the pulse as initially used, the minimum range was limited to about 1,000 feet.

A first revised version of the AI was flown in a Fairey Battle aircraft in June 1939. The maximum range was 12,000 feet and the minimum about 1,000 feet. In mock interceptions with a Hadley-Page Harrow aircraft, the minimum range seemed to be satisfactory. As in the Mk I, the operator's display used two CRTs for displaying combinations of range, azimuth, and elevation, and was easy to use.

In early August 1939, a successful demonstration was given to Air Marshal Dowding, and the system was then designated AI Mk II. Met-Vick and Pye were given contracts for building sets to be fitted into 30 Bristol Blenheim night-fighter aircraft for training purposes.

Based on nighttime tests by the Royal Aircraft Establishment, 1,000 feet appeared to be a satisfactory minimum range; Larnder, however, in one of the first OR studies, insisted that it must be nearer 500 feet. A considerable debate on this followed, between Bowen and Rowe (now the Bawdsey Superintendent).

The ASV system, evolving under Bowen along with the AI system, had all of the same basic improvements. The ASV would be used in two types of missions: sea search and homing. For searching, there was consideration of the ASV as a side-looking system. For experimental purposes, both transmitting and receiving antennas were placed on one side of an Anson, and flight tests over the English Channel were made in mid-May 1938. These clearly showed several vessels and a map of the Isle of Wight.

In examining various configurations for the ASV, Bowen considered using a rotating antenna, allowing a sweep over an angle from the front or side of the aircraft. An experimental unit was built using a moving dipole antenna with the output passing through a slip joint. A new type CRT display was developed, giving a pie-shaped presentation (what came to be known as a "B" scope). The obtained range was very small, likely due to losses at the slip joint, and the project was abandoned. This was, however, the first rotating antenna used in a British aircraft.

It was eventually decided that a single antenna configuration would be used for both types of missions. Since homing required a forward-looking system, the transmitting antenna was placed in the aircraft nose, and receiving antennas were arranged on both sides of the fuselage.

Elevation angle was not applicable (all targets were on the surface, and height of the aircraft came from the altimeter), so only target azimuth needed to be determined and only one CRT was necessary. The receiving antennas were split into two lobes for determining azimuth,

In all of the developments of airborne RDFs, the potential benefits of microwave sets had always been in the background. During 1939, Bowen's group made an attempt to examine this with the development of a 30-cm (1.0-GHz) pulsed RDF set with 2B250 valves from Western Electric. In August, a transmitter and receiver were completed, and tests were run using two-foot parabolic dishes (limited by the requirement to fit in the nose of an aircraft). The results were disappointing; the range was much less than that obtained with the then-existing AI sets, and the project was discontinued.

The Start of Army RDF

From the initiation of RDF work at Orfordness, the Air Ministry kept the British Army and the Royal Navy generally informed. For details, however, there had to be personal contacts. In the summer of 1935, Colonel Peter Worlledge of the Royal Engineer and Signals Board met with Watson Watt and was briefed on the RDF equipment and techniques. He immediately recognized that this could be of great value to the Army. His report, "The Proposed Method of Aeroplane Detection and Its Prospects," led the Signals Experimental Establishment (SEE) to become involved. (The SEE was at that time located at Woolwich Commons in London.)

Recognizing the advantage of being closely associated with existing RDF work, the SEE set up an "Army Cell" at the Bawdsey Research Center in October 1936. While generally under Watson Watt as the Station's Superintendent, the Cell was led by Dr. E. Talbot Paris and his assistant, Dr. Albert B. Wood. The staff included W. A. S. Butement and P. E. Pollard who, as previously described, had demonstrated a radio-based detection apparatus at the SEE in 1931.

Wood, who was on loan from the Admiralty, was an expert in acoustics and has been working at the Army's Air Defense Experimental Establishment (ADEE). He had also done considerable research in polarization of electromagnetic waves, and had pointed out that RDF signals should be polarized based on intended targets – horizontal polarization for targets with dominant length features (such as aircraft and ships) and vertical polarization for targets with tall features (such as towers and high buildings).

Working in a new facility on the Bawdsey Manor grounds, the Army Cell's work would emphasize two general types of RDF equipment: gun laying (GL) systems for assisting anti-aircraft guns and searchlights, and coastal defense (CD) systems for directing coastal artillery and defense of Army bases overseas. The Army initially called these activities "Cuckoo."

The Cell's initial work centered on a project led by Pollard. This was a GL system that also had an early-warning function. (At that time, "early warning" was the minute or so before the slow-moving bombers were in hearing and/or visible range, and "gun laying" was information for aiming seachlights and anti-aircraft guns on the target.) This system was code-named Mobile Radio Unit (MRU), the same name used for a truck-mounted communications set developed by the SEE during World War I, and was designed as a smaller, mobile version of a CH station.

The MRU operated at 13 meters (23 MHz) with a power of 300 kW. A single 105-foot tower supported a transmitting antenna, as well as two receiving antennas set orthogonally for estimating the signal bearing. In February 1937, a developmental unit detected an aircraft at 60 miles range.

MRU Truck

While the original MRU was being tested, development of an improved version was started. Code named the Transportable Radio Unit (TRU) and officially designated GL Mk I, the wavelength was changed to 5 m (60 MHz) to reduce the antenna size. The 50-kW transmitter was fitted into a large enclosure on a 3-ton truck and the receiver and display were on a separate trailer. A third vehicle carried a power generator. In October 1937, a prototype unit began field testing in Kent, and the first full target-tracking was achieved the following January.

Production on 400 deployable GL Mk I sets was started in June 1938 by Metropolitan-Vickers and A. C. Cossor. The first overseas system was sent to Malta in March 1939; others went into service at El Dhaba in Egypt in early 1940. Seventeen sets were sent to France with the British Expeditionary Force, but were destroyed at the Dunkirk evacuation.

The Air Ministry also adopted this system as a mobile auxiliary to the CH system. Further responsibility for units with this application was transferred to the Air Ministry in early 1938, and, as such, it was designated as AMES Type 9 Mk 1; Donald H. Preist was the project leader. It was also recognized that it could be used as a gap-filler and emergency substitute for the CH stations. In this application, it was designated AMES Type 9(T), and the first units were installed in the fall of 1939. In August, a unit temporarily placed outside Dundee, Scotland, spotted the *Graf Zeppelin* at a range of 90 miles on a reconnaissance flight.

In an improvement, the wavelength of the AMES Type 9 equipment (but not the GL Mk I system) was changed to 7.5 meters (40 MHz). Rotatable Yagi antennas, one each for transmitting and receiving, were supported one above the other on 65-foot field masts. The rotating antennas gave the system an "all-round looking" capability and also allowed tracking of targets. Earlier problems in transferring the signals from rotating antennas were solved using a device originally developed for loop antennas on radio direction finders. Designated AMES 9 Mk 2, this was the first British RDF system using rotating Yagi antennas.

During 1937, with Bowen's encouragement, Butement did research in UHF technology; like others, however, he found the lack of power from existing sources severely limited any applications. In early 1938, he began the development of a CD system based on the evolving 1.5-meter (200-MHz) AI/ASV sets. The transmitter had a 400-Hz pulse rate, a 2-μs pulse width, and 50-kW power (later increased to 150 kW). Although many of the transmitter and receiver AI/ASV components were used, the system would not be airborne so there were no limitations on antenna size.

After experimenting with Yagi and other arrays, a large dipole array, 10-feet high and 24-feet wide, was developed, giving much narrower beams and higher gain. The broadside array could be rotated at a speed around 1.5 revolutions per minute, either clockwise or counter-clockwise. Primary credit for introducing beamed RDF systems in Great Britain must be given to Butement. As a part of this development, Butement formulated the first – at least in Great Britain – mathematical relationship that would later become well known as the "radar range equation."

Alan Butement

William Alan Stewart Butement (1904-1990) was born and raised in Masterton, New Zealand, and went to England to study at the University College, London. Here he attended lectures by E. V. Appleton and received the B.Sc. degree in 1924. He was employed as a Scientific Officer with the Signals Experimental Establishment at Commons, London. At the end of WWII, Butement migrated to Australia, where he became the first Chief Scientist for Defense. He was awarded the D.Sc. degree from Adelaide University in 1960.

To improve the directional accuracy, lobe-switching was used in the transmitting array. In lobe-switching, two dipoles with slightly different orientations were placed at each array position, with the input to one given a slight time delay. A rapid, motor-driven switch alternated the power between the dipoles. With the persistence of the CRT screen, the two reflected signals showed up simultaneously as two distinct peaks. The relative heights indicated the aiming of the overall array; when they were equal, the aiming was directly toward the target. This technique, originally developed for air navigation, gave excellent angular accuracy.

With its narrow beam and lobe-switching, by May 1939 the CD Mk II RDF could detect aircraft flying as low as 500 feet and at a range of 25 miles. With an antenna 60 feet above sea level, it could determine the range of a 2,000-ton ship at 24 miles and with an angular accuracy of as little as a quarter of a degree. Orders were placed in June with Met-Vick and Cossor for 60 CD Mk II systems.

When the Air Ministry operations at Bawdsey were relocated to Dundee, Scotland, in early September 1939, the Army Cell joined with the Air Defense Experimental Establishment (ADEE), then at Biggen Hill, in moving to Christchurch, in Dorset, on the south coast of England.

Academic Connections

As the war became imminent and research work at Bawdsey enlarged, more highly qualified personnel were needed. With Tizard's assistance in setting up the meetings, Watson Watt visited Oxford, Cambridge, and other universities in recruiting attempts. While this was generally successful, the most significant accomplishment was obtaining the services Dr. Ernst Rutherford, 1908 Nobel Laureate, in soliciting leading researchers to become active in RDF development.

In the spring of 1939, a number of the leaders from the Cavendish Laboratory at Cambridge visited Bawdsey and were introduced to the RDF secrets. Following this, arrangements were made for about 80 of the top physicists from a variety of universities to visit Bawdsey, CH stations, and related government facilities for a month. There they would examine the existing work, particularly areas where they already had interest and expertise, and then return to their laboratories to engage in related research.

Wilfred B. Lewis (1908-1987), a physics graduate from Oxford University and a leading electronics specialist at Cavendish, joined the Bawdsey staff in July 1939 to coordinate this effort, as well as serve as Assistant Superintendent for Research. From

Wilfred Lewis

that point on, Lewis would have considerable influence on the Air Ministry's RDF work

This "learning" activity got underway on August 14, but Great Britain declared war on Germany two weeks later and, as described below, all of the Bawdsey research was relocated to safer points. However, this activity with university researchers continued. Many were formed into small teams and assigned to existing CH stations to learn the equipment and contribute to improvements. From then and throughout the war, university laboratories and members of their staffs attached to the government operations were a continual source of new ideas and technologies, as well as advanced engineering development for "crash" hardware projects.

Dr. A. C. Bernard Lovell, then of Manchester University, who would later lead one of the Air Ministry's most important airborne radar developments, was among these "visiting" physicists. In *Echoes of War: The Story of H2S Radar*, Lovell describes his entry into this activity:

> In August 1939, like many other young people pursuing academic research in universities, I was suddenly plunged into an entirely different kind of research and development. The three days that we spent at Bawdsey Manor revealed to me a different order of science and technology compared with that of my university experience. Even more electrifying were the visits that followed to the headquarters of the Fighter Command.

The ties of RDF development to physicists and physics laboratories came primarily because, in those times, it was necessary for persons engaged in experimental research to design, and often build, their own equipment. Consequently, most physicists had a good knowledge of electronics. Professor Edward Appleton was an excellent example. Educated in physics, his experimental work, for which he received the Nobel Prize in Physics, centered on problems in atmospheric science. Nevertheless, he was considered a high authority in radio electronics and was voted Vice President of the Institute of Radio Engineers.

The emphasis on physics and physicists, however, did have drawbacks in the development of RDF. Electrical engineers had the best knowledge of electronic components and commercial manufacturing practices in radio. The absence of such inputs was often noted as contributing to delays in the ultimate fabrication and installation of RDF

The lack of a high level of engineering on early equipment, however, was also likely the result of Watson Watt's instructional rule:

> Give them the third best to go on; the second best comes too late, and the best never comes.

Some time later, when examining what he felt was a very badly engineered piece of equipment, Hanbury Brown added to this:

> But don't give them the fourth best because it encourages them to throw the whole thing out!

THE ROYAL NAVY GOES IT ALONE

Like the War Office (Army), the Admiralty (Navy) had followed the Air Ministry's RDF developments from the start. In July 1935, Charles S. Wright, the Admiralty's Director of Scientific Research, and George Shearing, the Chief Scientist at the Experimental Department of His Majesty's Signal School (HMSS), were invited to visit Orfordness. (Shearing's department had provided the NT46 valves used in the Orfordness transmitter.) On July 24, they witnessed the tracking of a three-aircraft formation at a distance of 40 miles.

The Experimental Department, formed in 1919, was located at the Naval Barracks in Portsmouth. As a military organization, it was led by a naval officer designated the Experimental Commander, but most of the staff of engineers and scientists were civilian employees. This operation had a recognized history of research and development in direction-finding systems and high-frequency communications, perhaps leading the world in this area during the early 1930s.

Following published accomplishments from the commercial and government laboratories in other countries, the Department had initiated research in the generation and applications of microwaves. In 1928, Leonard S. B. Alder had prepared a report, "Methods and Means for Determining Positions, Directions, or Distances of Objects by Wireless Waves." Unfortunately, this concept was never reduced to practice.

The Department was well known for valve (vacuum tube) development, led by Robert W. Sutton; it was from this organization that Bowen had obtained the NT46 silica triodes used in his first RDF transmitter. With this background and having excellent laboratories at Portsmouth, it was logical that the Admiralty initiated its own RDF development at the HMSS Experimental Department.

Portsmouth, in Hampshire on the south coast of England, had been the major location of the Royal Navy for centuries. It is on the east side of the Portsmouth Harbor entrance, with the Solent estuary and the Isle of Wight to the south. The city had a population of about 200,000 in the late 1930s, a large portion being employed as civilians supporting the various Navy activities and as workers in the nearby shipyards.

The Admiralty's decision was a great disappointment to Watson Watt; he had wanted the Navy to join in having a central RDF research operation. The Admiralty, however, justified this through the argument that the designers must understand the special conditions of radio operations at sea – effects of gunfire, interaction with shipboard electronics, and interference from vessel structures, all of which they had already encountered in developing and operating their high-frequency direction-finders and communications radios. Although Watson Watt attempted to keep funding from going to the Navy's effort, an appropriation was made and HMSS started the work in September 1935.

Scientific officer R. F. Yeo was assigned to lead the new effort. He first went to Orfordness for six weeks to become familiar with RDF development. Then, during the following 18 months, he led a small team experimenting with pulse transmitters and receivers operating between 75 MHz (4 m) and 1.2 GHz (25 cm). Researcher O. A. Ratsey attempted to develop new types of VHF generators, and some testing was done by E. M. Collin and C. F. Bareford using a low-power magnetron. All of the work was under the utmost secrecy; it could not be discussed with other scientists and engineers at Portsmouth, and for some time there was almost no further contact with the RDF staff at Bawdsey.

A 75-MHz range-only set, this bring believed to be the lowest frequency for which antennas could be fitted to the masthead, was eventually developed and designated Type 79X. Tests were done using a training ship, but the activity had to be kept hidden even from the crew.

In August 1937, Captain Arthur J. I. Murray was appointed to command HM Signal School, and the conditions for RDF development radically changed. The work was placed under Cecil E. Horton, who had earlier led the highly successful direction-finding and high-frequency research. Horton opened the discussions with other members of the HMSS staff and also brought in consultants from universities and commercial firms. A field station in Nutbourne, 12 miles to the east, was set up under Alfred W. Ross to conduct antenna research.

John D. S. Rawlinson (1900-1990), a highly respected, long-time member of the staff, was made responsible for improving the Type 79X. To increase the efficiency, he decreased the frequency to 43 MHz (7 m). This was in the frequency band used in BBC's television transmissions; thus, following Bowen, commercial firms were turned to for receivers. Designated Type 79Y, it had separate, stationary transmitting and receiving antennas. Shore-based testing of this, and most future systems, was conducted at Southsea Castle, an area south of Portsmouth overlooking the Solent.

Prototypes of the Type 79Y air-warning system were successfully tested at sea in early 1938. The detection range on aircraft was between 30 and 50 miles, depending on height. The systems were then placed into service in August on the cruiser HMS *Sheffield* and in October on the battleship HMS *Rodney*. These were the first vessels in the Royal Navy with RDF systems.

In 1939, the Type 79Y was improved to become the Type 79Z. This contained a new type valve and had an increased power. Forty of these systems were placed into production in August 1939, with the designation simply Type 79, still operating at 43 MHz (7 m).

Type 79 Antennas

At the start of the war in September 1939, a Type 79 system was installed at Fort Wallington at the north end of Portsmouth Harbor. The first German raid on Portsmouth occurred on July 11, 1940, and this RDF system gave a warning of the attack.

During 1938, the Type 79 system was further improved and designated Type 279, a gun-control/air-warning system. This had two large antennas that rotated in synchronization on different masts. Intended for heavier vessels (battleships, cruisers, and carriers), the beam coverage was large and was almost useless against low-flying aircraft; nevertheless, it was put into production in early 1939. In 1940, it was modified to become Type 279M, using a single antenna.

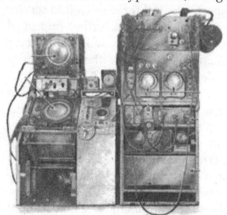

Type 281 Receiver & Display

For smaller cruisers, particularly those used as anti-aircraft ships, the Army's GL Mk I system was modified to operate at 85 MHz (3.5 m), thus reducing the antenna size. Designated Type 280, it had a very low power and was used only on an interim basis.

Type 281 was then developed for cruisers and battleships. Based on the Type 280, it also operated at 85 MHz but had a peak power of 1,000 kW at a pulse width of 2.5 μs or 350 kW at 15 μs. This was tested in 1940 on the light cruiser HMS *Didi*, and then first placed into service early the next year on the battleship

HMS *Prince of Wales*. The Type 281 sets were likely the most widely used RDF systems of the Royal Navy throughout the war.

When first used in combat on March 27, 1941, a Type 281 on the battleship HMS *Valiant* found the damaged Italian cruiser R.M.I. *Pola* at night off the coast of Greece. Unnoticed, the British were able to open fire, destroying the R.M.I. *Pola*, as well as two attending cruisers. In the ensuing battle, a force of British and Australian warships essentially annihilated the *Regia Marina Italiana* (Royal Italian Navy).

While the initial systems were reasonably satisfactory for sea search, air search needed a narrower beam. With limitations on antenna size, this required a higher frequency. For this, in August 1940 an ASV from the Air Ministry was modified for testing as Type 86, operating at 1.5 m (200 MHz). Then designated as Type 286, it used fixed antennas, one for transmitting and two for receiving, skewed in opposite directions.

By comparing the strength from the two receiving antennas, some bearing information could be obtained; however, for scanning, the ship itself was required to move around. The range was about 10 miles for large ships and 3.5 miles for surfaced submarines. Type 286W was the submarine version, with antennas mounted on frames on each side of the periscope.

The Type 286 sets were improved in mid-1941 with hand-driven steerable antennas and designated Type 286P. Somewhat later, Type 291M came out with a single power-driven antenna and a PPI display. There were also Type 291U and 291W variants; these were used on coastal vessels and submarines, respectively. All three went into service in late 1941, but were often replaced by Type 267, a further improved series of 1.5-m sets. (Note: sets were usually not sequentially numbered.)

John Coales

Simultaneously with the previously described developments, John F. Coales (1908-1999) was pursuing systems at higher frequencies. Coales was another long-time member of the research staff, having earlier worked in their direction-finding projects. During 1937, he developed a 25-cm (1.2 GHz) set using a klystron-like oscillator built by their valve laboratory. Near the end of the year, trials of the equipment were made, and it was concluded that with limited power then available at these frequencies, such sets could not meet the range requirements.

In early 1938, Coales changed to developing a 50-cm (600-MHz) system. H. C. Calpine was responsible for the transmitter and modulator.

This used a Western Electric 316A Doorknob tube, providing 25 kW output. C. F. Bareford developed the receiver and an A-scan display. Coales, assisted by W. F. Drury, developed the antenna; this involved a rotating pair of Yagi arrays and incorporated lobe-switching. By the end of the year, the assembly was completed. Designated Type 282, it underwent sea trials using the HMS *Sardonyx* beginning in June 1939. The set could locate low-flying aircraft at over 2.5 miles, and ships at twice this range.

At the outbreak of the war, the 50-cm set was moved to the anti-aircraft range at Eastney and mounted on a pom-pom Mark II director. Successful trials took place in February 1940, and approval for further development was given. In April, contracts were given for building 200 Type 282 sets.

Application of the Type 282 as a rangefinder for main armament directors brought the requirement for an entirely new type of antenna. For this, R. V. Alfred designed a large, cylindrical parabolic reflector, 3-meters long with 12 half-wave dipoles along the optical axis. This was also implemented using a 3-fold Yagi Array. Separate antennas were used for transmitting and receiving. Designated Type 285, trials in June 1940 on the HMS *Nelson* showed that this set could follow a destroyer out to about 15 miles. Put into production by eight different contractors, the first sets were installed on the HMS *Southdown* and the HMS *King George V* in November.

Type 282 Series Controls

Yagi Array for Type 285

Types 282 and 285 were particularly used with the Bofors 40-mm guns. In addition to these, there was also a Type 284 main gunnery director for anti-armament on large ships. The last non-microwave set to be developed at Portsmouth was still another 50-cm set, the Type 283. This was a blind-AA, barrage-fire system for large ships that became operational in 1942.

Coales was justifiably proud of these 50-cm systems; all of them were successful and used throughout the war.

THE WAR THROUGH EARLY 1941

During the spring of 1939, Rowe had led the senior staff of the Bawdsey Research Station in considering a move to a safer site – further removed from the coast and potential German bombers. Various sites were discussed, with factors such as availability of existing laboratory space, rail connections to London, and proximity of airfields. None of the suggested sites, however, were visited.

During July, the German *Graf Zeppelin II*, now operated by the German Air Force (*Luftwaffe*), made two flights near Bawdsey, then followed with a third flight in early August. It was assumed that these were reconnaissance flights to determine the nature of radio transmissions from the towers. (It was later found that the assumption was correct, but the Germans did not pick up any of the signals – they had used very-high-frequency receivers, corresponding to the frequencies of their own systems.)

After the third *Zeppelin* flight, everyone was convinced that Bawdsey would be an early target. No existing facilities appeared to be appropriate, so Watson Watt made a decision on a temporary evacuation site – it would be at his *alma mater*, the University College, Dundee, in Scotland. There was no airport at Dundee, but Bowen and the airborne group could use Scone Airfield, a civil facility in Perth, about 18 miles to the southwest. Watson Watt would make arrangements at both of these facilities for the eventual move.

On September 1, 1939, Hitler's forces invaded Poland. Rowe has described the feelings at Bawdsey:
> All arrangements had been made for the journey north and only word from Headquarters was awaited. In our conceit, it was felt probable that the enemy would bomb the Manor in the first hours of the war and perhaps even before war was declared.

Having defense pacts with Poland, France and Great Britain declared war on Germany two days later. A special train was brought to a rural station near Bawdsey on September 2. That night, in essentially total darkness, the staff loaded it with boxed materials and most of the heavy laboratory equipment, and it moved out to Dundee the next day. Similarly, 30 aircraft at Martelsham Heath (some Blenheims in the process of being fitted with AI sets) flew the research equipment and ground crews to the airport at Perth. Staff members primarily used their own automobiles for the move and also carried special equipment.

Early RDF in Great Britain

The part of Bawdsey Research Station responsible for the operation and development of CH station did not make the move to Dundee. It was successfully argued by the Chief of this group, C. H. Dewhurst, that operating in Scotland would make continued construction and maintenance very difficult, primarily because of transportation and communication limitations. This activity, therefore, was moved to Leighton Buzzard, a town some 40 miles north of London and near an RAF communications center.

The CH station at Bawdsey, by then in full operation, remained in place. It is interesting to note that during the ensuing war neither this CH station nor Bawdsey Manor were ever bombed, and Martelsham Airfield was hardly touched. The village of Orfordness, however, was severely damaged during German bombings, with many civilians killed.

As the main Bawdsey operation moved to Dundee, the Army Cell was made a part of the Air Defense Experimental Establishment (ADEE), and the combined operation relocated to a facility just being completed at Christchurch on the south coast of England.

To put the British RDF developments into perspective, the war itself in this period must be considered. When Poland surrendered to German and Russian troops at the end of September 1939, Germany settled back to prepare for the invasion of Western Europe. This led to what was called the Phony War, with little land action for over half a year. Neither was there any serious air activity over Great Britain.

Britain and France had superior naval forces, and initially tried, without success, to blockade Germany. In turn, the German Navy (*Kriegsmarine*) initiated submarine and surface warship attacks on merchant ships along the routes between Great Britain and the U.S. and Canada. This became known as the Battle of the Atlantic. When Italy entered the war in mid-1940, its submarines joined the German U-boats. Although officially neutral, the U.S. came to Britain's aid, sending 50 destroyers in exchange for leases on military basis.

In December 1939, three British cruisers severely damaged the German pocket battleship, *Admiral Graf Spee*, and it was scuttled off the coast of Uruguay. Through unofficial examinations of the vessel, the British found the first hard evidence of German RDF existence.

In January 1940, the Russians attacked Finland, and the ensuing defense against the vastly superior force is one of the most heroic stands in history. Here a tragic occurrence should be noted. Some of the first AI-equipped Blenheims were for a London squadron of the Auxiliary Air Force. The pilots were largely businessmen men – storeowners, bankers, and the like – serving under Lord Carlow as the commanding officer. Realizing that the aircraft were poorly armed and lacked armor protection, Lord Carlow had the entire squadron modified at his own expense. In February, the Air Ministry ordered Lord Carlow's squadron to strip off the RDF sets and go to the defense of Finland. The Blenheims were totally unsuited to operate in the severe winter of that country, and the squadron suffered almost a total loss, including Lord Carlow.

To establish air bases, Germany invaded Norway and Denmark in April 1940. At the beginning of May, Germany finally went on the long-awaited offensive, with the panzer divisions sweeping through Holland and Belgium, then into France. The previous fall, shortly after the start of the war, the British Expeditionary Force had been sent to the Franco-Belgian border. This had eventually built to ten infantry divisions, a tank brigade, and an RAF detachment with some 300 aircraft. After the German invasion, the Expeditionary Force suffered heavy losses and started withdrawing.

Neville Chamberlain resigned and on May 10, 1940, Winston Churchill was appointed Prime Minister and Minister of Defence. Starting May 26, some 330,000 British, Belgium, and French troops started a week-long evacuation from Dunkirk, France, mainly using civilian ships and leaving behind most of their equipment; this included several of the Army's MRU field units. Defense production worked at a feverish pitch, with a concentration on fighter plans and bombers. The U.S. provided large quantities of anti-aircraft guns and other equipment.

Fascist Italy, under Benito Mussolini, entered the war on June 10, 1940, and immediately began air attacks on southern France. With no other alternatives, France signed an armistice on June 22. This allowed German occupation of parts of the country, particularly opening their airfields on the coast to the *Luftwaffe*.

It was the plan of Field Marshall Hermann Göring to draw the RAF into being destroyed; then Britain could be invaded. In a movement called *Betriebseelöwe* (Operation Sealion), landing barges and troops were assembled at Dutch, Belgian, and French ports.

The Battle of Britain began on July 10, 1940. Over the next several weeks, the *Luftwaffe* conducted daytime bombing throughout the British Isle, concentrating on shipping facilities, transport arteries, production facilities, and military installations, but with much of it on civilian areas to weaken morale. The attacks increased to some 1,000 planes daily, with London as the primary target.

On August 21, the RAF stemmed the brunt of the German air assault by downing a large number of the attacking planes, and then repeated this in the days following. The final turning point was on September 15; the *Luftwaffe* sent 1,300 bombers escorted by 500 fighters to probe defenses along the entire eastern coast. The Germans lost 74 aircraft, a record number in a single day, and never again attempted massive daylight raids over Britain.

The Germans turned to a different type of attack – nighttime with fewer plans but concentrated on single cities and aided by their *X-Gerät* navigational system. Called the Blitz, this was a strategic bombing campaign conducted from September 1940 through May 1941, targeting populated areas, factories, and dockyards.

Some 22 cities suffered attacks, including London, Portsmouth, Birmingham, Manchester, Liverpool, and also Glasgow, Scotland, and Belfast, Northern Ireland. One of the most severe attacks took place on November 14 when 515 planes razed the industrial city of Coventry, destroying or severely damaging about 50,000 buildings and homes. For a total of 57 consecutive nights, London was bombed, including a tremendous incendiary attack on December 29. The RAF retaliated with similar attacks on Bremen in Germany.

Hitler's intention was to break the morale of the British people so that they would pressure Churchill into negotiating. The bombing, however, had the opposite effect, bringing the English people together to face a common enemy. Encouraged by Churchill's frequent public appearances and radio speeches, the people became determined to hold out indefinitely against the Nazi onslaught. "Business as usual," could be seen everywhere written in chalk on boarded-up shop windows.

Despite the Axis air and sea blockade of the British Isles, the still-neutral U.S. accelerated its shipments of food and defense supplies. On March 11, 1941, President Franklin Roosevelt signed the Lend-Lease Act, giving legality to the massive supply of military equipment.

The RAF inflicted increasingly heavy losses on the *Luftwaffe*. Using the Chain Home, the British Fighter Command was able to track and plot the course of German bombers, allowing fighter planes to be dispatched to attack the incoming bombers at the best possible position. As a result, the *Luftwaffe* never gained air supremacy over England.

The Germans recognized that they were not winning through the air war. Hitler gave up or postponed the planned invasion of the British Isles, and the Blitz ended in May 1941, when he transferred the *Luftwaffe* to Eastern Europe in preparation for the invasion of the USSR. During the Blitz, air attacks had killed more than 40,000 civilians, and in London alone left about 375,000 persons homeless.

DUNDEE AND THE AMRE

When the research equipment and personnel from Bawdsey arrived at Dundee, in the southeast corner of Scotland, they encountered a disaster. Although Watson Watt had obtained consent of the University College officials for research activities to be moved there, they were totally unaware of the extent of Bawdsey operations and had only two small laboratories available for the 200 or so relocating personnel and their equipment.

The space problem was eventually solved, but it set the activities back many months. The second floor of the Teachers' Training College and sections of several other buildings were taken over and placed under security. The personnel mainly found residency in the "town-and-gown" area of the West End of Dundee.

AMRE Facilities at Dundee

The operation at Dundee was named the Air Ministry Research Establishment (AMRE). Rowe continued as the Superintendent, with Lewis as the Assistant Superintendent for Research. Essentially none of the other university researchers took up work at Dundee.

Dundee was recognized as a temporary location while a permanent site was being selected and facilities constructed. This process, however, took longer than expected. With restricted space, during the next year much of the project activity was done at the newly associated university laboratories or through contracting with firms.

As described later, in late 1939, the Air Ministry asked the Army's newly formed operation at Christchurch to modify a number of their CD Mk II systems to be used to augment CH stations in detecting low-flying aircraft. These were called Chain Home Low (CHL) sets, and by the following February, 18 sets were ready for installation. At that time, John A. Ratcliffe, who had led the initial effort, joined the AMRE for further development of this system, now designated AMES 2.

Air Ministry installation crews performed the "heavy" work on the CHL, but the two antennas and lines were handled by a special AMRE team. Led by Dr. Dennis Taylor, much of this was done at the 200-foot level on ice-covered towers in gale-force winter winds. Included were Dr. William H.

CHL Tower

Penley and John C. Duckworth, scientists who would later continue with the development of similar equipment.

Work at the AMRE continued on IFF development, now being led by Dr. Frederick C. Williams, who was earlier on the Engineering Department faculty at Manchester University and well known for his circuit designs. The original IFF swept the frequencies used by CH stations. With the addition of CHL, frequencies for these sets had to be included. A mechanical device was incorporated to sweep the receiver-transmitter across the necessary bands. The result was IFF Mk II; this design was completed and turned over to Ferranti for production.

There was no question that RDF operation at higher frequencies would have many benefits. With shorter wavelengths, antennas could be smaller – especially important for airborne systems and, to an extent, for shipboard use. With this and other benefits in mind, in late 1939 the Air Ministry contracted with EMI in Hayes and GEC Research Laboratories in Wemberly to jointly pursue work leading to a microwave RDF system.

Rowe envisioned that this could result in a transmitter with a searchlight-type beam, capable of sweeping an observational field. Work was already in process on a CRT display that could show targets in a space 360 degrees around the antenna. One of the first accomplishments at GEC was in developing the Micropup valve (vacuum tube), eventually capable of producing several kilowatts power at up to 600 MHz (50 cm). This was considered a further upgrade of the AI set, and the project was placed under Bowen.

Although not extensive, the *Luftwaffe* did make sporadic bombings in the first part of the war. In mid-March 1940, they bombed the naval base at Scapa Flow in the northeastern Orkney Islands, their first major air action against Great Britain. A site near Swanage, a seaside town in Dorset on the southern coast of England, had been selected as the "permanent" location of the Air Ministry's RDF development, and many of the needed buildings were already under construction. Realizing that the Scapa Flow bombing could have well been Dundee, the move to Swanage was accelerated.

ST. ATHAN AND AIRBORNE RDF

When the airborne group had relocated to Scotland on September 3, 1939, 29 of the starting 30 aircraft made it safely to Perth, a few miles from the AMRE in Dundee. The fate of the missing aircraft was never determined, but it was assumed to have gone down in the North Sea. There was a suspicion, however, that it might have made its way to Germany, carrying some of the highly secret equipment.

The situation at Perth's Scone Airfield was somewhat the same as that at the University College. The private airfield was already being used to train RAF pilots. Although the owner agreed to take them on, the small amount of unused space was swamped by the planes, equipment, and support personnel arriving from Martelsham Air Base. Operations, nevertheless, got underway in a few days; makeshift hangars were used to complete the first Blenheim AI installations in October.

A building at the airfield was made available for completing the design of the 200 MHz (1.5 m) ASV. The prototype system was much like the AI Mk II, but used a VT90 valve in the transmitter; this was a new high-frequency valve developed by the GEC Research Laboratory at Wembley. When the prototype was tested at Gosport Way on the southern coast, it was only 20 feet above sea level but detected many vessels on the English Channel.

Near the end of October 1939, a more suitable place for Bowen's airborne group was found; they moved to a very large RAF training and maintenance air base opened a year earlier at the village of St. Athan near Cardiff in South Wales. Here the effort concentrated on improving AI and ASV sets, as well as assembling sets and installing AIs in Bristol Blenheim and ASVs in Lockheed Hudson aircraft. St. Athan was a long way from Dundee, and this complicated the reporting to AMRE and the interaction with the primary research staff.

The living quarters were satisfactory – wooden huts interconnected by passages, all with central heating. The working conditions, however, were horrible. Essentially all of the space made available was in large hangar-type buildings without heat. Bowen described their attempts to work in heavy clothing and gloves in the searing winter wind:

> Here was one of the most sophisticated defense developments being introduced to the Royal Air Force, and it was being done under conditions that would have produced a riot in a prison farm.

Nevertheless, almost immediately the group was turning out RDF-equipped planes at the rate of one per day.

A major problem in installing the equipment concerned the RF cables. The internal transmission of high-frequency signals over even a small distance – such as from the transmitter to the antenna – required the use of co-axial cable. At that time, this cable was hand-made, threading the internal wire through small plastic beads, then inserting this into a breaded-wire sheath. Where numerous bends were needed, such as in an aircraft, the cable was easily damaged and often failed.

Shortly after Bowen moved to St. Athan, the ICI Company approached him for advice on how their electrical insulator material –

called polyethene – might be used in RF applications. Reginald O. Gibson and Eric W. Fawcett had developed this material in 1933, but its uses had mainly been in coating underwater telephone lines. After testing the RF characteristics, Bowen recognized that this could be used to continuously support the central wire in co-axial cable.

ICI quickly began manufacturing such a cable, resulting in great improvements in installing the airborne RDF equipment. It was also adopted throughout Great Britain in many types of applications. In the intervening years, Fawcett had joined DuPont in the U.S., leading to this firm releasing the same material under the name polyethylene.

The prototype ASV set, designated ASV Mk I, was fitted into a Lockheed Hudson aircraft in December 1939. It was an immediate success, detecting a large ship at a range of about 20 miles and coastlines at 30 miles or more. Flying at altitudes between 1,000 and 6,000 feet, it detected British submarines at up to 3 miles in rough seas and 6 miles in calm waters. Based on these tests, 300 ASV Mk I sets were put under contract for aircraft of the Coastal Command. The E. K. Cole Company was brought in to build transmitters, and receivers were from Pye.

As the ASV sets became available, Bowen's team started their installation on aircraft of the Coastal Command, with Hanbury Brown and Keith Wood handling the land-based aircraft. John W. S. Pringle, a zoologist recently added from Cambridge, went to Pembroke Dock on the southwestern tip of Wales to coordinate installations on Catalinas, the long-range flying boats. There, Squadron Leader Sidney Lugg made a brilliant suggestion that was followed up by Pringle: place an IFF set in the landing field's control tower and allow the ASV to use it for homing.

With only small modifications to accommodate the ASV frequencies, this system was quickly adopted. Called "Mother" for obvious reasons, ASV-equipped aircraft could easily use this beacon to obtain bearing and range to their base from 50 or more miles out. This development was credited with saving many lives throughout the war.

In his continuing poor relationship with Rowe, Bowen had been criticized for the lack of engineering on the airborne systems, and he and his group were discouraged. Having been told of the airborne sets and their importance, however, King George VI and Queen Elizabeth braved the snow and ice to visit St. Athan on February 9, 1940. Bowen described the visit:

> He [the King] passed down the line and had a pleasant word for every member of the group. We were honored by the visit and it gave an enormous boost to the group that had been feeling the cold, cold winds of adversity in more senses than one.

In general, the initial AI and ASV equipment had, indeed, been poorly engineered. Bernard Lovell noted this in his 1991 book, *Echoes of War*. This was not true for the CH station equipment because that hardware had received final engineering and fabrication by firms highly experienced in similar equipment. The airborne equipment, hurriedly designed in the research laboratories, did not have this advantage. Engineering on such equipment means far more than simply performing as expected; such things as simple as placement in the aircraft, interaction with other equipment, human factors/ease of operation, component reliability, and maintenance accessibility could not have been expected in the conditions of their initial development.

Such was the case with the ASV Mk I. Consequently, in mid-February 1940, the Air Ministry transferred responsibility for ASV improvements to the Royal Aircraft Establishment at Farnborough. Touch and Hanbury Brown went with this move, and thus began the break-up of Bowen's highly productive airborne RDF group.

ASV Antennas on Wellington

At Farnborough, the set was improved to become the ASV Mk II, one of the best and most-used airborne systems of the war. For interrogation, this set had Yagi antennas on both sides of the aircraft, giving lobes covering a swatch 25 miles wide. When a target was found, the pilot switched to a forward-looking pattern for final guidance.

In addition to outfitting aircraft with AI sets, Bowen's remaining group at St. Athan continued their development. Lewis devised a means for improving the pulse shape. A new valve from GEC, the E1130 Micropup, was added, increasing the power output to over 5 kW. These changes resulted in the set giving a good target signal at near four miles. This improved set was designated AI Mk III.

In April 1940, Cole and Pye were given contracts for 3,000 each of AI Mk III and ASV Mk II sets. The initiation of bombing by the Germans, however, led to a priority being given for AI sets, delaying delivery of ASVs until later in the year.

While pilots are most often recognized, certain aircraft – such as the Blenheim – also had a radar operator. The glamour of being a fighter pilot kept those positions filled, but it was not so with the radar operators, positions requiring skill and intelligence for adequate training. The shortage was solved by waiving the RAF prohibition against such personnel wearing eye glasses. When this was done, hundreds of well-qualified persons volunteered, including a large number of school teachers.

The first kill of a German aircraft by an AI-equipped Bristol Blenheim IV was on July 22, 1940. Initially tracked by a CH station, an unsuspecting Dornier night bomber was shot down over the sea near Bangor. During the next several months, however, the number of successful nighttime intercepts was small. The Blenheim was fast enough to overtake most bombers and but did not have good fire power; also, the then-used AI Mk II did not have sufficient sensitivity.

Blenheim IV with AI Radar

As previously noted, EMI had an outstanding group developing television. This was led by Alan D. Blumlein (1903-1942), perhaps Britain's most versatile electronics engineer. As the war started, EMI changed its positiom on accepting government contracts; at their own expense, they had Blumlein and Dr. Eric L. C. White develop an early-warning set similar to the Army's GL Mk I. While this set did not meet any existing requirement, it opened the door for EMI to participate in RDF work.

Alan Blumlein

Rowe, without Bowen's knowledge, gave a contract to EMI in December 1939 for Blumlein and White to further improve the AI. (This resulted from the continuing debate between Rowe and Bowen concerning the AI's minimum range.) Through EMI's effort, the minimum range was reduced to about 500 feet. The system incorporating these and other improvements already made on the Mk III became AI Mk IV in August 1940, and Cole and Pye switched their AI production to this version.

A new night-fighter, the Bristol Beaufighter, started coming off the production line in August 1940. With a powerful engine providing speeds greater than the German bombers, and, bristling with cannons and machine guns, it was a formidable

Bristol Beaufighter

machine. They were sent to St. Athan to be equipped with the AI Mk IV.

Beaufighters were immediately put into action. Before the end of 1940, they were being used very successfully. However, one was soon shot down over occupied France and, since the pilot was listed as captured, it was assumed that the AI set might have survived and been examined by the Germans. Somewhat later, one of the German airborne sets was saved from a crash in Great Britain, and the display was almost a duplicate of that in the AI.

Bowen had turned his attention to the emerging microwave AI project and in June 1940, joined this activity at Swanage. The following September, he left the airborne group to be a member of a technology-exchange mission led by Sir Tizard to America. Most of the remaining personnel of his airborne group, including all of those engaged in research on new equipment, joined the operations at Swanage.

ASV and Leigh Light on Wellington

By mid-1941, the ASV Mk II was finally being produced. Squadron Leader Humphrey de Verde Leigh suggested that this unit be used to turn on a powerful searchlight mounted on the aircraft, changing night into day and greatly increasing submarine kills. The 22-million candlepower, 24-inch searchlight was produced by Savage & Parsons. Implemented in June 1942, the ASV/Leigh-Light combination soon wreaked havoc on surfaced U-Boats, reducing the losses in Allied ships from 600,000 to 200,000 tons per month.

Leigh Light Application

CHRISTCHURCH AND THE ADEE/ADRDE

During WWI, the Army had established a Searchlight Investigation Group to improve the performance of searchlights against air raids. At the end of the war, this became the Searchlight Establishment, then in 1922 was combined with related activities and named the Air Defense Experimental Establishment (ADEE). Located at Biggen Hill, the ADEE was involved with the improvement of searchlights and the related equipment needed to find aircraft.

Early RDF in Great Britain

In the early 1930s, it appeared that the only practical detection method was acoustic – hearing the noise from aircraft engines. Major (Dr.) W. S. Tucker, Director of Acoustic Research at the ADEE, began the development of a "sound mirror" system, composed of a 200-foot concrete sound collector supported by a number of 30-foot bowl-type collectors. With the first demonstration of an RDF system at Bawdsey, the acoustic project was discontinued. General work in acoustics, however, continued at the ADEE; this included experiments toward developing an acoustical fuse for projectiles.

A related organization was the Signals Experimental Establishment (SEE), established in 1916 at Woolwich Commons in London. Operated by the Ministry of Supply, here was concentrated the Army's research and development of radio communication equipment, including jamming hardware and direction-finding sets. As previously described, it was at the SEE in 1931 that W. A. S. Butement and P. E. Pollard proposed a Coastal Defense Apparatus using ultra-short radio waves reflected from ships. Although an apparatus was constructed and demonstrated, the work was not continued. Then as RDF came into being, Butement and Pollard were assigned to the Army Cell when it was formed at Bawdsey in October 1936.

As the war loomed closer, the ADEE greatly increased in its activities and, believing that Biggen Hill would likely be bombed, it was decided that a move to safer location would soon be necessary. Christchurch in Dorset on the south coast was selected.

Christchurch is both a town and a borough. Christchurch town has existed in one form or another since the Iron Age when the area between the River Avon and the River Stour was used as a trading point and safe mooring for visiting ships. In 1938, the large Somerford Grange, located about three miles northeast of the town, was purchased by the Ministry of Defense, and construction of wooden buildings for the ADEE began. This was directly adjacent to the Christchurch Airfield, a small, previously private field that had been taken over by the RAF in 1936 but remained relatively unimproved.

At the start of the war in early September 1939, the movement from Biggin Hill to Christchurch began. At this same time, the Army Cell at Bawdsey, then with about 20 persons, also relocated to Christchurch and joined the ADEE. For the RDF group, a special facility was built on the high cliffs at nearby Steamer Point. Looking out over the Solent (the surrounding waters) and to the Isle of Wight in the distance, Steamer Point was an excellent location for testing the Coastal Defense (CD) and Gun Laying (GL) systems.

The main facilities were not completed when the move began; thus, the Searchlight and Acoustics Sections remained at Biggin Hill for some time. When the move was completed, the combined operation had some 200 personnel and continued to be called the Air Defense Experimental Establishment (ADEE).

Earlier in March, Dr. D. H. Black had been appointed Superintendent of the ADEE, and he continued in this position at Christchurch. Dr. Talbot Paris, the Technical Director of the original Army Cell remained the leader of the RDF Group. At the time of the move, Butement had been named an Assistant Director of Scientific Research, and continued to lead the CD research activity. Pollard remained responsible for the GL development.

Air Defense Experimental Establishment (ADEE) at Christchurch

There was an urgent need to improve the effectiveness of the anti-aircraft guns, and the work, started earlier at Biggen Hill, toward an acoustical fuse continued. The acoustics authority, Dr. Albert B. Wood, served as a consultant. Tests were made using projectiles and rockets with microphones imbedded in tail fins, but with only about 15 percent detonation rate.

With his background in radio, in October 1939, Butement turned to this technology as a potential solution. One method would place a highly compact RDF set on the projectile, setting off the detonation when close proximity to the target was attained. An analysis, made by Butement with assistance from Dr. Edward S. Shire and Amherst Felix H. Thompson, indicated that high-frequency radiation from an oscillator in the projectile could be reflected with a Doppler shift from the moving target, reenter the circuit in the projectile, and interact with the original oscillation to generate a beat signal that could be detected. The projectile would be detonated when the Doppler signal reached some magnitude, which would be when the projectile was close to the target.

In addition to the huge problem of packaging such a device in a small projectile, there was the question of the vacuum tubes surviving the acceleration forces at firing. Intelligence reports gave some hope to solving these problems; it was found that the Germans had developed a tube-based electronic package for projectiles that survived the launch shock.

Unfortunately, the demands on personnel and funds at the start of the war were such that little more was done at that time. In less than a year, however, the concept and circuit had been adopted in the United States as the VT (variable-time) Fuse, the most-manufactured electronic device of the war. In later years, Butement said that he considered the proximity fuse as his most significant accomplishment.

When the Bawsey Research Station was evacuated, the CD system remained with the CH station. Admiral Sir James Somerville asked for a demonstration test of the CD in detecting submarines. After a successful demonstration, Sommerville asked to be supplied three sets to "close the Fair Isle Channel." Watson Watt asked if John Cockcroft's team at Cambridge could modify Butement's CD equipment for this application.

At that time, John Douglas Cockcroft (1897-1961) was one of the most respected researchers in England. Educated in electrical engineering, experienced in industry, and with a graduate degree in mathematics, he joined Cambridge's Cavendish Laboratory in 1924. There he and Ernest T. S. Walton developed a linear accelerator that, for the first time, disintegrated an atomic nucleus. Later, for this accomplishment, they shared the 1951 Nobel Prize in Physics. In 1934, Cockcroft was placed in charge of the Royal Society Mond Laboratory at Cambridge.

John Cockcroft

Cockcroft took on Watson Watt's assignment and, assisted by John A. Ratcliffe and others, started building the Chain Home U-boat (CHU) systems at Cambridge. After a short time, however, they realized the need for working with Butement and other experienced designers and moved to the ADEE at Christchurch. There, using a CD transmitter, they soon had the sets working. During the 1939-40 winter months – with high gales and snow – the team selected sites and installed the CHU sets for the Admiralty. The equipment detected submarines broadside at 25 miles and aircraft about three times this distance.

At the same time that the CHU project was underway, it was recognized that the modified CD (Mk II) could be very useful in detecting aircraft at low altitudes. The Air Ministry asked for five additional sets to be installed for this purpose at locations along the east coast. Watson Watt then recommended that these be broadly adopted in the CH network, giving the stations a capability for detecting low-flying aircraft as well as seeing behind the line (the two major shortcomings of the original stations).

This recommendation had been immediately accepted by the RAF, and in November a number of the CD systems currently under manufacturing contract were designated for this application. Known as Chain Home Low (CHL), one unit would be placed at most CH stations, mounted on the transmitting towers at the 200-foot level or on a separate mast.

With the nation now at war, it was imperative that this capability be added quickly. Although German air activity over Great Britain did not start immediately, there was mine-laying in the surrounding seas by low-flying planes, and these were undetectable by the existing CH facilities.

For the necessary equipment integration and checkout, Ratcliffe assembled a team at the ADEE and took on the responsibility. As an emergency measure, several CH stations were equipped with hastily assembled CD sets to serve the CHL function. These were initially placed at ground level on cliffs overlooking the sea. Separate antennas were needed for transmitting and receiving. The antenna rotational drive mechanisms were not ready, so they initially had to be aimed by hand.

Initial CHL Station

By the end of February 1940, 18 CHL sets were completed by the ADEE and turned over to the AMRE at Dundee for installation (described earlier). Further responsibility for the systems was then placed with the AMRE, and Ratcliffe joined them for ongoing CHL development.

Development of the 5-m (60-MHz) Gun Laying equipment continued under Pollard at Christchurch. It was still necessary to have separate transmitter and receiving vans. A wooden structure with dipole antennas at both ends extended on each side of the receiver van, allowing improved determination of azimuth by rotating the van and also through using lobe-switching.

A height-finding (elevation) capability, developed by Leslie H. Bedford of Cossor, was added. This used two movable dipoles on a tower and a reflecting wire-mesh ground plane. Reception from the two antennas was compared using a goniometer (as in the CH system).

In late 1939, the improved system became the GL Mk II, and in June 1940 was put into production by Metropolitan-Vickers and A. C. Cossor. Some of the original GL Mk I sets were retrofitted with the height-finder; these were designated GL E.F. Over the next three years, about 1,700 GL Mk II sets were put into service, including over 200 supplied to the Soviet Union.

The GL Mk II was noted for its mechanical construction and reliable electronics, but it did not provide any significant improvements in blind-firing capability. This, however, was likely due to inadequacies in fire directors using the RDF data. Patrick M. S. Blackett, who had been an original member of the Tizard Committee in 1934, was asked to organize a team and analyze the performance of the new GL system.

Building on the operational research techniques initiated earlier by Harold Larnder at the Bawdsey Research Station, the team found the data to indicate that anti-aircraft units with the GL Mk II averaged 4,100 rounds fired per hit, compared with 18,500 for the original GL Mk I. This corresponded to an average error of about two degrees in predicting the direction. The need for a proximity fuse, such as that designed by Butement, was clear.

From this study, a permanent team – informally called Blackett's Circus – was kept together to perform operational research on many of the wartime activities in Great Britain. After the United States entered the war, President Roosevelt sent a personal envoy to Great Britain to examine this process, and he returned to implement it in America as operations research. (For his earlier academic accomplishments, in 1948 Blackett received the Nobel Prize in Physics.)

To improve the accuracy of fire control, the extensive work on 1.5-m (200-MHz) systems was examined for the GL systems. From this, work on the Searchlight Control (SLC) RDF was initiated, with SCL Mk 1

Searchlight Control RDF

emerging in April 1941, followed with Mk 2 containing minor improvements. About 100 total of the two versions were built but saw essentially no service.

There had also been work by Shire at the ADEE on a 50-cm (600-GHz) GL system. By March 1940, a set operating on a rooftop at Steamer Point ranged aircraft up to five miles, and in June picked up echoes of enemy ships near Cherbourg on the coast of France. This work, as well as the introduction of improved SLCs, was put aside with the advent of the multi-cavity magnetron and microwave systems.

In early 1940, John Cockcroft joined the ADEE an Assistant Scientific Officer. Later in the year, he accompanied Sir Tizard to America on a technology exchange mission (see Chapter 3). Upon his return in December, Cockcroft was appointed the Chief Superintendent of the ADEE at Christchurch. In mid-1941, the operation was reformed into the Air Defense Research and Development Establishment (ADRDE).

SWANAGE AND THE TRE

Swanage is a picturesque holiday town in Dorset on the southern coast of England. Christchurch is about 16 miles to the northeast, and Portsmouth another 30 miles further eastward, then the locations of Army and Navy RDF research facilities. About 200 people were involved in the move in May 1940, and this made problems for not only the research space but also adequate housing, particularly for the married personnel. However, the few visitors to a holiday town in wartime meant landlords were pleased to have tenants.

Old Swanage

The new operations were initially named the Ministry of Aircraft Production Research Establishment (MAPRE), then after several months changed to Telecommunications Research Establishment (TRE), a name that would remain throughout the war. (Hereinafter, all activities after the move to Swanage will be called TRE.)

The TRE facilities were not set up in Swanage itself, but primarily adjacent to Worth Matravers, a small village about three miles to the west on Purbeck Peninsula (often called Isle of Purbeck). This location was chosen because it had a flat cliff-top site, good for testing radar, and was then further away from German-occupied territory. There were no existing government facilities in Worth Matravers, so a crash building project was started. The facilities were still being completed when the first personnel and equipment arrived in mid-1940.

The facilities were laid out in five sites, A through E, all enclosed by security fences. A CH station for experimental work and operator training was built at Site A. This had two 240-foot towers for supporting antennas – a steel mast for the transmitting and a wooden mast for the receiving.

Site B was the main facility; it contained a number of 20- by 50-foot laboratory buildings, many with blast-protecting ramparts. There was a large central building for services, including a cafeteria. Site C had a few buildings used mainly for microwave experimentation, and included a large open area for testing; it also served as a landing site for small aircraft.

An experimental CHL station was set up on a cliff-side ledge overlooking the sea at Site D. The RAF had an operational CH station at Site E, a short distance away from the research facilities. Bicycles were the main mode of transportation, between sites as well outside locations.

Telecommunications Research Establishment (TRE) – Worth Matravers

Regular operational stations were relatively close: one of the original 20 CH stations was at Ventnor on the Isle of Wight south of Portsmouth, and in late 1940 the first Ground Controlled Interception (GCI) station was built at Sopley, a few miles north of Christchurch.

There was no airfield close to Swanage, but there was one at Christchurch. Although this field was barely useable, the Air Ministry directed that it would be used by the TRE airborne group. The Army ADEE had relocated to Christchurch in the fall of 1939, but had no use for the airfield. It had been acquired by the RAF from a private owner in 1936, but almost no improvements had been done.

Actually the airfield was four grass-covered fields side-by-side in an "L" configuration, all surrounded by trees. The two side areas were 3,000 feet in length; but not suitable for landing the larger aircraft such as Blenheims, Beaufighters, and Mosquitoes. A vacant building was converted to serve as the electronics workshop and soon a variety of aircraft were parked around the airfield and under the trees.

Within a short time, a large hangar, supporting facilities, offices, and two laboratories were constructed along the north side of the field. Under Group Captain Frank C. Griffiths, the Telecommunications Flying Unit (TFU) at Christchurch Airfield would be used by the TRE for flight testing of experimental equipment for over a year. In 1941, Hurn RAF Airfield would open; this was somewhat nearer to Swanage and the TFU transferred there.

Integration of factory-produced AI and ASV sets into aircraft remained at the RAF St. Athan Airfield. As more equipment and planes were delivered, installation was expanded into other RAF facilities where the work was performed by both military personnel and contractors.

Rowe was named the Chief Superintendent of the TRE, and lower-level superintendents were responsible for specialty groups. The Sunday Soviets meeting continued to be held, now with an even greater variety of participants. Staff members were expected to work six days each week, with staggered off-days. To accommodate this irregular schedule, some churches in the area had services on Friday night at least once a month.

Prime Minister Winston Churchill, in a report to his cabinet in September 1940, identified the key significance of technology to the war effort:

> Our supreme effort must be to gain overwhelming mastery of the air . . . we must regard the whole sphere of RDF . . . as ranking with the Air Force of which it is an essential part. The multiplication of the high-class scientific personnel . . . should be the very spear-point of our thought and effort.

The first significant development at the new facilities was an improvement for the CHL antennas. Congreve J. Banwell had started a project concerning a common antenna while at Dundee, and it was completed early in the summer of 1940 at the TRE. For this, Banwell and Roland J. Lees developed a transmit-receive circuit (using a pair of spark gaps and quarter-wave lines) that sufficiently suppressed the transmitter pulse going into the receiver, allowing a single antenna for both transmitting and receiving. The spark gap, exposed to the atmosphere, had a limited life. To correct this, a gas-filled device was later developed by N. L. Harris at the GEC Research Laboratory.

The AI development activities of the TRE also grew. The laboratories were soon filled with test gear and experimental circuits for transmitters and aerial designs that would permit installation into the cramped space of an aircraft. While the principles of airborne RDF were understood, the issues of weight, size, and power supply all still had to be resolved.

With the assistance of Blumlein and White of EMI, the AI Mk V and Mk VI evolved, the latter to accommodate a single-seat fighter. The improvements included changes in the display system. For the Mk VI, Drs. F. C. Williams and F. J. U. Ritson developed an automatic strobe that locked onto the target, giving the pilot a spot indicator of the target position. The Mk VI was the last version of the meter-wave AI sets

While Rowe believed in the potential merits of microwave RDF sets for AI, he did not feel that such systems could be developed in time to affect the war. Nevertheless, in May 1940, he brought in Philip I. Dee (1904-1983), a leading physicist at Cavendish Laboratories, to head the microwave efforts. Dee brought with him a number of other researchers, including Dr. Herbert W. B. Skinner, who served as his deputy. Their laboratory was initially in Building 40 at Site C.

By the fall, TRE required additional space away from the main facility. In September 1940, two large estates, each with several buildings, were acquired: Durnford House, a preparatory school in Worth Matravers, and Leeson House, a former girls' school in the village of Langton Matravers, between Worth Matravers and Swanage. The administrative staff, including Rowe, moved into Durnford, along with some research operations.

Later in the fall, Forres School in Swanage itself was taken over for setting up a variety of RDF training programs under John A. Ratcliffe. An activity called Post-Design Services (PDS) would also be established by Ratcliffe at Forres.

There were still significant activities in the meter-wavelength ground RDF systems, particularly the CHL sets. Dr. William H. Penley managed the team engaged in these activities; Dr. Carl H. Westcott served as the scientific advisor and led in many important technical developments.

Dee and his microwave group were in what was called TRE Leeson. Their work would continue but be somewhat "hidden," and the facility

TRE Leeson House

was very suitable for this – near to other TRE activities but separated. There was a spectacular view from Leeson House across Swanage Bay and to the Isle of Wight, some 40 miles away; this geography was later used in testing as microwave systems gained power.

The Ground-Based Intercept

At the start of the war, RAF fighter pilots were given target locations, then flew to where they could either visually see the target or locate it with the AI equipment. Even with its improvements, the AI remained limited in maximum range by ground returns. Thus, the fighters often missed finding their targets, particularly at night or in bad weather.

As a solution, Hanbury Brown came up with a new type of system. Since the CHL could follow bombers on both sides of the CH area of protection, he proposed using this technology in widely distributed stations, tracking both the target and fighter aircraft and with operators using VHF radio-telephones to guide pilots to the target vicinity. This was called Ground Controlled Interception (GCI), and stations were developed as one of the first major projects started at TRE.

By December 1940, three GCI installations were completed and undergoing tests. A further 12 mobile units were ordered for immediate use. Watson Watt reflected the widespread optimism:

> The Beaufighter had ample performance and adequate armament to deal with many of the German bombers then being used. . . . Our hopes began to rise that at long last we had a combination of airborne [RDF] and ground control that was adequate to handle an interception.

The GCI stations used essentially the same type transmitter and receiver as the CHL, operating at 1.5 m (200 MHz). It was necessary, however, for the antennas to be significantly different. A station would need to keep both the target and fighter aircraft within its coverage and operate over essentially a hemisphere. For this, the first experimental stations used separate transmitting and receiving antennas, with the pointing apparatus positioned by the operator.

The receiving antenna had two dipole arrays, one above the other, allowing the elevation angle, and thus height, to be estimated. Later, a single antenna was used in the GCI for both transmitting and receiving, with use being made of the previously mentioned circuit developed by Banwell and Lees for keeping the transmitter pulse out of the receiver.

Since the CHL and hence the GCI, were based on systems developed by the Army, continuing assistance was given by the ADEE – later renamed ADRDE – at Christchurch.

The first GCI station became operational in January 1941. With a radio call sign of "Starlight," it was located near the village of Sopley in Hampshire, a few miles north of Christchurch. Although designed at the TRE, the station equipment was built by the ADEE. This initially had the antenna manually turned by airmen peddling in response to bell-code signals from the operator, similar to signaling from the bridge to the engine room on ships.

GCI Mobile Antennas

The PPI display was added, giving the operator a map showing the relative positions of the two aircraft being followed. By the spring of 1941, GCI was an operational reality with 12 stations. These systems permitted use of new tactics that increased enemy bomber losses from 0.5 percent to about 7 percent, and the first phase of massive night bombings by the *Luftwaffe* ended by May.

GCI Sopley Control Room

Sopley, as one of the first GCI stations, was visited by a succession of Air Ministry officials and government leaders. Included were Prime Minister Churchill and King George VI on May 7, 1941. While the King watched behind the GCI controller, Beaufighter ace John Cunningham successfully intercepted an enemy aircraft. It is popular legend that the King stepped outside after the interception and saw flames from the enemy aircraft as it fell.

A major shortcoming of the initial GCI was the lack of precision height-finding, a long-standing problem in the RDF development. C. J. Banwell, who had been working this problem since Bawdsey, devised a system for obtaining this information. He used an arrangement of antenna elements along a 200-foot tower, giving a fan-shaped beam about 1-degree wide. Using a phase shifter, this could be moved up and down in elevation; the position giving maximum signal could then be translated to target height.

Somewhat later, Banwell also used his transmit-receive isolation device to allow a common antenna in the height-finder set; this was called the Variable Elevation Beam (VEB) height-finder system. There was limited production, with VEBs put into service at several CH sites.

The work then progressed to applying the same technique for use on the CHL and related GCI stations. Under Penley, a 50-cm (600-MHz) VEB height-finder using antennas on 20-foot poles was developed for the CHL stations. This system was designated AMES Type 20, Decimetric Height Finder, but saw limited application because of the advent of microwave height-finding systems.

Penley also directed the development of a 50-cm (600 MHz) VEB system for the GCI. Here use was made of an antenna with a wire-mesh reflector 10-feet wide by 30-feet high designed by Leslie H. Bedford of Cossor. This gave a fan-shaped beam about 2-degees wide that could be physically moved slowly up and down, as well as scanned 5 degrees to each side at 7 scans/second. This was designated AMES Type 16, Fighter Direction Station.

Over the next years, a wide variety of GCI stations were developed. For protection, some of these were placed underground (in a room called "the well") with only a camouflaged antenna exposed. While the norm was for a station to track one target and fighter at a time, some operators were able to handle two interceptions. This led to some stations having two PPI displays operating from one receiver. Like the wartime RAF airfields, GCI stations eventually covered Great Britain.

The PPI – A Map on the CRT Display

When the cathode-ray tube was first used by Watson Watt in his lightning instrumentation at Slough, it had a horizontal trace for time and showed signal inputs on a vertical scale – an "X-Y display." Distance was determined by comparing displays from two separated stations, This was modified by J. F. Herd when he applied the input from one station to the X-trace and the input from a second station to the Y-trace, allowing the combined signals to indicate the direction of a common signal (a lightning strike).

As Bowen was planning the AI and ASV systems at Bawdsey, he suggested that a common display showing azimuth and elevation angles would be beneficial, somewhat like Herd had done. As RDF systems with rotatable antennas were developed, this concept was proposed to allow the display to map the target location.

In planning the CHL, this display was again considered. In this display, the time base would rotate in synchronism with the antenna direction, and the signal from the target was shown on this revolving line. This was called the "radial time base" (RTB). Some work on this was done by the CHL team, but, since the initial system used two antennas and they were hand-rotated, the display work had little priority.

Rowe and Lewis visualized the RTB being used with a station having a common antenna for transmitting and receiving, and with a sharp beam rotating like that of a lighthouse. While at Dundee, Geoffrey W. A. Dummer had been assigned to lead the display development. A 50-cm (600-MHz) transmitter and receiver with a rotating antenna was being worked jointly by GEC and EMI, and this was intended to be used for testing the display. The RTB development made good progress, but there was little progress on the RDF equipment.

For the display, it was necessary to have developed a new type of CRT with a long-persistent screen, allowing the target spot image to remain through a complete rotation in the display. This was eventually accomplished by using two different phosphors – one of low persistency to quickly indicate the target with a blue flash, and a second one with an orange afterglow that was activated by the initial flash. To emphasize the pattern-sustaining glow, a filter passing the orange color was placed in front of the CRT face.

After the move to Swanage, the 600-MHz RDF equipment was still not completed, but the display was ready for testing. Dummer then coupled the display with a 200-MHz CHL set that had a revolving antenna, and very satisfactory demonstrations were made during the summer of 1940. In September, a demonstration was given to Sir Philip Joubert, Assistant Chief of Air Staff of the Coastal Command, who was very impressed and ordered it to be adopted. The name of the display was soon changed to plan-position indicator (PPI), a term used for a similar display (but with an inferior screen) in the U.S.

To incorporate this into equipment already being installed, such as the CHL and GCI, would require considerable modifications. However, France had been overrun, and the Germans held positions near to Great Britain. Then all efforts at the TRE were directed toward defense against invasion. Most of the staff stopped research and went to the CHL stations being set up to ensure that they gave the best possible performance. They converted a few of them for single turntable operation (one antenna atop the other), put in technical improvements, and checked the performance. Some were found to be in poor locations, and for these they selected better sites for allowing good detection.

Improved Identification

IFF Mk I and II had met with reasonable success, but improvements in their operating reliability drove continuing developments. The great importance in this reliability is illustrated in an event told by Rowe in *One Story of Radar*:

> Each of two pilots had felt uncertainty regarding the identity of the other and were discussing the incident. One said 'You were lucky, I was about to open fire'. The other said 'You were luckier; I did'.

The airborne transmitters and receivers of the initial systems swept through all frequencies of ground RDFs, resending on the same frequency any interrogating pulse and giving a visual indication on the receiver screen. As the number of transmission bands increased, this sweep became more difficult and time-consuming. In addition, Mk II would sometimes respond to German systems. A new technique was needed.

Dr. B. Vivian Bowden from Cambridge joined the TRE to lead the IFF Group. Working out of a lodge at the main gates to the Leeson House, they tried a variety of new techniques. Eventually, Dr. F. C. Williams solved all of the problems with an entirely new IFF technique.

This new system used a separate transmitter sweeping across 157-187 MHz every 2.9 seconds and co-located with the ground RDF transmitter. The receiver-transmitter on the interrogated aircraft responded with a pulse of changeable coded length, with aircraft distress indicated by a very long pulse. Designated IFF Mk III, this system was put into British production by Ferranti in early 1941. Operationally, there was a problem in that the pilot needed to change the response code at a precise time, sometimes when he was engaged in combat.

Ground-Based Navigation Systems

A number of systems had been developed in the 1920s and '30s that used radio signals for navigation. These, however, were only effective over relatively short distances. In 1937, Robert J. Dippy at Bawdsey proposed a radio navigation system using coordinated transmissions from three or more radio stations. When received by an aircraft, the transmissions could be used to pinpoint the aircraft's position. Dippy's system required the transmission of precisely synchronized pulses, and a low-level project was started for the development.

Central to the concept was the fact that all points where the time difference between any two of the stations could be graphed was a hyperbola. With a transmitter at the locus of the hyperbola, the distance between the transmitter and the receiver could be calculated. By using three transmitters – one the master and the others slaves that repeated the signal from the master – the hyperbolas overlapped and the receiver position was given from where the hyperbolas intersected. The system required the transmission of precisely synchronized pulses.

When the TRE was established, Dippy was placed in charge of a group to speed the development of his hyperbolic navigation system. With the code name "G" or "GEE" – actually short for "Grid" – the system operated in the short-wave frequency range 20 to 86 MHz.

The initial GEE system, designated AMES 7000, had a major limitation in that it required line-of-sight between the ground-based transmitter and airborne receivers, and was thus limited in range to about 350 miles. This, however, covered the Netherlands and the Ruhr Valley industrial district of Germany. The aircraft position could be established

GEE Receiver on Bomber

to within an elliptical area of some one by six miles. The GEE system was first used in bombing runs by the RAF in March 1942.

Oboe was another ground-based navigation system for bombers. It was conceived by Alec H. Reeves and developed with the assistance of

Oboe Ground Station

Dr. Frank E. Jones at about the same time as GEE. This used two widely spaced, precisely located, ground stations designated "Cat" and "Mouse," and a transponder on the aircraft. Pulse-coded signals transmitted from Cat would be used in the planes to navigate in a circular arc that passed directly over the target. (Reeves had earlier patented pulse-code modulation when he was with the Standard Telephone Research Laboratory.) Mouse, an RDF set, determined an aircraft's ground speed and gave a coded signal when the drop point was reached.

The initial Mark I Oboe system, also designated AMES Type 9000, used 100-MHz transmitters. Like GEE, the range was limited to some 350 miles because of the need that a line-of-sight be maintained between the aircraft and both ground stations. The accuracy, however, was very good; bombs released from an aircraft flying at 30,000 feet would usually strike within about 300 feet of the selected target. Sometimes called by the users "through-the-clouds bombing," Oboe first became operational in December 1942.

COUNTERMEASURES AND INTELLIGENCE

In September 1939, Dr. R. V. (Reginald Victor) Jones (1911-1997) became the first civilian scientist at the Air Ministry headquarters. In a short while, he was named Assistant Director of Intelligence (Science), and his accomplishments were such that he is called the Father of Science and Technology Intelligence. Jones graduated in 1932 from Oxford with a First Class Honors in physics, then did research at the Clareton Laboratory to complete his Ph.D. degree. He initially worked at the Royal Aircraft Establishment (RAE), Farnborough, where he was instrumental in developing the very successful VHF radiotelephone system use by the RAF. Through this, he gained great expertise in the analysis of radio signals.

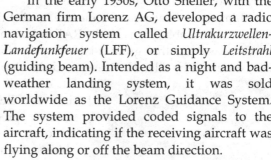

R. V. Jones

In the early 1930s, Otto Sheller, with the German firm Lorenz AG, developed a radio navigation system called *Ultrakurzwellen-Landefunkfeuer* (LFF), or simply *Leitstrahl* (guiding beam). Intended as a night and bad-weather landing system, it was sold worldwide as the Lorenz Guidance System. The system provided coded signals to the aircraft, indicating if the receiving aircraft was flying along or off the beam direction.

As Germany started night-time bombing of Great Britain in 1940, the Lorenz System was modified to have two separated transmitters with their beams crossing, the second providing bomb-release positioning. Described in Chapter 5, this was called *X-Leitstrahlbake* (Direction Beacon) and nicknamed *Knickebein* (Bent Leg). From interrogations of downed German airmen and examination of radios in wreckage, Jones found how *Knickebein* functioned. At the TRE, a Counter Measures Group was set up under Dr. Robert Cockburn to develop means of countering the beam system. Before joining the TRE,

Cockburn had worked with Jones at the RAE, developing the RAF's VHF radiotelephone system.

The German beams had been code-named "Headaches," so the countermeasure transmitters were called "Aspirins." Electrodiathermy sets, used in hospitals for electrically cauterizing wounds, were pressed into service as temporary broadband jammers. With antennas attached, they transmitted radio noise over a wide range of frequencies and disrupted *Knickebein* transmissions.

The Battle of the Beams followed and the *Knickebein* system was soon replaced by a more sophisticated modification of the Lorenz System – the *X-Geraet* (Device) using several beams – that was used in the incendiary bombing of Coventry. Again, countermeasures transmitters (called "Bromides") developed by Cockburn's team soon neutralized the system.

Near the end of 1940, an even more sophisticated system, *Y-Geraet*, began operating. In this, the bombers picked up the navigation beam and retransmitted it back to the originating station. The moving aircraft generated a Doppler shift in the reradiated signal, allowing the ground controller to determine the aircraft speed and, in turn, its flight position. The controller could then signal when to drop the bombs, thus eliminating the need for the second *X-Geraet* beam.

While *Y-Geraet* was being tested, it was discovered by the British and the high-power BBC television station in London was pressed into service as a countermeasure. Picking up the reradiated signal, it was rebroadcast by the BBC station at the same frequency, thus jamming the system. This was in place by February 1941, and used successfully the first time *Y-Geraet* operated in combat.

In May 1942, when Germany turned from massive air attacks on Great Britain, they set up standard navigation beacons in France, Germany, and Norway to assist in small-scale attacks. The TRE then developed fake beacons, called "me-cons" to confuse the *Luftwaffe* pilots. They had considerable success, even turning around two bombers and guiding them to land on RAF bases by mistake.

As the Chain Home RDF system was being developed, the possibility of German electronic countermeasures (ECM) was considered, and electronic counter-countermeasures (ECCM) techniques were incorporated in the design. The system could easily switch frequencies, the simplest ECCM for jamming.

Before the war, the open literature indicated that the Germans had developed a *Funkmessgerät für Untersuchung* (radio measuring device for reconnaissance), but little detailed information was available. Further, few British officials believed that the Germans had invented their version

Radar Origins Worldwide

of RDF. In December 1939, the British Navy forced the German pocket battleship *Graf Spee* to be scuttled in shallow waters off Montevideo, Argentina. L. H. Bainbridge-Bell from the AMRE was sent to Montevideo and joined sightseers in climbing over the abandoned warship. In his report, he said that he examined what he believed to be an antenna array for a gun-ranging unit. This was the first hard evidence that Germany had RDF systems.

As time passed, signals believed to be from German RDFs were picked up, and these were used by Jones for identifying characteristics of the equipment. Although there was no organized search for such signals, Jones conducted impromptu searches wherever possible. As an example, Derek J. Garrard from the TRE was to drive from Swanage to London for a meeting with Jones, and was asked to carry a variety of receivers in his car. Along the southern coast, Garrard picked up an unusual signal on 375 MHz. When this was reported to Jones, he correlated it in time with the shelling of British ships by shore guns. Message intercepts at that same time had mentioned "aiming using *Seetakt*," thus tying together a set's name, application, and operating frequency. Garrard later joined the Intelligence Staff of the Air Ministry and assisted Jones in analyzing German systems throughout the war.

Considerable information was gained through high-level message intelligence, code-named Ultra, produced by cryptographic operations at Bletchley Park. (This activity was located in Buckinghamshire on the main highway between Oxford and Cambridge, primary sources for the cryptographers.) Listeners at intercept stations around Great Britain took down German radio messages – usually in Morse code – by hand, and these were delivered by motorcyclists to Bletchley Park. There, analysts used electro-mechanical Enigma machines, captured codebooks, and mathematical ingenuity for deciphering the coded messages, believed by the Germans throughout the war to be "highly secure." (A German Enigma machine and instructions on its operation had been secretly provided to Great Britain in 1939 by Polish cryptographers.) Decoded messages gave the names for many of the systems, such as *Freya* (125 MHz), *Würzburg* (560 MHz), and *Mammut*, a giant phased array of 16 *Freyas*.

Bletchley Park

As Great Britain started large-scale bombing of Germany, the attrition of bombers as they entered the European mainland indicated that the defending fighter planes were being warned by some large, distributed system similar to the Chain Home. Jones, using signal intercepts and intelligence reports, mapped this as a system of *Würzburg* and *Freya* sets spaced every few miles along the coast facing Great Britain. The British called this the "Kammhuber Line," after the developer Brigadier General Josef Kammhuber.

To establish an effective ECM for the Line, Jones needed more specific information about the *Würzburg* system. Extensive aerial reconnaissance was made of a cliff-top installation near Bruneval on the coast of France. After considerable analysis of intelligence and debate at top levels, Operation Biting (popularly called the Bruneval Raid) was authorized by the Prime Minister. Colonel (Dr.) Basil F. J. Schonland, a South African who had developed their RDF and who was then head of the British Army's operational research activities, put together a full plan for the raid, including a rehearsal exercise. Donald H. Preist, a civilian expert from the TRE, provided technical advice.

Site Photo – Radar Lower Left

On February 27, 1942, a raiding party of 116 men parachuted into Bruneval. The party was composed of RAF personnel and crack troops from the Army's Brigade of Guards. Included was RAF Flight-Sergeant C. W. H. Cox, an RDF operator. The Royal Navy positioned several landing crafts offshore; Preist was aboard, operating a receiver that monitored signals from the *Würzburg*.

Preist with Hardware

The raiding troops met considerable resistance from the Germans guarding the site, but were able to photograph the equipment, take out key electronic units (selected by Cox), and capture a *Würzburg* technician. After retreating from the cliff tops onto the beach, they were evacuated by the Navy. British losses were two killed, six wounded, and eight captured. Five Germans were killed and two taken prisoner.

The photographs and electronic units were of great benefit to Jones's intelligence staff as well as Cockburn's ECM designers at the TRE. Those examining the equipment marveled at how well it was engineered. Much of it was in modules, an architecture that greatly facilitated maintenance but not used in British systems at that time.

From the captured materials, an operating *Würzburg* was reproduced in a few weeks. At the RAE, Farnborough, a design flaw in the receiver was found, leading to the development of a jammer that worked very effectively on future flights over the Ruhr Valley.

In the British bombers, transmitters built for radio amateurs were initially used for jamming German ground systems. To know when ECM should be used – when they were being "illuminated" – amateur radio receivers were also used, particularly the American-built Hallicrafter S-27 covering 27-145 MHz.

Bibliography and References for Chapter 2

Appleton, E. V.; "On two methods of ionospheric investigation," *Proc. Phys. Soc.*, vol. 45, p. 673, 1933

Bowen, E. G.; *Radar Days*, Inst. of Physics Pub., 1987

Bowen, E. G.; "The development of airborne radar in Great Britain 1935-1945," in *Radar Development to 1945*, ed. by Russell Burns; Peter Peregrinus, 1988

Brown, Louis; *A Radar History of World War II – Technical and Military Imperatives*, Inst. of Physics Pub., 1999

Carter, R. H. A.; "A personal reminiscence: RDF and IFF," in *Radar Development to 1945*, ed. by Russell Burns; Peter Peregrinus, 1988

Churchill, Winston S.; *The Hinge of Fate*, vol. IV of *The Second World War*; Cassell, 1951

Coales, J. F., and J. D. S. Rawlinson; "The Development of Naval Radar 1935-1945," *J. Naval Science*, vol. 13, nos. 2-3, 1987

Cockcroft, J. D.; "Memories of radar research," *IEE Proceedings*, vol. 132, Oct., p. 327, 1985

Ditton Park Archive; "History of Radio Research at Ditton Park"; http://www.dittonpark-archive.rl.ac.uk/histTime.html

Dobbs, J., and J. Ludlow; "The C.H. radiolocation transmitters," *J. of the IEE*, vol. 93, pt. IIA (Radiolocation), p. 1007, 1946

Guerlac, Henry E.; *Radar in World War II*, vol. 8 of *The History of Modern Physics 1800-1950*, Amer. Inst. of Physics, 1987

Howse, Derek; *Radar at Sea: the Royal Navy in World War II*, Naval Institute Press, 1993

Jefferson, S.; "Radar at Worth Matravers and Swanage," The Purbeck Radar Museum Trust, 1990

Jones, R. V.; *Most Secret War: British Scientific Intelligence 1939-1945*, Hamish Hamilton, 1978

Kinsley, Gordon; *Orfordness – Secret Site*, Terence Dalton, 1983; *Bawdsey – Birth of the Beam*; Terence Dalton, 1983

Larnder, Harold; "The Origin of Operational Research," *Operations Research*, vol. 32, no. 2, p. 465, 1984

Latham, Colin, and Anne Stobbs, (Eds.); *Radar – A Wartime Miracle*, Sutton Publishing, 1996

Latham, Colin, and Anne Stobbs, (Eds.); *The Birth of British Radar - The Memoirs of Arnold "Skip" Wilkins*, Speedwell, 2006

Lovell, Sir. Bernard; *Echoes of War*, Adam Hilger, 1991

Merchant, P. A., and K. M. Heron; "Post Office Equipment for Radar," *Post Office Electrical Engineering Journal*, vol. 38, Jan. 1946

Millar, George; *The Bruneval Raid – Flashpoint of the Radar War*, Doubleday & Company, 1975

Neale, B. T.; "CH – the first operational radar," in *Radar Development to 1945*, ed. by Russell Burns, Peter Peregrinus, 1988

Penley, W. H., and R. G. Batt; "The British Radar Story 1935-1945," The Purbeck Radar Museum Trust, 1997

Penley, W. H.; "Penley Radar Archives—the Early Development of Radar in the UK," 2006; http://www.penleyradararchives.org.uk

Putley, E.; "Ground control interception," in *Radar Development to 1946*, ed by Russell Burns, Peter Peregrinus, 1988

Rowe, A. P.; *One Story of Radar*, Cambridge University Press, 1948

Rowlinson, Frank; *Contributions to Victory*; Metropolitan-Victors Electric Co., 1947

Soulounac, A; "The story of IFF (Identification Friend or Foe)," *Proc. IEE*, vol.132, pt. A, p. 435, Oct. 1985

Swords, S. S.; *Technical History of the Beginnings of Radar*, Peter Peregrinus, 1986

"Tesla, at 78, Bares New Death Beam: Death-Ray Machine Described," *New York Sun*, July 11, 1934

Tomlin, D. H.; "The origins and development of UK army radar to 1946," in *Radar Development to 1945*, ed. by Russell Burns, Peter Peregrinus, 1988

Tomlin, Donald H.; "From Searchlights to Radar – The Story of Anti-Aircraft and Coastal Defense Development 1917-1953," IEE Conference on the History of Electrical Engineering, p. 110, July 1988

Watson Watt, R. A., et al: "The return of radio-waves from the middle atmosphere." *Nature*, vol. 137, p. 866, 1937

Watson Watt, Robert A. "State of RDF Research," Report to CSSAD, Orfordness Research Station, Sept. 9, 1935

Watson Watt, Robert A.: "Detection and Location of Aircraft by Radio Means." (Secret), Radio Research Station, Slough, February 27, 1935

Watson-Watt, Sir. Robert; *Three Steps to Victory*, Odhams Press, 1957; *The Pulse of Radar*, Dial Press, 1959

Whitehead, J. (James) Rennie; "Memories of a Baffin," 1995; http://www3.sympatico.ca/drrennie/memoirs.html

Wilkins, A. F.; "The Story of Radar," *Research*, vol. 6, p. 434, Nov. 1953

Chapter 3
EARLY RADIO-ECHOING IN AMERICA

Between the turn of the century and the end of World War I, much of the development of radio technology for the military in America was accomplished by industry. In the 1920s, the situation changed; the radio industry became almost totally devoted to consumer electronics. Consequently, although with limited budgets, the U.S. Army and Navy had to develop their own radio research and development capabilities.

With the arrival of the Great Depression in 1929, industry again sought contract work from the Government. But the Army and Navy were also affected financially and tended to hold onto the scarce funds, sustaining their still-fragile developmental capabilities. With any ordinary innovations readily offered by industry, the emphasis within the government military organizations naturally turned to development of items with special benefit to themselves.

One such activity was what was often called radio-echo detection – the use of radio signals to detect the presence of targets. This was soon enlarged to include the distance or range to the target and sometimes given the name radio direction-finding. This had a strong relationship to undersea acoustical technologies that began prior to WWI and were still being developed. Recognizing that such applications had significant military connections and should therefore be kept classified, these projects became even more closely held. Eventually, the encompassing name radio detection and ranging was adopted, with the acronym RADAR used for cover, soon followed by simply the name radar.

MILITARY RESEARCH LABORATORIES

In the United States, two research laboratories were central to the evolution of radio into radar: the Naval Research Laboratory (NRL) and the Signal Corps Laboratories (SCL). Both of these organizations were completely dedicated to furthering military technologies and had broad activities, but in the latter half of the 1930s, work in radar dominated. Several industries, especially the Bell Telephone Laboratories (BTL) and the research laboratories of RCA and General Electric, also made significant contributions, but almost exclusively in components and technologies, rather than system development.

Radar Origins Worldwide

Naval Research Laboratory

In the period before World War I, research supporting military technology, particularly in radio communications, had been relatively unorganized in the U.S. This situation started to change when America's foremost inventor, Thomas A. Edison, made a far-reaching proposal. In an interview published in the May 30, 1915, issue of *The New York Times Magazine*, Edison said that the Navy should have an "inventions factory," modeled on those then existing in progressive American industries.

Secretary of the Navy Josephus Daniels was impressed with Edison's suggestion and asked him to take the lead in recruiting a technical advisory group to assist in the endeavor. With typical enthusiasm, Edison gathered 24 of the best luminaries in the scientific-engineering community and formed the Naval Consulting Board (NCB) of the United States, whose charge was to identify and shepherd state-of-the-art military invention.

Acting on the recommendation of the NCB, Congress in 1916 appropriated funds for establishing a central research laboratory as a unit directly under the Secretary of the Navy. The onset of World War I delayed further progress on this development, but the NCB itself had a very active role in a number of wartime programs, including the emergence of underwater sound techniques and hardware.

It was 1920 before construction of the Naval Research Laboratory (NRL) started. After more delays, it was officially opened on July 2, 1923. Thomas Edison had been very vocal in insisting that the

Naval Research Laboratory 1923

Laboratory should be run by civilians, not Navy officials. Although most of his recommendations concerning the NRL were accepted, this one was not. In protest of placing the NRL under the direction of a naval officer, America's best-known inventor resigned from the NCB and did not attend the opening ceremony. A bust of Edison is at the entrance of the present NRL main building, in recognition of his founding of this prestigious research organization.

Located on a 27-acre campus along the east bank of the Potomac River at Bellevue in the District of Columbia, the NRL began life as a small-scale invention factory much as Edison and the NCB had envisioned it should. It had a few doctoral-level scientists, but most of the technical staff were engineers whose work was aimed at inventing new devices, rather than adding to what Edison believed was "an already enormous heap of untapped basic scientific understanding."

As a unit directly under the Secretary of the Navy, the NRL's top management was composed of naval officers. Captain Edwin L. Bennett was the first official Director, but the NRL was only one of his many responsibilities and his office was at the Navy Department, not the Bellevue campus. It was Captain Edgar G. Oberlin, the Assistant Director, who was the hands-on manager, making most of the executive decisions during his tenure that lasted until 1931.

The NRL absorbed most of the Navy's existing research and development activities, mainly forming the Radio Division. These included the Navy Radio Laboratory at the National Bureau of Standards, the Washington Radio Test Shop, and the Aircraft Radio Laboratory. Most of the initial personnel in this Division came from these three operations, including the Superintendent, Dr. A. Hoyt Taylor, who previously headed the Aircraft Radio Laboratory.

The Sound Division was another initial activity at the NRL. Dr. Harvey C. Hayes, who had previously led the underwater sonics research and development at the Experimental Station at Annapolis, was named Superintendent. A short time later, the Heat and Light Division was established with Dr. Edward O. Hulburt as the Superintendent. Hulburt had worked for the Army Signal Corps in Paris during the war. All three of the divisions had a major common interest in electronics.

The new buildings at the NRL would have been essentially empty except for 34 train-car loads of war surplus machinery and equipment that was provided by the Army. Also, many of the initial personnel brought with them their own research equipment. Congress gave an initial operating allocation of only $100,000, a sum that did not even cover the staff salaries and overhead. Thus, any Navy bureau needing work done by the NRL had to pay from its own budgets.

One of the initial five buildings on the NRL campus was designed to be a training facility. Activities here began in 1924 with the opening of the Radio Materiel School (RMS). This was developed to provide high-level maintenance training on the fleet communications equipment.

While primarily for highly select enlisted personnel, it also included a radio engineering program for warrant officers, as well as special courses for regular officers.

The initial 18 professional members of the Radio Division were all well known to Hoyt Taylor, the Superintendent. Essentially all were transferred from the merged organizations. Taylor himself was ideally suited to the new appointment.

Albert Hoyt Taylor (1874–1961), a native of Chicago, graduated from Northeastern University and became an instructor at the University of Wisconsin. He later studied at the University of Goettingen, Germany, where in 1908 he was awarded the doctorate in electromagnetics. Returning to America, he accepted the position as head of the Physics Department, University of North Dakota.

Albert Hoyt Taylor

As a Lieutenant in the Naval Reserve, Taylor was called to active duty in 1917 and assigned as District Communications Officer, 9th Naval District, Great Lakes, Illinois. On his own initiative, he developed a research project in underground antennas.

After Great Lakes, Taylor was sent to the Marconi Wireless Station (then operated by the Navy) at Belmont, New Jersey, and headed the Trans-Atlantic Communications Systems along the East coast. Following this he was assigned as Director of the Aircraft Radio Laboratory at the Anacostia Naval Air Station in the District of Columbia. Eventually promoted to Commander, Taylor returned to civilian status in 1919 but remained as the Director at the Laboratory until joining the NRL when it opened.

Taylor carefully selected the initial key personnel for the NRL Radio Division. Dr. J. M. Miller was named Assistant Superintendent and Director of the Precision Measurements group. Other initial directors and their groups were W. B. Burgess, Direction Finders; T. M. Davis, Receivers; L. A. Gebhard, Transmitters; C. B. Mirick, Aircraft Radios; and L. C. Young, General Research.

Burgess had worked in developing direction finders at the NBS Navy Radio Laboratory. Davis, formerly a Gunner's Mate, had assisted in making major contributions to receivers at the Radio Test Shop. Mirick came from the Aircraft Radio Laboratory. Both Gebhard and Young had been members of Taylor's team at Great Lakes, then followed him to Belmar and the Aircraft Radio Laboratory. Gebhard would eventually succeed Taylor as superintendent of the Radio Division. To this day,

Taylor and Gebhard remain connected through the intersection of NRL campus streets named for them.

Leo Clifford Young (1891–1981) was closely tied to Taylor in all of his Navy work. These two men had a strong friendship through their love for amateur radio, Taylor with call letters 9YN and Young with W3WV. In contrast to Taylor, Young's formal education stopped with a high-school diploma. He was, however, highly self-educated in radio technology. He grew up on a farm near Van Wert, Ohio, and built his first receiver when he was 14 years old. After high school, he worked as a railroad telegrapher for a number of years. Young was an avid amateur radio operator, and in 1913 set up the central control station for the Navy-Amateur Network. A member of the Navy Reserve, he was called up at the start of the war and assigned to the Great Lakes Naval Radio Station, where Hoyt Taylor was the officer-in-charge. Young followed Taylor in his other wartime assignments, and then joined him as a primary assistant when the NRL was formed.

Leo Young

Science writer Ivan Amato, in his 1998 book, *Pushing the Horizon,* which heralds the first 75 years of the Naval Research Laboratory, described the technical environment:

> In these first decades of the 20th Century, the anatomy of technology was more self revealing. Knobs turned, metal touched metal, dials swung from number to number, parts moved. . . . It was with their tools in hand and radio parts all asunder that many in the Radio Division had developed an intuitive relationship with the unseen electromagnetic waves.

Most of the initial tasks of the NRL Radio Division centered on new transmitters and receivers. Early developments included high-frequency transmitters producing 10-kW output power with piezoelectric-crystal control up to about 10 MHz, and the Navy's first high-frequency receiver, the RG, covering 1-20 MHz.

The Fleet was preparing for a cruise to Australia and New Zealand in early 1925. To examine the merits of high-frequency communications, a global experiment was planned. With cooperation of the American Radio Relay League (ARRL), amateurs worldwide would attempt communicating with the Fleet flagship, the USS *Seattle,* that was equipped with a HF transmitter operating at 5.7 MHz and an RG

Louis Gebhard with HF Transmitter

receiver. The same type receivers were also at the NRL and at San Francisco, Honolulu, and Balboa. The tests were very successful, with good Morse-code communications between the *Seattle* and Washington even when the ship was in Melbourne, almost 10,000 miles away.

Following these tests, immediate plans were made to equip ships and shore stations with transmitters capable of up to 18 MHz. Within a few years, this led to the U.S. Fleet's having far more reliable communications capability than any other fleet in the world.

It had been observed that communications between 2 and 4 MHz were better during the night but those between 4 and 12 MHz were better in daylight hours. Taylor arranged with a number of geographically distributed amateurs to gain data on these phenomena. These experiments resulted in the discovery of a "skip distance" – a zone of silence that varied with frequency. The existing theory of reflections due to the ionosphere (then called the Kennelly-Heaviside Layer) did not explain this. In 1925, Taylor teamed with Hulburt of the Heat and Light Division to develop a more generalized mathematical account of propagation based on refraction in the ionosphere.

The resulting paper, "The Propagation of Radio Waves Over The Earth," was published in *Physical Review, vol. 27* (1926). Hulburt later credited this work with putting the NRL on the scientific map because its significance went beyond military communication and into the realm of basic science. Much later, a panel from the American Physical Society would rate this work as one of the 100 most important applied research papers of the century.

The observation of signal interference made in 1922 by Hoyt Taylor and Leo Young at the Aircraft Radio Laboratory (described in Chapter 1 under Radar Precursors), was again made at the NRL in mid-1930. In tests being made on a receiving antenna, variations in received signal strength were noted when an aircraft flew over the transmission path. This was reported by Taylor as an aircraft detection technique. Although no funding was provided at that time for development, this serendipitous discovery would ultimately lead to radar and have a profound effect on the conduct of warfare.

During the early 1930s, unfunded research on signal interference as a potential for aircraft detection continued at a low level. Recognizing the advantages of higher frequencies for this radio application led to investigations with microwaves, but the lack of transmitters at these wavelengths soon brought this to a halt.

With the onset of the Great Depression, the Navy's appropriations were cut to the core. The Radio Division of the Bureau of Engineering had been funding more than half of the NRL work. As the financial conditioned worsened, the Radio Division recommended that the work not only be further curtailed but that the NRL be taken out from the Office of the Secretary of the Navy and placed in the Bureau of Engineering.

Captain Oberlin, now Director of the NRL, made a number of attempts to sway the Secretary of the Navy, but these were in vain. In November 1931, the Secretary ordered the change. Oberlin then bypassed the Navy channels, appealing to the Naval Appropriations Subcommittee of the House of Representatives. The Navy's position remained unchanged, and Oberlin was relieved of his command in early 1932.

Overall, the operations at the NRL were hardly affected by the Great Depression. The staff was only slightly reduced, and all of the key personnel were retained. Later, every civilian employee in the Government took a temporary pay cut. Some work in the Radio Division during this period included development of the decade frequency synthesizer and incorporation of the Dow electron-coupled circuit, devised to eliminate crystal controls in high-frequency transmitters.

By 1934, the economic situation had improved. Further, Rear Admiral Harold G. Bowen was appointed Chief of the Bureau of Engineering and was personally interested enough in the NRL to place it directly under his authority in the Bureau. The NRL received more funding for research, and in a short while, work on radio-based detection, and later ranging, was central to the NRL Radio Division. Other related activities did include a project led by Carlos Mirick involving the development of radio-controlled aircraft (America's first drones) and research in high-frequency communications.

Through the years, the activities of the NRL Sound Division were closely allied with those of the Radio Division, each centering on improvements in electronics. The respective Superintendents, Dr. Harvey Hayes and Dr. Hoyt Taylor, were the leading scientists in the Laboratory and worked together in promoting the NRL. In fact, Hayes was Taylor's main supporter in obtaining the first appropriated funding for radar research. Although the Sound Division had systems that used water for acoustic-signal propagation and the Radio Division used electromagnetic waves, equipment for both involved transmitters, receivers, signal processors, and displays.

Signal Corps Laboratories

At the beginning of World War I in 1917, the Army Signal Corps had established a training facility for signal troops in east-central New Jersey. Shortly after opening, it was named Camp Vail, after Alfred M. Vail, an inventor associated with Samuel F. B. Morse. Later that year, the Army determined that a need existed for a research laboratory devoted to radio and electronics, and established the Signal Corps Radio Laboratories at Camp Vail.

Col. Squier at Radio

Under the direction of Col. (Dr.) George O. Squier, the Radio Laboratories centered on the standardization of vacuum tubes and the testing of equipment manufactured for the Army by commercial firms. Experimentation was also being done on radio communication with aircraft, and on aircraft detection using sound. Squier personally made a major contribution to communications by developing multiplexing, for which he was elected to the National Academy of Science in 1919.

After the end of the war, aviation communication was transferred to the Signal Corps Aircraft Radio Laboratory at Wright Field in Dayton, Ohio. The Radio Laboratories continued at a low level, centering on design and testing of radio sets and field-wire equipment. The facility survived as an Army installation when all Signal Corps schools, both officer and enlisted, were moved to Camp Vail. The initial curriculum in the officers' division included courses in radio, telegraph, and telephone engineering, as well as signal organization and supply. For enlisted personnel, there were courses in radio, electricity, photography, meteorology, gas engines, and motor vehicle operation. In the early 1920s, the facility was elevated to Fort Vail. The training operation there was named the Signal School and expanded to accommodate the ROTC and National Guard, as well as personnel from other services.

In 1925, Fort Vail was renamed Fort Monmouth. Although overshadowed by the Signal School and at a reduced scale due to budget restrictions, the Radio Laboratory remained an important activity at Fort Monmouth. Developments included a variety of radios for voice and Morse-code communications. Coupling capabilities in electronics and meteorology, in 1929 the Laboratory developed and launched the first radio-equipped weather balloon.

Going into the 1930s, continuing decline in economic conditions led to the consolidation of the Signal Corps' widespread laboratories. The Electrical and Meteorological Laboratories and the Signal Corps Laboratory at the NBS were moved from Washington, D.C., to Fort Monmouth. The Subaqueous Sound Ranging Laboratory was transferred there from Fort H. G. Wright, N.Y. The Aircraft Radio Laboratory, however, remained at Wright Field. On June 30, 1930, the consolidated operations became the Signal Corps Laboratories (SCL). The initial SCL had a personnel strength of 5 officers, 12 enlisted men, and 53 civilians. Major (Dr.) William R. Blair was named Director.

This Laboratory was responsible for the Army's ground radio and wire communication development and for improvement of the meteorological service. The next year, the SCL was also made responsible for research in the detection of aircraft by acoustics and electromagnetic radiation. While the number of personnel was inadequate for major work in these many and diverse areas, it must be noted that Blair, the Director, was personally knowledgeable in all of them.

SCL Sound Detector

William R. Blair (1874-1962) was awarded the Ph.D. degree in 1906 from the University of Chicago. His dissertation involved experimental studies of microwave reflections, including those from non-metallic surfaces. After graduation, he took a position with the U.S. Weather Bureau as an aerological specialist. There he prepared a major report, "Meteorology and Aeronautics," for the NACA (National Advisory Committee for Aeronautics, predecessor of NASA) that was widely circulated as a basic handbook. The theoretical portions of the report were published in the *Proceedings of the Philosophical Society*, vol. 58 (1917).

William Blair

When World War I began, Blair was commissioned as a Major in the Aviation Section of the Army Signal Corps Reserves. Following the war, he remained in the Army as a meteorologist and participated in planning the first round-the-world airplane flight in 1924. While attending the Command and General Staff College, he made a study of acoustical direction-finding for antiaircraft artillery, and soon realized that this could better be done using electromagnetic waves. In 1926, Blair was assigned as the Chief of Research and Engineering under the Chief Signal Officer. Eventually promoted to Colonel, Blair served as Director of the SCL from 1930 until his retirement in 1938.

During the 1920s, the Army Ordnance Corps at Frankfort Arsenal had made tests in detecting infrared emitted from airplane engines or reflected by their surfaces. When the SCL was formed, this work was transferred to that Laboratory. Carrying this forward, in 1931 Blair initiated Project 88, "Position Finding by Means of Light." Here "light" was used in the general sense of electromagnetic radiation, including infrared and the very-short radio waves with line-of-sight transmission characteristics. Lieutenant (Dr.) Harold A. Zahl was Blair's primary assistant in these studies.

Initially, emphasis was placed on special devices with high-gain amplification for detecting reflected infrared from a searchlight. Civilian physicist S. Herbert Anderson led this effort. In August 1932, this equipment was used to track a blimp at a distance of over a mile. Further pursuit of active detection techniques was then abandoned because of the limit of infrared energy available from searchlight sources.

Although Anderson and Zahl continued research in the passive detection of infrared emitted from aircraft motors, Blair became convinced that practical detection systems would involve reflected radio signals. No doubt he was influenced by his earlier doctoral research, and was also aware of work in this that had been done at the NRL. In December 1930, representatives of the SCL had been briefed at the NRL on the beat-interference phenomena.

EMERGENCE OF RADAR

The early 1930s were a financially difficult period for research in America's armed services. Only a short time after the American financial crash and still several years before the rise of Hitler, the nation's domestic economic worries were pressing and any foreign menace seemed remote. The Great Depression resulted in further reductions to military budgets, which had never been large following World War I. Nevertheless, a few dedicated researchers at the NRL and the SCL initiated some of the most important developments of the 20th century – those related to radar.

In the period from 1933 to 1936, both the NRL and the SCL pursued research toward radio detection of targets. It was not until near the end of this period, however, that any significant interaction between the two organizations occurred. At that time, arrangements were made for mutual visits and exchanges of technical information, mainly to avoid duplication of effort. This also allowed them to better compare their work with that being done in industry.

Some work in what was sometimes called "radio-echo detection" was also being conducted at a very low level in research laboratories of RCA and General Electric as well as the Bell Telephone Laboratories. These activities were primarily outgrowths of their communications research and mainly in the microwave region.

In October 1936, Lt. Colonel Louis E. Bender of the Office of the Army Chief Signal Officer visited various industrial laboratories to assess the state of their research applicable to radio-based detection. He found that, while they had made progress in some of the necessary components, none of these firms was prepared to take on the development of a complete aircraft detector with a reasonable assurance of meeting the requirements at an early date. His findings were summarized as follows:

> Comparison of the work done by commercial concerns applicable to this problem with that done by the Laboratories of the Army and Navy leads to the conclusion that the latter are further advanced, showing more practical results, and definitely more promising.

Navy Radar

An event of long-term significance to the NRL occurred on June 24, 1930. Lawrence A. Hyland was determining the characteristics of an aircraft's receiving antenna. A 32.8-MHz (9.14-m) transmitter was at the NRL, and an aircraft with the receiving antenna was about 2 miles away at Bolling Field. The antenna was a 15-foot wire strung between the cockpit and tail, and the aircraft was rotated to obtain the receiving pattern. Hyland observed that the signal strength delivered to the receiver fluctuated – a beat effect – when an airplane flew overhead.

This phenomenon was shown to Carlos Mirick, Director of Aircraft Radios. Recognizing that this was similar to phenomena that Hoyt Taylor and Leo Young had observed in 1922, Mirick told them of the discovery. Taylor was very interested and authorized preparation of a request to the NRL Director asking for funds. This request was passed on to the Radio Division of the Bureau of Engineering; here the Division Chief, Comdr. Stewart A. Manahan, not only denied the request but also stated that he could see no possible naval use of such a project.

Detection by Interference Beat. Captain Edgar G. Oberlin, the NRL Executive Director, found the interference phenomena interesting and authorized some discretionary funds for confirming experiments. Over the following months, however, Young and Hyland mainly continued the work on their own time. The phenomenon was observed using portable receivers at distances as much as 10 miles and frequencies up to 65 MHz.

On November 5, 1930, Taylor submitted to the Bureau of Engineering an 11-page, detailed report entitled, "Radio-Echo Signals from Moving Objects." Not receiving a response, in January the Acting Director of the NRL followed with a strong endorsement letter:

> The Director considers subject matter of the utmost importance and of great promise in the detection of surface ships and aircraft. No estimate of its limitations and practical value can be made until it has been developed. However, it appears to have great promise and its use applicable and valuable in air defense, in defense areas for both surface and aircraft and for the fleet in scouting the line.

In just days after receiving the letter, the Bureau assigned the NRL a problem designated as W5-2, with a Confidential security classification. This assignment specified that the Laboratory "investigate use of radio to detect the presence of enemy vessels and aircraft." This, however was not accompanied by specific funding, and thus amounted to an unfunded mandate.

In December 1930, detection by the beat phenomena had been demonstrated to representatives of the Army's SCL, Coast Artillery Board, and Air Corps. In early 1931, the USS *Los Angeles*, ZR3, the Navy's first dirigible, was visiting the east coast, and arrangements were made for it to fly over the NRL. Using a local broadcasting station as the transmitting source, that recorded excellent beat signals on an oscillograph.

According to Hyland, the observation results were sent to Comdr. Manahan, the Radio Division Chief at the Bureau of Engineering, hoping to influence a change in his position. In response, Manahan personally came to the NRL and berated Hyland for continuing the observations; this ultimately led to Hyland's resigning. (Lawrence "Pat" Hyland later had a highly successful career in industry, ultimately becoming the President and CEO of Hughes Aircraft.)

Despite the altercation with Hyland, the NRL Director allowed the research to continue under a low level of discretionary funding. Meanwhile, Taylor's report had very slowly made its way through the bureaucracy in Washington. In January 1932, it was finally forwarded by the Secretary of the Navy to the Secretary of War (Army) with the following suggestion:

> Certain phases of the problem appear to be of more concern to the Army than to the Navy. For example, a system of transmitters and associated receivers might be set up about a defense area to test its effectiveness in detecting the passage of hostile craft into the area. Such a development might be carried forward more appropriately by the Army.

This was, in turn, forwarded to the SCL with the comment that "the subject is of extreme interest and warrants further thought." The response of the SCL Director, William Blair, was that he was already aware of the Navy experiments and "the possibilities of this method of finding airplanes are being considered."

Research in interference or beat of signals continued at the NRL on a part-time basis. A system with several 60-MHz (5-m) stations was conceived by Young and Hyland for the protection of an area around Washington. Sufficient components of the system were set up to prove that the presence of aircraft could be detected when they were within 50 miles of the center of the protected area. Taylor called this a system of "radio screens."

This system involved separated transmitting and receiving stations, and was thus not adaptable to shipboard use. Based on the previously noted earlier memorandum from the Secretary of Navy, the NRL shared this with the SCL. Like other projects of this era, however, the lack of funding soon brought this development to a close at the NRL, and the SCL did not take it up.

In June 1933, a patent application was filed by the Bureau of Engineering in the names of Taylor, Young, and Hyland for using the beat-interference method in a system for "detecting objects by radio." The patent was granted in November 1934. (Note that this was before the Daventry experiment in Great Britain on the same phenomena.)

Based on microwave work being done at the BTL, in 1933 NRL researchers turned their attention to reflections of ultra-high-frequency radio waves. Some basic laboratory experiments were made, but it was soon realized that the power at these frequencies of existing and potential transmitter tubes from American and foreign sources would be a major obstacle. The microwave project was then abandoned.

Pulsed Systems. In 1934, Young returned to the pulsed-transmission technique that he and Louis Gebhard had developed for Gregory Breit and Merle Tuve of the Carnegie Institution in their ionospheric measurements research. Young had earlier discounted the possibility of this technique for target ranging because of the continuous nature of the ionosphere as contrasted with small reflecting areas of airplanes and ships. Now, however, he realized that pulsed techniques were the mainstay of work in the NRL's Sound Division. There, the Director, Harvey Hayes, and his associates had developed a technique in which an echo from an underwater sonic pulse could be detected and shown on a cathode-ray screen to measure the distance to a relatively small target.

With a new technique having a strong potential for solving the radio detecting and ranging problem, the activity was placed under Robert M. Page, charged with turning Young's concepts into reality.

Robert Page

Robert Morris Page (1903–1992) was raised on a farm near Minneapolis, Minnesota. He had joined the NRL in 1927, just after receiving an undergraduate degree in physics from Hamline University, a small Methodist-supported school in Saint Paul. Although he had no formal background in engineering, building radios had been his hobby since childhood, and he quickly gained the confidence of Taylor by providing very creative solutions to a wide variety of problems. Years later, Page served as the NRL Director of Research and ultimately received 65 patents, including 40 related to radar.

Page, assisted by LeVern R. Philpott, Robert C. Guthrie, and Arthur A. Varela, first gave attention to the transmitter. In a short while, they had designed and built a 60-MHz (5-m) transmitter that could emit 10-μs pulses with a wait-time of 90 μs between pulses. This allowed the initial pulse to radiate out, then be reflected and received during the dead period. With the velocity of radiation at 300,000 kilometers per second (186,000 miles per second), the range could be determined from the transmit-to-receive time interval.

The Receiver Section provided a super-heterodyne set that was modified to increase the bandwidth and also to reduce "paralysis" caused by the strong transmitter pulse. The output was monitored using a standard oscilloscope. Separate dipole antennas with tuned reflectors were used for both transmitting and receiving; these were spaced about 175 feet (10 wavelengths) and aimed up the Potomac River.

This "breadboard" system was tested in late December 1934 and detected an airplane at a distance of one mile as it flew up and down the river. Although the detection range was small and the indications on the oscilloscope monitor were almost indistinct, it demonstrated the basic concept. Based on this, Page, Taylor, and Young are generally credited with initiating radar development in the United States.

The next year was spent on improving the receiver to solve the problems peculiar to reception of microsecond pulses and with the receiver adjacent to the transmitter. The pulsing reflected signal required the receiver to have an output bandwidth far beyond that of conventional receivers, about the same as that of video amplifiers in recently emerging television sets.

The new equipment operated at 28.6 MHz (10.5 m), allowing the use of an existing large antenna built for that frequency. The transmitter, giving 3-kW pulse power at 5-7-μs pulse width and about 1800 PRF, was built by Guthrie. Drawing on the sonic system from the Sound Division, a cathode-ray tube with a radial time-base was developed to display the range to the reflecting target (it was improved and patented by Page as the Plan-Position Indicator or PPI.)

First Experimental Pulsed Radar

In considering the time required for the development, it is noted that Page, as well as others on the team, made this progress despite the lack of financial support specifically for the project. They worked on the project "between" more conventional radio problems that had direct funding. Moreover, the Bureau of Engineering took no action toward securing needed funds. In early 1935, after receiving permission from the NRL Director, Taylor and Hayes (from the Sound Division) made a direct appeal to the Senate Subcommittee for Naval Appropriations. This resulted in a special earmarked appropriation of $100,000.

Finally, with adequate funding, the improved system was completed in late 1935. Tests during the following months showed a detection range up to 25 miles. The following June, the system was demonstrated to officials from the Bureau of Engineering, and an NRL report was submitted detailing the accomplishments. Based on this report, Rear Admiral Harold G. Bowen, Chief of the Bureau, directed the NRL to "give the highest possible priority to the development of shipboard equipment," and also wrote:

> It is requested, upon receipt of this letter, that the subject problem be placed in a Secret status, that all personnel now

cognizant of the problem be cautioned against disclosing it to others, and that the number of persons to be informed of further developments in connection therewith be limited to an irreducible minimum.

A 200-MHz (1.5-m) system was then developed by Varela, allowing a smaller antenna that would be necessary for shipboard use. The system also used a duplexer, suggested by Young and developed by Page and Varela, for switching a single antenna between transmit and receive modes. This ingenious device, which connected the transmitter and disconnected the receiver and vice versa during each cycle of the transmitter pulse, was made from two quarter-wave lines operating with what was called resistance inversion. Similarly operating devices would later become common to all radar equipment.

Testing Aboard the USS Leary

In April 1937, Admiral William D. Leahy, Chief of Naval Operations, witnessed a demonstration of the equipment at the NRL and became convinced of its capabilities. That same month, the first sea-borne testing was conducted. The equipment was temporarily installed on the USS *Leary*, an old destroyer, with a Yagi antenna mounted on a gun barrel for sweeping the field of view. Planes at ranges up to 18 miles were located, with the distance limited by the low power available from the transmitter tube. The technology was fully disclosed to the BTL in July; this was the first industry brought in on the development.

In 1938, the Chief of Naval Operations directed that radio detection equipment be placed in the fleet for operational purposes. The NRL further improved its 200-MHz (1.5-m) system and designated it XAF. Page developed the ring-oscillator circuit, allowing several tubes (in even numbers) to function in parallel. This transmitter delivered 15-kW, 5-μs pulses to a 20-by-23-foot, stacked-dipole "bedspring" antenna. H. E. Reppart developed a high-resolution, single-line target display (later called an "A" Scope) with 2-mile marker pips. In laboratory tests, the XAF was successful in detecting planes at ranges up to 100 miles.

In a parallel activity, the basic principles of radio detection and ranging equipment had been disclosed to RCA in March 1938, and the firm was given a contract for the development of a 385-MHz (0.779-m), 1-kW peak power experimental system. The preliminary equipment was designated CXZ. Although not satisfactorily developed, the CXZ was tested on the battleship USS *Texas* during December 1938.

First Fielded Naval Systems. The NRL had succeeded in developing a radio-based detection and ranging system that would shortly revolutionize U.S. naval warfare. Despite this, they had difficulty in obtaining sufficient funds to rapidly pursue this and similar projects. In 1939, the Bureau of Engineering obtained only $25,000, exclusive of salaries, for NRL electronics research.

The XAF was installed on the battleship USS *New York* for sea trials starting in January 1939, and became the first operational radio detection and ranging set in the U.S. fleet. During a 3-month period, it routinely detected ships at 10 miles and aircraft at 48 miles. It was

XAF Antenna on USS New York

used for navigational purposes, for spotting shot falls, and, amazingly, for tracking in-flight projectiles.

After testing both the CXZ designed by RCA and the XAF from the NRL, the Navy decided that the XAF would be more reliable. In his report on the tests, Rear Adm. Alfred W. Johnson, Commander of the Atlantic Squadron, stated, "The XAF equipment is one of the most important military developments since the advent of radio itself. The development of equipment is such as to make it now a permanent installation in cruisers and carriers."

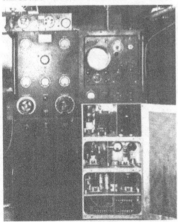

XAF Transmitter and Receiver

In May 1939, the principles of the XAF were disclosed to RCA and Western Electric, with competitive bids requested from the two firms for the further development and production of the system. The contract was awarded to RCA. Five of these search systems, designated the CXAM, were delivered in May 1940. One was placed aboard the USS *California*, a battleship that was sunk in the Japanese attack on Pearl Harbor. Others were on the aircraft carrier USS *Yorktown* and cruisers USS *Chicago*, USS *Northhampton*, and USS *Pensacola*.

Cruiser with FA Radar

Based on success at the BTL from internally funded work on higher-frequency technology, the Navy issued a contract to Western Electric for a 500-MHz (0.6-m) shipboard fire-control radar initially designated CXAS and later FA); and deliveries started in June 1940. Even with an output of only 2 kW, the FA suffered badly from a short life of the transmitter tubes.

The acronym RADAR was coined from <u>Ra</u>dio <u>D</u>etection <u>A</u>nd <u>R</u>anging. The origin is attributed to two U.S. naval officers, Lt. Commander S. M. Tucker and Captain F. R. Furth. In November 1940, the Chief of Naval Operations directed the use of the word as non-classified for reference to the then-secret project. The acronym quickly became the name "radar." For some time, this name was not publicly used and, even in official documents, "detection" was often replaced by "direction-finding."

The CXAM was further refined into the SK system and produced by General Electric starting in late 1941. Improvements included a 15- by 15-foot, rotating antenna (called the "flying bedspring") and a Plan Position Indicator (PPI), a map-like CRT display developed by Page at the NRL. With a 330-kW output, the SK could detect aircraft to 150 miles. The SK remained the early-warning radar for aircraft carriers, battleships, and cruisers throughout the war.

Derivatives of the SK included the SC, produced by General Electric, with a 15- by 5-foot antenna and primarily

SK "Flying Bedspring" Antenna

for destroyers, and the SA, built by RCA, with an even smaller antenna for destroyer escorts and other smaller vessels. About 500 sets in this series were built for the U.S. Navy. All of these systems were also used on Allies' ships. An airborne version, the ASA, intended for the Navy's R4D (C-47) aircraft, was developed, but the antenna was too large for practical mounting.

Aircraft safety drove the development of radar altimeters. Such systems needed very little transmitted power (the earth is a *large* target).

SK Controls with PPI

Designated AYA, pulsed system operating at about 500 MHz (0.6 m) was investigated by the NRL for this application. The project was turned over to the Army's SCL and in early 1940 the equipment was put into production by RCA as the SCR-518. These units ultimately gave accurate measurements to a relative altitude of 40,000 feet.

SD Radar Mast

The NRL also developed the SD, an aircraft-detection radar for use by surfaced submarines. Operating at 114 MHz (2.63 m) and with 100-kW output, it used a periscope-type antenna. The system allowed detection, but not bearing, of approaching aircraft; bearing, however, was relatively unimportant since crash-diving was the normal defense. Production of this radar was started by RCA in 1941.

One of the first applications of Operations Research in the U.S. Navy was in determining the best operating technique for the SD system. This was a trade study, comparing probabilities of target detection versus detection by the enemy of the submarine through intercept of the radar signal. It was decided to use the radar in most circumstances.

As radar came into being as a detection device, there was no means of identifying the targets as to friendly or enemy, particularly those airborne and under poor visibility conditions. In 1937, the NRL addressed this problem and developed the first recognition technique, later called Identification Friend or Foe (IFF). Designated the XAE, this consisted of a unit on the aircraft that, when in the vicinity of friendly ships, transmitted a series of coded signals. On shipboard, a Yagi antenna on a hand-held mount was aimed at the aircraft and, upon receiving and decoding the signal, would flash a light, signifying the approach of a friendly aircraft. From that start, the NRL took the lead in what was then called recognition systems. Included were the ABD and ABE systems, incorporated in the airborne radars.

Wartime Naval Research Laboratory

Army Radar

Upon becoming Director of the Signal Corps Laboratories (SCL) at Fort Monmouth, New Jersey, in 1930, Dr. (Major, later Colonel) William R. Blair initiated low-level research in detection by target-infrared and radio signals from engine ignition. Some success was made in the infrared detection, but little was accomplished using radio. Starting in 1932, progress at the Naval Research Laboratory on radio interference for target-detection was passed on to the Army, but it does not appear that any of this information was used by Blair.

Microwaves and Doppler Techniques. During 1933, the SCL undertook a systematic survey of what was then known throughout the world about the methods of generating, modulating, and detecting radio signals less than one meter in wavelength. Such radio waves, in 1933, constituted a "no man's land" between the useful radio spectrum on the one hand and the optical spectrum on the other.

The SCL's first definitive efforts in radio-based target detection started in 1934 when the Chief of the Army Signal Corps, after seeing a microwave demonstration by RCA, suggested that radio-echo techniques be investigated. The SCL called this technique radio position-finding (RPF). New sub-projects were set up with the titles of "radio-optics," "radio-optical transmission systems," and "detection and position finding of aircraft by radio-optical rays."

Emphasis was placed on assessing capabilities of the existing microwave tubes. Tests were made with the German-built Hollmann tube, giving a 5-watt output at 50 cm (600 MHz). The only available sample of this tube was X-rayed and duplicated at the SCL. Ranges were determined for 9-cm (3-GHz) magnetron equipment brought to the SCL by Drs. Irving Wolff and Ernest Linder of RCA. Laboratory experiments were also carried out with RCA Acorn tubes, producing wavelengths as short as 45 cm (660 MHz) in a Hartley oscillator circuit.

During 1934 and 1935, tests of microwave RPF equipment resulted in Doppler-shifted signals being obtained, initially at only a few hundred feet distance and later over several miles. These tests involved a bi-static arrangement, with the transmitter at one end of the line of transmission and the receiver at the other, and the reflecting target passing through or near the path. The development state of this Doppler-beat detector was summarized by Blair in his 1935 annual report on the project:

> To date the distances at which reflected signals can be detected with radio-optical equipment are not great enough to be of value. However, with improvements in the radiated power of transmitter and sensitivity of the receiver, this method of position finding may well reach a state of usefulness.

To house the activities of the SCl, Squier Hall was constructed in 1935. The facility was named to honor Maj. General (Dr.) George O. Squier, founder of the SCL and Chief Signal Officer during WWI. Lt. Colonel Roger B. Colton was assigned as the SCL Executive Officer.

Squier Hall – Fort Monmouth

Pulsed Detection Technique. In a 1934 report concerning Doppler-based detection, Blair had noted that the SCL might investigate another technique:

> Consideration is now being given to the scheme of projecting an interrupted sequence of trains of oscillations against the target and attempting to detect the echoes during the interstices between the projections.

This suggestion for using a pulsed transmission possibly came from his knowledge of the pulsed transmitter supplied by the NRL to Breit and Tuve of the Carnegie Institution for determining the height of the ionosphere.

In 1936, a modest project in pulsed microwave transmission was started by W. Delmar Hershberger, SCL's Chief Engineer at that time.

Lacking success with microwaves, Hershberger visited the NRL (where he had earlier worked) and saw their meter-wave pulsed set. Back at the SCL, he and Robert H. Noyes built an experimental set with a 75-watt, 110-MHz (2.73-m) pulsed transmitter and a receiver patterned on the one at the NRL. A request for project funding was turned down by the War Department, but, with the backing of the Chief Signal Officer, Maj. Gen. James B. Allison, $75,000 for support was diverted from a previous appropriation for a communication project.

Hershberger left in October 1936, and Paul E. Watson (later Lt. Colonel) took over as the SCL Chief Engineer and led the project. A field setup near the coast was made with the transmitter and receiver separated by a mile. On December 14, the experimental set detected at up to 7 miles range aircraft flying in and out of New York.

The Radio Position Finding (RPF) Section was formed for this activity, with Ralph I. Cole heading receiver work and William S. Marks leading transmitter improvements. The receivers – one each for separate azimuth and elevation antenna – were made less susceptible to the transmitter pulse. These receiving antennas, plus the transmitting antenna, were made of large arrays of dipole wires on wooden frames. The system output was used to aim a searchlight. The first demonstration of the full set was made on the night of May 26, 1937. An unlighted bomber was detected and then illuminated by the searchlight. The observers included the Secretary of War, Henry A. Woodring; he was so impressed that the next day orders were given for the full development of the system.

With strong support from General Allison, a special Congressional appropriation of $250,000 was obtained. The frequency was increased to 200 MHz (1.5 m). The transmitter used 16 tubes in a ring-oscillator circuit (developed at the NRL), producing about 75-kW peak power. Colton wanted lobe-switching for the receiving antennas, and Major James C. Moore was assigned to head the resulting complex electrical and mechanical design. Engineers from Western Electric and Westinghouse were brought in to assist in the overall development.

Testing at Fort Hancock

First Fielded Army Systems. For better security and more space, the RPF Section was moved to Fort Hancock. This was an isolated location on Sandy Hook, a sand-bar peninsula reaching into the New York Harbor. During 1938, Blair's health failed, and the position of SCL Director was taken over by Roger Colton, who was then promoted to Colonel.

Early Radio-Echoing in America

Roger Baldwin Colton (1887–1978) was raised in Jonesboro, North Carolina. He earned a B.S. degree in physics from Yale, and in 1910 was commissioned in the Army Coast Artillery. After several years of service, he was sent to MIT, where, in 1920, he received the M.S. degree, also in physics. Later, he graduated from the Command & General Staff School and the Army War College.

Transferring to the Signal Corps as a Major in 1930, Colton was attached to the SCL and participated in the early infrared and microwave research. Here be became a life-long advocate of radio-based detection systems and leading promoter of this technology within the Army. Working closely with Edwin H. Armstrong, Colton is also credited with developing the Army's first tactical FM radio communication set.

Roger Colton

After succeeding Blair as the Director of the SCL, Colton remained until September 1944, when he transferred to the Army Air Forces. He was awarded the Legion of Merit and the Distinguished Service Medals for his work at the SCL. Colton retired in 1946 as a Major General.

Colton arranged for the demonstration of a prototype system in late November 1938. Designated SCR-268, this was primarily intended for aiming searchlights associated with anti-aircraft guns; the radar allowed coarse pointing of a thermal detector, and this then aimed the searchlight. The night demonstration was for the Coast Artillery Board and was conducted at Fort Monroe, just off the coast near Hampton, Virginia.

SCR-268 – Army's First Fielded Radar

This was almost a failure because the target, a Martin B-10 bomber at 20,000 feet, was blown off course and flew miles out over the Atlantic. After a long return flight, it came above an opening in the clouds and, to the delight of the observers, was immediately illuminated by the radar-directed searchlight.

Production of 212 SCR-268 sets was started by Western Electric in 1939, and it entered service in early 1941; about 3,100 sets were eventually built. Later, the PPI was added and the system was designated SCR-516, a low-altitude early-warning radar.

Another observer at the May 1937 test had been Brig. Gen. Henry H. "Hap" Arnold, then Assistant Chief of Staff of the Army Air Corps. This led to a request from the Air Corps for a simpler, longer-range, early-warning system. In parallel with the completion of the SCR-268, a new project led by Dr. (Major, later Lieutenant Colonel) Harold A. Zahl got underway. Good funding and a high priority were received; thus, development was quickly completed.

Harold Zahl

Harold A. Zahl (1905–1983) was born in Chatsworth, Illinois, the son of a Methodist minister. He graduated from North Central College in the suburbs of Chicago in 1927, and then attended the University of Iowa where he earned the M.A. degree in 1929 and the Ph.D. degree in 1931, both in solid-state physics. Upon completing his doctorate, he joined the SCL staff and was also commissioned in the U.S. Army. In successive years, his work in infrared and radio-detection technologies was widely recognized. Between 1948 and 1966, Zahl, then a civilian, served as the SCL Director of Research.

Operation at 106 MHz (2.83 m) was selected. A single water-cooled WL-530 tube provided 8-kW (100-kW pulsed) power in a conventional circuit. Other simplifications centered on the antenna, including elimination of lobe-switching and adding a duplexer developed by Zahl. Overall, there was a sacrifice in accuracy, but thus was balanced by ease in maintenance and greater range (up to 240 miles).

SCR-270 and SCR-271 at Fort Hancock

There were two configurations – the SCR-270 (mobile) and the SCR-271 (fixed-site). Westinghouse received the production contract, and started deliveries near the end of 1940. Eventually, about 800 sets were built; they were the Army's early-warning radars throughout the war.

Early Radio-Echoing in America

The Army deployed five of the first SCR-270 sets around the island of Oahu in Hawaii. On the morning of December 7, 1941, one of these radars was being operated by Privates Joseph Lockard and George Elliot. At 7:02, a cluster of blips appeared on the screen at a range of 136 miles due north. They checked the movement for 18 minutes, first thinking there was something wrong with the radar, then passed the observation on to the Aircraft Warning System, just being established by the Army and Navy at Fort Shafter. The Duty Officer dismissed it as "nothing unusual," believing the detection to be from a flight of U.S. bombers known to be approaching from the mainland. The alarm went unheeded. Lockard and Elliot continued tracking until 7:39 when the planes were only 20 miles away. Sixteen minutes later, the Japanese hit Pearl Harbor.

SCR-270/271 Console

Taking over an earlier project of the NRL, the Radio Laboratory developed the SCR-518 pulsed-radar altimeter for the Army Air Forces. Operating at 518 MHz (0.579 m), this system was produced by RCA starting in 1940. The final system weighed less than 30 pounds and was accurate to about 42,000 feet above ground. The Laboratory was also involved in an early version of a portable, radar-based instrument landing system, eventually designated the SCS-51.

Testing at Twin Lights

Evans Signal Laboratory. During 1941, the Laboratory relocated again, this time to Camp Evans, a site a few miles south of Fort Monmouth. This site included the original facility of the Marconi Belmont Station, and a central building commonly called the Marconi Hotel became the headquarters. Outdoor testing of hardware was often done at Twin Lights, a lighthouse station between Camp Evans and Fort Hancock.

Evans Headquarters -- Marconi Hotel

Although the SCL initiated its radar research using microwaves, it never returned to developing sets in this wavelength region during wartime activities. The Evans Laboratory of the SCL did, however, push the frequencies higher, primarily through Harold Zahl's development in 1939 of the VT-158, a tube generating 240-kW pulse-power at up to 600 MHz (0.5 m). This was actually four triodes and their associated circuit tightly packaged in one glass envelope.

Zahl's VT-158

Following the surprise bombing of Pearl Harbor, there was a crash program to obtain radars to protect the Panama Canal Zone from a similar attack. To detect low-flying aircraft at a range allowing sufficient warning, a high-frequency radar for picket ships 100-miles offshore was needed. Dr. (Captain) John W. Marchetti led a 20-person team in using the VT-158 to adapt SCR-268s for this application. The special project was completed in a few weeks, with the first set installed on the M.S. *Nordic*.

John Marchetti

SCR-268 Below Deck on Nordic

Marchetti's team then went on to convert this into the AN/TP-3, a light-weight, transportable system and the last major radar fully developed by the SCL. The set could be assembled and placed into operation by a small crew in 30 minutes. During the war, the AN/TPS-3 was used for early warning at beachheads, isolated areas, and captured air bases. Zenith manufactured about 900 of these systems. Dr. Marchetti later became the first director of the SCL Cambridge Research Center.

Early Radio-Echoing in America

AN/TPS-3 Radar

Wartime Evans Signal Laboratory

Typical Development Facilities at Evans Signal Laboratory

Bibliography and References for Chapter 3

Allison, D. K.; "New eye for the Navy: The origin of radar at the Naval Research Laboratory," Report 8466, Naval Research Laboratory, Sept. 28, 1981

Amato, Ivan; *Pushing the Horizon: Seventy-Five Years of High Stakes Science and Technology at the Naval Research Laboratory*, Special Publication, Naval Research Laboratory, 1998

Brown, Louis; *A Radar History of World War II - Technical and Military Imperatives*, Inst. of Physics Pub., 1999

Bureau of Naval Personnel, "Radar Operator's Manual," Training Manual 01090, Department of the Navy, 1945

Colton, Roger B.; "Radar in the United States Army," *Proc. IRE*, vol. 33, p. 749, 1947

Chiles. James R.; "The Road to Radar," *Invention & Technology Magazine*, vol. 2, issue 3, Spring, p. 24, 1987

Davis, Harry M.; "History of the Signal Corps Development of US Army Radar Equipment, Part II," Office of the Chief Signal Officer, 1945

Drury, Alfred T.; "War History of The Naval Research Laboratory," in the series, *U.S. Naval Administrative Histories of World War II*, Department of the Navy, 1946

Fort Monmouth Historical Office; *A Concise History of the U. S. Army Communications-Electronics Life Cycle Management Command and Fort Monmouth, New Jersey*, CE LCMC Historical Office, 2005; electronic version, http://www.monmouth.army.mil/historian/pub.php

Friedman, Norman; *Naval Radar*, Naval Institute Press, 1981

Gebhard, Louis A.; "Evolution of Naval Radio-Electronics and Contributions of the Naval Research Laboratory," Report 8300, Naval Research Laboratory, 1979

Goebel, Gregory V.; "The Wizard War: WW2 & The Origins of Radar," a book-length document; http://www.vectorsite.net/ttwiz.html

Guerlac, Henry E.; *Radar in World War II*, vol. 8 of *The History of Modern Physics 1800-1950*, Amer. Inst. of Physics, 1987

Howeth, Linwood S.; *History of Communications-Electronics in the United States Navy*, Government Printing Office, 1963; elecyronic version, http://earlyradiohistory.us/1963hw.htm

Hyland, L., A. H. Taylor, and L. C. Young; "System for detecting objects by radio," U.S. Patent No. 1981884, 27 Nov. 1934

Hyland, L. A.; "A personal reminiscence: the beginning of radar 1930-1934," in *Radar Development to 1945*, edited by Russell Burns, Peter Peregrinus Ltd., 1988

Orr, William I.; "The secret tube that changed the war [Zahl's tube]," *Popular Electronics*, p. 57, March 1964

Page, R. M.; "Early history of radar in the US Navy," in *Radar Development to 1945*, ed. by Russell Burns; Peter Peregrinus Ltd., 1988

Page, R. M.; "Monostatic Radar," *IEEE Trans. AES*, No. AES-13, No. 2, p.557, Sept. 1977

Page, Robert Morris; *The Origin of Radar*, Anchor Books, 1962

"Radar: A Report on Science at War," Office of Scientific Research and Development, distributed by Office of War Information, 15 Aug. 1945; http://www.ibiblio.org/hyperwar/USN/ref/Radar-OSRD/index.html

Radar R&D Sub-Committee; "Operational Characteristics of [U.S.] Radar Classified by Tactical Applications," FTP 217, Joint Committee on New Weapons and Equipment, 1 August 1943; http://www.history.navy.mil/library/online/radar.htm

Schooley, Allen H. "Pulse Radar History," *Proc. IRE*, vol. 37, p. 404, April 1949

Skolnik, Merrill I.; "Detecting Radar's Development," *The Scientist*, vol. 1, 19 October, p. 22, 1987

Skolnik, Merrill I.; "Fifty Years of Radar," *Proc IEEE, Special Issue on Radar*, vol. 73, p. 182, 1985

Swords, S. S.; *Technical History of the Beginnings of Radar*, Peter Peregrinus Ltd, 1986

Taylor, A. H. and Hulburt, E. O; "The Propagation of Radio Waves Over the Earth," *Physical Review*, vol. 27, p. 189, February 1926

Taylor, A. H.; "Radio echo signals from moving objects," Memorandum E4222, Naval Research Laboratory, Nov. 1930

Taylor, A. Hoyt; *Radio Reminiscences: A Half Century*, Naval Research Laboratory, 1948, 1960

Terrett, Dulany; *The Signal Corps: The Emergency (to December 1941)*, 4th edition, Government Printing Office, 2002

Tuve, M. A., and Dahl, O.; "A transmitting modulating device for the study of the Kennelly-Heaviside layer by the echo method," *Proc. IEE*, vol. 16, p. 794, 1928

Vieweger A. L.; "Radar in the Signal Corps," *IRE Trans Mil. Elect.*, MIL-4, p. 555, Oct. 1960

Zahl, Lt. Col. Harold A., and Major John W. Marchetti; "Radar on 50 Centimeters," *Electronics*, January, p. 98, 1946

Zahl, Harold; "One Hundred Years of Research," *IRE Trans. on Mil. Elect.*, vol. Mil-4, Oct., p. 397, 1960

Chapter 4
MICROWAVES BRIDGE THE ATLANTIC

Beyond the very high frequency (VHF) portion of the electromagnetic spectrum, there is the ultra high frequency (UHF) portion. Here the frequency is typically stated in gigahertz (GHz) and the wavelength in centimeters (cm), giving rise to the name microwaves. The one basic equation stated in Chapter 1 still holds, but with conversions in frequency and wavelength, this becomes: f (GHz) = $30/\lambda$ (cm). A half-wave dipole antenna for 1 GHz would be only 15 cm or about 5.9 inches long. Herein is a major reason for desiring UHF equipment: smaller antennas.

The size factor is also at play in other components. For example, as the frequency is increased, the required spacing between electrodes in vacuum tubes must become smaller. In the microwave region, this is bad, not good. Even in high-frequency tubes (of that day) such as the Acorn and its stronger brother the Doorknob, interelectrode spacing limited their operation to far less than a GHz.

A technique known as velocity modulation can be used to overcome the interelectrode limitations. In 1920, Drs. Heinrich G. Barkhausen and Karl Kurz in Germany invented a velocity-modulation tube that could operate at up to centimeter wavelengths. Although widely used in experimental transmitters and receivers, the device could not produce significant power. In the mid-1930s, the Standard Telephone Company in Great Britain developed an 18-cm (1.9-GHz) system using Barkhausen-Kurz tubes and giant parabolic dishes that allowed communications across the English Channel, but it was closed because it could not compete with submarine cables.

The klystron was another velocity-modulation device capable of producing microwaves of relatively low power. In 1935, Drs. Oskar E. and Agnesa A. Heil published a paper in Germany on the basic concept. In 1937, likely without knowledge of this paper, brothers Russell H. and Sigurd F. Varian at the Lawrence Livermore Laboratory, Stanford University, developed a similar tube, the klystron.

The klystron received much attention after its public release in early 1939. Although initially developed as a possible high-power RF driver for a cyclotron (particle accelerator), it was quickly adopted in a smaller version as a microwave transmitter for experimentation in other applications. While its output frequency could not be readily changed, it could be made to produce low-power signals at an adjacent frequency and thus could be used (with questionable results) as a local oscillator in experimental receivers.

In 1920, the same year as the Barkhausen-Kurz tube invention, Dr. Albert W. Hull, at General Electric in the U.S, developed a velocity-modulation device that used a magnetic field to control the electron flow, but it neither had high power not could operate in the microwave spectrum. Over the years, researchers in many countries developed various forms of the magnetron that operated at higher and higher frequencies, but still with relatively low output powers.

Eventually, magnetrons were developed with split (actually dual) and cavity anodes that produced power approaching 100 watts at centimeter wavelengths. Scientific papers and patents on these date from the early 1930s. A severe shortcoming to most of these was the lack of frequency stability. Philips Company in The Netherlands marketed a 15-cm (2-GHz) magnetron rated at 50 watts; this device, with a four-segment anode, had been developed by Klass Posthumus who also published a paper detailing the design of advanced magnetrons.

At Bawdsey in September 1938, Dr. Edward G. Bowen's group constructed a 30-cm (1.0-GHz) pulsed RDF system using Western Electric 2B250 triodes in both the transmitter and receiver. With two-foot parabolic dishes (limited by the requirement to fit in the nose of an aircraft), the range was much less than that obtained with the then-existing VHF AI systems; thus, the project was discontinued.

There was also research on microwave devices in the Experimental Department of HMSS at Portsmouth under their Ultra-Short-Wavelength program. To support microwave development, Sir Charles Wright, the Admiralty's Director of Scientific Research and also Chairman of the interservice (Army, Navy, and Air Force) Committee for the Coordination of Valve Development (CVD), placed contracts with Birmingham, Oxford, and Bristol Universities for developing 10-cm, high-power valves. Wright had also seen the 30-cm system at Bawdsey and had discussed with Bowen the need for microwave sets in airborne applications.

At Birmingham University, CVD funding went to a laboratory under Dr. Marcus Oliphant that had been established to investigate devices for producing microwaves. Marcus Laurence Elwin Oliphant (1901-2000) was originally from Australia and earned a First Class Honours degree in physics from Adeliede University in 1923. After teaching and performing research at Adeliede, in 1927 he received a scholarship to work under Professor Ernest Rutherford at Cavindish Laboratory of Cambridge University; there he earned his Ph.D. in 1929. After progressing to become Rutherford's deputy, he changed to Birmingham University in 1939.

Oliphant had been among the academic participants visiting the Bawdsey operations and had led a group of university researchers in an extended stay at a CH station on the Isle of Wight, thus gaining a good understanding of RDF operations. Initially, research at Birmingham centered on the Varian klystron, and by late 1939 they had built a unit that produced 400 watts of continuous power at about 10 cm (3 GHz).

Many of the personnel in the Birmingham laboratory were trained in nuclear physics. Oliphant explained this by saying: "By virtue of their training, the staff would not be inhibited by prior convictions about what could or could not be done."

Marcus Oliphant

About this same time, Robert A. Sutton of the Admiralty's valve laboratory at the Experimental Department found a way to slightly adjust the frequency of a klystron, resulting in a much lower power but producing a signal such that could be used as a local oscillator of a super-heterodyne receiver in detecting a 10-cm input pulse.

THE MULTI-CAVITY MAGNETRON

John Turton Randall (1905-1984) was born in Lancashire and earned B.Sc. and M.Sc. degrees in physics from the University of Manchester. He worked in the GEC Wimbley Laboratory, where he became known as the leading research authority on luminescent powders for fluorescent lamps. In 1937, he started doctoral studies at the University of Birmingham, but with the advent of the war, he joined Oliphant's microwave research laboratory.

Henry (Harry) Albert Howard Boot (1917-1983) was a native of Birmingham, England. Attending the University of Birmingham, he received the B.Sc. degree in physics and began graduate studies in that field (eventually earning the Ph.D. degree in 1941). As the war neared, he also changed to Oliphant's microwave laboratory.

Boot and Randall were among the academic group that, just before the war, visited the CH station on the Isle of Wight for several weeks. Upon returning to the Birmingham University, they were assigned by Oliphant to further improve the klystron tube. By November 1939, they concluded that there was little prospect of improvement and switched their efforts to applying the resonant-cavity configuration of the klystron to the magnetron. In Boot's words,

> We concentrated our thoughts on how we could combine the advantages of the klystron with what we believed to be the more favorable geometry of the magnetron.

For designing the resonant cavity, they returned to Hertz's wire loop and expanded it to a three-dimensional cavity. They calculated that the diameter should be 1.2 cm to resonate at 10 cm (3 GHz). A device was constructed from a copper block (to dissipate the anticipated heat) that had six cylindrical cavities arranged symmetrically in a circle around a tungsten-wire cathode. This device was sealed in a glass container with a continuous evacuator. It was surrounded by an external magnet and connected to a high-voltage source.

Boot & Randall Magnetron

In mid-February, power was applied to heat the cathode, and Boot noted that indications of a strong RF output were immediately found:

> The amount of power produced was, for the first time, capable of quite spectacular effects. It was uncomfortably hot to hold the hand near the output lead ... and a small arc sprayed into the surrounding air.

Boot and Randall kept their device to themselves for a few days, making small changes and trying to determine the output frequency and power. On February 21, 1940, they first demonstrated the resonant cavity magnetron to Oliphant and others at the University. It produced an estimated 400 watts of continuous power at a wavelength of 9.8 cm (3.1 GHz). (A fluorescent light tube was first used as the indicator of power; in earlier employment, Randall had developed the phosphor coatings used inside these lamps.)

It is logical to ask why researchers attempting to improve microwave devices were not aware of the published literature on higher-power magnetrons. Oliphant and his senior staff undoubtedly knew of these papers, but Boot and Randall, in the rush of wartime, apparently pursued their work without this advantage. They later said,

> Fortunately we did not have time to survey all the published papers on magnetrons or we would have become completely confused by the multiplicity of theories of operation.

Research on magnetrons was also being conducted at the GEC Research Laboratory. Here Dr. Eric C. S. Megaw had been working for

Microwaves Bridge the Atlanic

several years on magnetrons and klystrons under the Admiralty's Ultra-Short-Wavelength program. In November 1938, he had demonstrated at the Admiralty Signal School in Portsmouth a split-anode magnetron (GEC E821) that produced a pulse output of 1.5 kW at 37 cm (810 MHz).

With the advent of the Boots and Randall magnetron, a decision was made to turn this device over to GEC, combining the two efforts and developing a magnetron that might be mass produced for use in all defense applications. (Hereinafter, when a magnetron is mentioned without noting the type, it will be in reference to a resonant-cavity device.)

In May 1940, Megaw produced the E1188, a sealed-off version (the original device required connections to a vacuum pump) with about the same characteristics as the Boots and Randall magnetron. In this effort, Megaw was assisted by S. M. Duke, who worked at both GEC and Birmingham University.

The first version was followed in June by the E1189, a lighter-weight version capable of generating a pulsed output of 10 kW. This incorporated a cylindrical, oxide-coated cathode, an improvement found in a magnetron developed by Drs. Henri Gutton and Robert Warneck at the firm FSR in France. The FSR magnetron had been rushed to Great Britain just before the German occupation of France.

E1198/CV38

A short time later, GEC released the E1198 (also designated CV38) with eight cavities and working at 9.1 cm (3.3 GHz). It was about four inches in diameter. Twelve E1198 magnetrons were initially built. All three types of these first magnetrons required an external magnetic field of about 1,500 gauss and a pulsed, high-voltage source of near 10,000 volts.

Magnetron Without Cover

At the TRE in Swanage, Dr. Philip I. Dee led the microwave developments group, with Dr. Herbert W. B. Skinner serving as his Deputy. Believing that high-power microwave tubes would eventually be available (at this time, TRE had not been made aware of the new magnetron), they used the klystron to conduct research on the supporting components, with Skinner concentrating on crystal diodes for receivers and Dr. A. C. Bernard Lovell working on wave-guides, horn antennas, and covering domes. Oliphant's group had built a 10-cm set using klystrons, and in June this had been taken to Dee at Swanage for use in the component research.

In the experimental set, the klystron output was fed into a waveguide, then coupled to a horn antenna. While this radiated a well-formed beam, the length of the horn presented problems for potential airborne installations. As an alternative, Lovell tried a dipole antenna with a cylindrical parabola reflector. This produced a beam comparable to that from the horn, and was about a tenth the length – highly suited for an airborne microwave antenna.

Microwave receivers required a mixer diode for the front-end. Vacuum-tube diodes could not function at these wavelengths, and existing crystal detectors were very difficult to configure. At GEC, a silicon-tungsten crystal with a "cat's whisker" had been used in their 25-cm klystron system, but was not practical for mass production. To solve this problem, Skinner built a high-purity silicon crystal with a tungsten point contact, all in a wax encapsulation and then enclosed in a glass tube. This was the first solid-state microwave component.

The local oscillator was another vital front-end receiver component. This needed to produce a low-power signal 45 MHz removed from the primary signal (nominally 3 GHz). A special klystron made by Sutton at the Admiralty's valve laboratory was used in the experimental set, but could not be readily adjusted to produce the desired frequency. By October 1940, in his continued work on improving the klystron, Sutton had developed the reflex klystron, a low-power, tunable device. Also called the Sutton Klystron, this, with the magnetron, made microwave RDF systems possible. A major feature was that the resonant cavity was adjustable from outside the vacuum enclosure, facilitating the frequency tuning.

An initial E1189 magnetron from GEC was brought to the TRE on July 19, 1940, and Dee tasked Lovell to "wire it into a crude pulsed system as soon as possible." For this, the 10-cm klystron set from Oliphant was modified to accept the magnetron. A pulse-modulator for the magnetron was built by J. R. Atkinson and W. E. Burcham. The existing receiver, with an older low-power klystron from Sutton as the local oscillator, was modified by A. G. Ward to use Skinner's now-working crystal diode as the mixer; this fed a standard 45-MHz IF amplifier from an AI set. Both the transmitter and receiver were coupled to separate dipole antennas backed by 3-foot diameter parabolic reflectors.

The apparatus was set up outside for initial testing, and on August 8, small echoes were received off the ancient Norman chapel about a mile away at St. Aldhelm's Head. In testing the set on August 12, an aircraft several miles away flew into the beam and produced a firm echo, the first such using the new magnetron.

On August 13, the microwave apparatus was demonstrated to Watson Watt, Rowe, and Dee, and again detected aircraft echoes. Later in the day, Lovell asked one of his young engineers, Reginald G. Batt, to fit a sheet of metal to his bicycle and ride it along the cliff edge some distance away. This target was detected very clearly, although the return was exactly where the ground return would normally have shown, indicating another great advantage of microwave operations. Lovell and his team rushed back to the laboratory to tell Dee, leaving Batt still peddling his bicycle and unaware of this success.

Twelve months later, 10-cm systems were in operational use with the Royal Navy.

CONTINUED WAR AND ITS CONCLUSION

Similarly to the presentations in the previous chapters, the activities in British and American radar development from 1941 through the end of conflict in 1945 should be viewed in perspective with the ongoing war. This section provides an overview of this portion of the war – particularly that part after the entry of the United States.

With the failure of the German *Luftwaffe* to destroy the RAF, Hitler's planned invasion of the British Isles was dropped or postponed. Germany's attention then turned westward to the USSR. On June 22, 1941, German troops invaded the Soviet Union, intending to sweep a path to Moscow in a few months. The Russian military resistance, however, was much greater than anticipated, and in four months the Germans had only reached Leningrad (St. Petersburg). Here they faced severe supply problems and an unusually early winter; their momentum was lost, but they started a long and uneasy occupation of the city.

While Germany was pursuing the air war against Great Britain, Hitler's close ally, Benito Mussolini, sent Italian forces to attack British holdings in Africa. After overcoming Somalia, Kenya, and the Sudan, they invaded Egypt in September 1940, intending to secure for the Axis a strategic position on the Mediterranean and control of the Suez Canal. Although the Italian forces were vastly superior in numbers, they were routed in Egypt by the British and driven back into Libya. Hitler then sent in the *Afrika Korps* under General Erwin Rommel, and by May 1942, the British were pushed back to El-Alemein, Egypt.

The Mediterranean Fleet, composed of British and Australian warships, cornered the Italian Royal Navy (*Regia Marina*) off the coast of Greece on March 28, 1941. In the resulting Battle of Cape Matapan, the

Italian fleet was essentially destroyed and throughout the war never again ventured to sea together.

On December 7, 1941, Japanese warplanes attacked the U.S. military base in Pearl Harbor, Hawaii. On the following day, the U.S. declared war on Japan. On December 10, Germany and Italy declared war on the U.S., and Congress immediately responded with a declaration against them. The U.S. and Great Britain were finally official allies in World War II.

Attack on Pearl Harbor

During the early Pacific war, the Japanese captured Hong Kong, the Philippines, Singapore, and many important islands. In May 1942, the Japanese and American navies fought the Battle of the Coral Sea, northeast of Australia, with a clear victory for the U.S. This was followed in June by the Battle of Midway, resulting in the U.S. gaining superiority in the Pacific and dealing a naval blow to Japan from which it never fully recovered. In land battles, however, the Japanese dominated; by the end of 1942, they had taken Burma and were moving toward India. The Royal Navy was dispatched to successfully defend the Bay of Bengal.

From the start of the war, RAF bombing of German targets had been with relatively small formations, and, even after the "terror" bombing of London and other cities, German industries were the main targets. In March 1942, Great Britain abandoned this type of operation and started massive "carpet" bombing of cities. On March 30, 1,000 RAF planes conducted a nighttime raid on Cologne.

After the entry of the U.S. into the war with Germany, large numbers of planes and personnel were sent to Great Britain. Some small sorties by the U.S. Army Air Forces took place during late 1942, then the next year they supplemented the RAF with large-scale daytime bombing, initially targeting industry. This day and night strategy mainly continued throughout the war in Europe. Hamburg, Kassel, Pforzheim, Mainz, Berlin, and Dresden were largely destroyed; post-war estimates of civilian deaths from bombing are between 300,000 and 600,000.

In a conference at Washington, D.C., in June 1942, President Roosevelt, Prime Minister Churchill, and their respective top military staffs planned the invasion of North Africa. This invasion began on November 8, 1942, with Lt. Gen. Dwight D. Eisenhower leading combined British and American troops into Algeria; they quickly overcame the Vichy French forces.

Germany retaliated by sending large numbers of troops into Tunisia. British forces under Field Marshal Bernard L. Montgomery then attacked from the south while American troops did the same from the north. This two-pronged effort was highly effective, leading to the surrender of Axis forces in North Africa on May 12, 1943. The stage was set for an assault across the Mediterranean on Hitler's Fortress Europe.

The Allied invasion of Sicily began on July 10, with British General Harold Alexander in overall command. Lt. Gen. George S. Patton led the U.S. landing on the southern coast, while Montgomery led the combined British and Canadian forces invading the southeast coast. The Sicilian campaign ended in 38 days and led to the downfall of the ruling Fascist government.

On September 3, 1943, an armistice between Italy and the Allies was signed. This was also the date of the invasion of German-occupied Southern Italy by the Allies. At the end of November, Premier Joseph Stalin of the Soviet Union met with Churchill and Roosevelt in Tehran, Iran, to strategize the conclusion of the war. For the remainder of 1943, the progress north against the German forces in Italy was slow, difficult, and costly. It was not until June 4, 1944, that Rome was freed. The Germans, however, continued the war in parts of Italy for another year.

The successes in North Africa, Sicily, and Italy were accompanied by an increased air assault on German targets by British planes. On several occasions the RAF mounted raids involving 1,000 or more planes on key cities in the industrial Ruhr Valley. In 1943, the U.S. joined in this assault, with the British assigned night attacks and the U.S. mainly involved in daylight missions. Raids on Berlin started in the summer of 1943, and continued until the end of the war.

The Battle of the Atlantic – a name coined by Churchill – continuously pitted U-boats (*Unterseeboots*) and heavily armed ships of the German Navy (*Kriegsnarine*) against Allied convoys that were mainly going from North America to Great Britain and the Soviet Union. The convoys were protected by the British, Canadian, and U.S. navies and, to the extent of their flight duration, large aircraft and blimps equipped with radar and carrying depth bombs.

Extending over six years, this naval warfare involved thousands of ships in more than 100 convoy battles and at least 1,000 single-ship engagements. The advantage switched back and forth as both sides developed new weapons, tactics, and countermeasures. Over 2,600 Allied merchant ships were sunk. The Allies gradually dominated however, driving the German surface raiders from the ocean by early

1942, and decisively defeating the U-boats in a series of engagements the next year. Altogether, of near 1,200 *Kriegsnarine* submarines that put to sea, 65 percent were sunk or irrecoverably damaged. In 1945, new German submarines were introduced, but were too late to affect the course of the war.

Starting in 1943, massive amounts of weapons and supplies accompanied by huge numbers of troops were brought to Great Britain and accumulated at ports of demarcation. The long-awaited invasion of France began on June 6, 1944 (D-Day), with General Dwight D. Eisenhower designated as the Supreme Commander. The target was the coastline of Normandy between Cherbourg and Le Havre. Opposition by the German Army was great; it was almost a month before the vital port of Cherbourg was taken and put into use supplying the ground troops. The significant roadways into the interior were not secured until the end of August.

While the fighting in Normandy was taking place, the invasion of Southern France began on August 14, 1944. With the Allies pressing from the west and south, the Germans withdrew almost completely from France, with those occupying Paris surrendering to Free French troops under General Charles de Gaulle on August 25.

By mid-September, Allied armies were at the border of Germany but the progress was halted by major resistance. In mid-December, the Germans initiated a surprise counterattack into Belgium and Luxembourg. The "Battle of the Bulge" ensued. By the end of January 1945, this final thrust was stopped, with Germany losing 220,000 men, marking the last time that the *Wehrmacht* initiated a major offensive.

In this period, Germany started desperate attacks by unmanned flying weapons: the *Vergeltungswaffe 1* and 2 (the V-1 flying bomb or "buzz bomb" and the V-2 long-range rocket). Approximately 10,000 V-1s were fired at England between June 1944 and January 1945; over 2,400 reached London and killed about 6,200 people. Over 3,000 V-2s were launched against England, Belgium, and France between September 1944 and March 1945. England was the target of about 1,400 V-2s, mainly London, where nearly 2,800 persons were killed. While these new weapons caused considerable devastation, they had little strategic effect.

In Eastern Europe, the 27-month occupation of Stalingrad was finally broken by the Russians at the end of January 1943. This is considered by many military historians as the turning point of WWII, leading to other successful offensives against the badly weakened German Army. By the end of 1943, Soviet troops were at the Polish border, and during 1944 most of the eastern countries had been captured.

Microwaves Bridge the Atlanic

On February 4, 1945, Churchill, Roosevelt, and Stalin (the "Big Three") met in Yalta to discuss post-war plans. Shortly after this, the Allied armies on the western Germany border launched a powerful assault, and by early March had reached the Rhine. At the same time, the Soviet army began a major offensive into eastern Germany, and by April they were 30 miles from Berlin. On April 26, the American and Russian troops met about 70 miles south of Berlin.

Hitler committed suicide on April 30. In accord with agreements at the Tehran and Yalta Conferences, the USSR army was allowed to enter Berlin alone on May 2. The formal surrender of the German High Command to the Allies occurred on May 8, 1945 (V-E Day).

Churchill, Roosevelt, and Stalin in Yalta

The Pacific war was carried out under four Allied commands: U.S. General Douglas MacArthur in the southwest Pacific, U.S. Admiral Chester Nimitz in central Pacific operations, British Lord Louis Mountbatten in southeast Asia, and Generalissimo Chang Kai-Shek in mainland China. The Chinese did not officially declare war on Japan until December 9, 1941, four years after their conflict began; war was also declared against Germany and Italy at this same time.

In April 1942, U.S. intelligence read coded Japanese messages and found that a major naval force was being assembled for further seizures in the South Pacific. All available U.S. warships in the region were directed to intercept them, and the Battle of the Coral Sea ensued. Carrier-based planes gave the U.S. its first victory of the war.

In June, another Japanese task force was discovered approaching Midway Island. In a four-day battle, land- and carrier-based planes again wreaked devastation. The Japanese lost four aircraft carriers, two large cruisers, three destroyers, and 300 planes, compared with American losses of one aircraft carrier, a destroyer, and 150 planes. From that point, the Japanese could never muster sufficient naval strength to prevent Allied movements against their captured islands.

Starting in November 1943, Nimitz launched the land offensive with "island-hopping" attacks, eventually taking the Gilbert, Marshall, and Mariana Islands. Beginning in December, General MacArthur conducted a series of successful operations in New Guinea and the Admiralty Islands, and aimed toward the Philippines. The Japanese attempted an interception, and in a mid-June battle in the Philippines Sea lost three aircraft carriers and 400 planes.

The invasion of the Philippines started in October. In an all-out attempt to stop the invasion, the Japanese Navy assembled a strike force in the China Sea. Despite their initiation of *kamikaze* attacks, the ensuing battle in early February resulted in a complete Japanese defeat. The full liberation of the Philippines was declared on July 5, 1945.

The Japanese had taken the Burma Road early in the war, cutting the overland supply to the Chinese Army. An aerial route over the Himalayas (the "Hump") from India to China was helpful, but in 1944 the Japanese attempted to cut this by initiating an invasion of India through Burma. British and Indian troops eventually repelled this, but it was not until May 1945 that the Japanese were pushed out of Burma.

American submarines played a major role in the Pacific war. They sank about one-third of the Japanese combat ships and two-thirds of their merchant vessels. By the end of the war, Japan's transport fleet was largely destroyed, having a major effect on that country's ability to move military forces between combat zones and to replenish critical supplies.

The U.S. forces in the Pacific continued their island-hopping, preparing for invading the Japanese homeland. On February 19, Iwo Jima was stormed. Only 400 miles from Japan, this tiny island was vital as an airbase for bombers. Raids by B-29s started immediately, with continuous attacks resulting in heavy damage on industrial and military operations in major Japanese cities.

Next was Okinawa, considered by Japan as a home island. Starting on March 26, this involved 1,400 vessels in the largest amphibious assault of the Pacific war. While the Japanese Army was defending the land, the Japanese Navy launched more than 2,000 *kamikaze* aircraft and boat attacks on U.S. naval vessels supporting the invasion. America's concentration of naval, air, and ground fire was unmatched in warfare history, with over 98,000 tons of ammunition expended.

Before Okinawa was secured on July 2, America suffered 49,000 casualties, including 12,500 killed or missing. In the air, 763 U.S. planes were shot down. The Navy lost 36 ships sunk, and another 368 were damaged, with about 4,900 men killed. On the other side, it is estimated that 110,000 Japanese died in battle, 7,800 planes were downed, 10 ships sunk, and 150,000 Okinawans – one-third of the native population – perished.

The casualties in this tragic event are noted primarily because this was the precursor to the dropping of atomic bombs the next month on Hiroshima (August 6) and Nagasaki (August 9). This ended the war and stopped the planned invasion of Japan's homeland. Potential American casualties had been projected at many times the losses at Okinawa. The Japanese death toll from the two atomic bombings was approximately 214,000, significantly less than their total military and civilian Okinawan losses.

Japan officially surrendered on September 2, 1945 (V-J Day), thus bringing to a close World War II – the most costly, most destructive, and most widespread war in history.

EARLY MICROWAVES IN THE U.S.

In the United States, just as in Great Britain, it was recognized that detection sets operating in the microwave region would offer many advantages. As noted earlier, both the Naval Research Laboratory and the Signal Corps Laboratories had initially pursued UHF research, but soon abandoned it because of the lack of high-power output tubes at these wavelengths.

In American industries and government laboratories, there had been considerable work on microwave technology. Major accomplishments had particularly been made in components. The magnetron was originally developed – but not matured – by Dr. Albert W. Hull at General Electric. High-frequency tubes, including the Acorn, Doorknob, and Lighthouse, were already in use. Research on microwave "plumbing" (waveguides and horn antennas) was underway. A number of organizations, notably RCA Camden Laboratory, Bell Telephone Laboratories, and Signal Corps Laboratories, had built elementary detection sets using low-power magnetrons or Barkhausen-Kurz tubes.

Cyclotrons were being built at several universities for experimental atomic studies, and microwave sources were needed for these facilities. At the University of California at Berkeley, Dr. Ernest O. Lawrence, inventor of the cyclotron, operated the Radiation Laboratory. As a microwave source for this Laboratory, in 1937, brothers Russell H. and Sigurd F. Varian built on existing velocity-modulation technology to develop the klystron. This device produced microwaves at levels higher than any other sources of that day, but still at relatively low levels.

Early in the summer of 1940, Dr. Vannevar Bush (1890-1974), President of the Carnegie Institutution of Washington and also a highly respected Professor at MIT, had

Vannevar Bush

prepared a plan for forming a top-ranking group to have control over war-related scientific research. On June 15, President Roosevelt issued an Executive Order creating the National Defense Research Committee (NDRC), and naming Bush as the Chairman. The commission of the NDRC was as follows:

> To correlate and support scientific research on the mechanisms and devices of warfare except those related to problems of flight. . . . It shall aid and supplement the experimental and research facilities of the War and Navy Departments and may conduct research for the creation and improvements of instrumentation, methods and materials of warfare.

Included in the NDRC was a Microwave Committee, indicating the viewed importance of this to the preparedness effort. Alfred L. Loomis was named by Bush to head this Committee.

Alfred Loomis

Alfred Lee Loomis (1887-1985), educated as an attorney and very wealthy from years as a Wall Street tycoon, had an insatiable appetite for scientific research. He formed the Loomis Laboratories in the 1920s at Tuxedo Park, an exclusive residential area about 60 miles north of New York City. Here he personally paid for a cadre of highly qualified researchers to pursue studies in a wide variety of areas, including electromagnetics, physical optics, and ultrasonics – foundation stones of microwave radar and sonar. Loomis did personal research in the accuracy of timekeeping devices, coordinating with scientists at the Bell Telephone Laboratories.

As America mobilized for war, Loomis was anxious for his Laboratories to make significant contributions to these efforts. In early 1939, Dr. Karl T. Compton, President of MIT, suggested that Loomis Laboratories join in an MIT project on distance-finding by radio. Loomis was immediately intrigued by this and suggested using microwaves in an airplane-detector for anti-aircraft weapons that would "include a fairly simple computer which would control the gun directly." A team at Loomis Laboratories was assembled and, led by Frank D. Lewis, began research in microwave technology. Within a short time, a simple object-detection system using Doppler returns was demonstrated.

Microwave development in the United States was waiting for a high-power UHF source.

THE TIZARD MISSION

In mid-1940, Great Britain was in a desperate situation. There were essentially no allies for her support, and she was approaching the limit of productive capacity, particularly in electronics. Sir Henry Tizard made the unbelievably bold suggestion that Great Britain should offer her utmost secrets to the United States in exchange for assistance in technical and production matters.

As might be expected, there was considerable opposition to this suggestion. For example, Watson Watt publicly stated that both research and production capabilities of the United Kingdom were adequate, and America "had nothing to offer." Admiral Sir James Sommerville said, "Anything told to the American Navy went straight to Germany." Nevertheless, newly appointed Prime Minister Winston Churchill gave it his strong support and approached President Franklin Roosevelt on setting it up. Lord Lothian (Philip H. Kerr), British Ambassador to the U.S., made the official proposal on July 8, 1940, and it was approved by both governments before the end of the month.

It was agreed that Vannevar Bush and other members of the NDRC would be the primary representatives of the U.S. and serve as the main interface with their British counterparts.

In Great Britain, much attention was given to the selection of members of what was called the Tizard Mission. Those eventually chosen were Sir Henry Tizard, Mission Leader; John D. Cockcroft, Army Research; and Dr. Edward G. Bowen, Air Ministry Research. Arthur E. Woodward Nutt, an Air Ministry Official, would serve as the Mission Secretary. The Armed Services were represented by Brigadier F. C. Wallace, Army; Captain H. W. Faulkner, Navy; and Group Captain F. L. Pearce, Air Force. All of these military men had recent combat experience, as well as expertise in the various technologies.

All types of documentation were assembled in Great Britain for the visit. This included information on RDF technology, rockets, explosives, superchargers, gyroscopic gunsights, submarine detection devices, self-sealing fuel tanks, and even the initial basics of the jet engine and an atomic bomb. Among these treasures, however, nothing carried the all-pervasive importance of the resonant-cavity magnetron, Great Britain's most closely guarded secret.

These documents, as well as one of the 12 type E1198 magnetrons, were packed into a metal solicitor's "deed box" to be personally carried by E. G. Bowen, the youngest member of the team.

Pierce went ahead by air to make arrangements in Washington. Tizard also went earlier, traveling first by air to Ottawa, Canada, where he briefed Dr. C. J. Mackenzie, President of the National Research Council of Canada (NRCC), as well as Canadian Chiefs of Staff on the purpose of the Mission. While in Ottawa, he divulged the magnetron concept to Dr. John T. Henderson, Head of the NRCC Radio Division, and asked him to represent Canada on the Mission.

On August 23 in Washington, Tizard met with the U.S. Secretary of State, W. Franklin Knox, to establish the ground rules for the exchange. Three days later, he received an audience with President Roosevelt, who welcomed him but explained that political considerations prevented the U.S. from sharing details of the Norden bombsight. (Concerning this exception, it should be noted that Patrick Blackett, an original member of Tizard's 1934 Committee, had developed a bombsight for the Air Ministry, and this was in competition with the Norden unit for adoption in the U.S.) On August 28, Tizard first met with Vannevar Bush.

The other Mission members had taken the SS *Duchess of Richmond* (a Canadian liner just converted to a troop transport) from Liverpool, departing just as the port was being bombed. During the passage, Cockcroft calculated that the black box would stay afloat should the ship be torpedoed, so holes were drilled in each end. They arrived safely in Halifax, Nova Scotia, on September 6. The box was taken over by U.S. Army guards and transported by an armored vehicle to Washington. Cockcroft and Bowen went by air to Ottawa for meetings with the Canadian NRC, while the others flew directly to Washington.

An ASV Mk I set was to have been shipped on the transport with the Mission members, and then installed by Bowen at Ottawa on a Hudson aircraft of the Royal Canadian Air Force. It would then be flown to Washington for showing at the meeting. Unknown to Bowen and other Mission members, a last-minute decision had been made by the Air Ministry not to send the ASV, so this portion of the exchange had to be scrapped.

The Mission finally assembled in Washington, and official meetings started September 12 at the Army and Navy Headquarters. On the American side, the members of the exchange were a number of representatives from the NDRC and senior military officers of the U.S. Army and Navy. Also included were Vice Admiral Harold G. Bowen, Director of the Office of Naval Research, and Major General Joseph C. Mauborgne, the Army's Chief Signal Officer.

As these attendees were being selected, there had been significant reservations on the part of some members concerning General Mauborgne. It was known that in a recent meeting with the NDRC Chairman, Bauborgne had stated,

> When a nation goes to war, it is necessary to freeze all new development – developments should stop because production problems of the present day are so vast. You have to fight a war with what you have.

The initial meetings went on for about 10 days, with Tizard giving introductory comments, followed by one of the Mission members providing details. U.S. officials and specialists then did the same, and included visits to the Naval Research Laboratory (NRL) and the Signal Corps Laboratories (SCL) for demonstrations of radar systems (the name "radar" was then introduced into the British technical community). For the Navy, this was the CXAM, very similar to the CHL and operating on the same frequency band. The Army showed the SCR 270, the mobile air-warning radar, and the SCR 268, a searchlight director system.

Over the next several weeks, members of the Mission visited a number of facilities, collecting documents, equipment, and components that were sent back to England by diplomatic couriers. In general, it was concluded that both the U.S. Army and Navy had made significant progress in developing lower-frequency radars and their components, but that the British had a substantial lead in airborne sets.

The High-Power Magnetron

A low-power microwave radar using a klystron was shown at the NRL, and presentations were made on activities at the BTL, RCA, General Electric, and MIT on microwave devices. However, the klystron – already well known to the Mission members – and some low-power split-anode magnetrons were the only microwave sources being used. During these discussions, there had been no specific mention by the Mission members of the resonant-cavity magnetron. It had been "hinted" that means for generating high powers at centimeter wavelengths had been discovered in Great Britain, but how this was done had not been disclosed.

On the evening of September 19, there was a special meeting involving J. D. Cockcroft and E. G. Bowen of the Committee; Dr. John Henderson, the representative from Canada; Dr. Karl Compton, President of MIT; Admiral Bowan; Carroll Wilson (Bush's *alter ego* and personal assistant at the NDRC); and Alfred Loomis, Chairman of the Microwave Committee of the NDRC. The meeting took place at Loomis's apartment in Washington.

At that time, the resonant-cavity magnetron was first disclosed outside of Great Britain. E. G. Bowen described the situation:

We quietly produced the magnetron and those present at the meeting were shaken to learn that it could produce a full 10 kilowatts of pulse power at a wavelength of 10 centimeters.

Everyone was cautioned that even the existence of the device must be held at the highest levels of secrecy. It was noted, however, that a major purpose of the disclosure to America was to elicit support for the continued development and mass manufacturing of the magnetron Loomis suggested that a follow-up meeting be held at his laboratory in Tuxedo Park, New York.

The follow-up meeting was held on September 28 and 29 at Tuxedo Park. Bowen and Cockcroft again showed the magnetron, this time discussing its details and showing the drawings. Those present were Alfred Loomis; Carroll Wilson; Dr. Edward Bowles, Professor of Electrical Engineering at MIT; and Dr. Hugh Willis, Director of Research at Sperry Gyroscope Company. Both Bowles and Willis were members of the NDRC Microwave Section, and Bowen noted their reaction:

> The atmosphere was electric – they found it hard to believe that such a small device could produce so much power, and that what laid on the table in front of us might prove to be the salvation of the Allied cause.

It was immediately agreed that the Microwave Committee would sponsor production of the magnetron and the BTL would be asked to be the manufacturer.

Tizard was brought up to date, and readily agreed to turning the magnetron over to Loomis's Committee and for BTL to produce it. He then returned to England, leaving Bowen to continue with the transfer of magnetron technology. After a meeting between BTL executives and vacuum-tube specialists in New York, the consensus was that production of the device would be straightforward and could easily be done in their tube department.

On October 6, Bowen visited the BTL laboratory in Whippany, New Jersey, to demonstrate the magnetron in operation. The laboratory had sources for producing the necessary external magnetic field and pulsed high-voltage. When these were connected and the filament turned on, an inch-long glow discharge immediately came from the magnetron's output terminal. Based on his past observations, Bowen estimated that the output was about 15 kW.

The BTL Tube Department had already received its instructions: if the device produced anything near what had been promised, they were to immediately manufacture 30 exact copies for the NDRC. Although the power exceeded expectations, a problem arose. It was necessary to X-ray the existing magnetron, and the BTL scientists found that it had eight cavities, not the six described by Bowen. This was quickly resolved; the 12 improved magnetrons had been made by Megaw with a mixture of numbers of cavities, and, not knowing this, Bowen had brought an eight-cavity version but had documentation for the six-cavity device.

Since the eight-cavity magnetron had been proven to work at the BTL, this version was selected for manufacturing as type CV41. They had their first tube in two weeks, and the full 30 by the end of a month. Meanwhile in Great Britain, the six-cavity improved version was put into production as type CV38 in a hastily arranged facility at the Cavindish Laboratory, leading to a difference in the early magnetrons used in the two nations.

In his book *Scientists against Time*, Dr. James P. Baxter, III, official NDRC historian, stated:

> When the members of the Tizard Mission brought one [magnetron] to America in 1940, they carried the most valuable cargo ever brought to our shores.

Other Radar-Related Secrets

While the magnetron-related activities were taking place, there were parallel discussions and visits concerning other matters. Radar topics included identification friend-or-foe, electronic countermeasures, the plan-position indicator, crystal diodes for detectors, and transmit-receive antenna switches. Both sides had developments in each of these and they were compared, seeking to find the best for common use.

The problem of sending the ASV set to the U.S. was resolved, and an ASV Mk II, an AI Mk IV, and an IFF Mk II were shipped to the Naval Research Laboratory (NRL) for demonstration under Bowen's guidance. The ASV was installed in a PBY flying boat, and first operated over land in late November, then, with Bowen as the operator, over the sea on December 2, 1940. The tests were a complete success; the U.S. had been introduced to airborne radar.

At the same time that the ASV was being demonstrated, the AI was sent to Wright Field at Dayton, Ohio, for examination by the Army Air Corps. Bowen, tied up with the ASV demonstration and the initiation of a central laboratory at MIT, left this up to a representative of the British Air Command. It was installed in a Douglas A20 Havoc, a recently introduced fast attack bomber. Again, the testing went very well.

The third system sent to the U.S. was the IFF Mk II. This was installed in a Navy fighter aircraft by the NRL for testing in comparison with a system that they had under development. The tests favored the

British system, and this was adopted by the Navy as the ABD and ABE (different sets for various radars), and by the Army as the SCR-535.

To examine first-hand the radar (RDF) activities in Great Britain, President Roosevelt authorized the Army Signal Corps to commission a number of physicists and engineers and allow them to visit the secret operations in England. It took more time than expected to recruit and prepare the first group, so it was not until September 1941 that 35 such persons made their way overseas. Their mission was highly secret, but they became intimately involved in defense activities; two were killed in action. An anti-aircraft battalion commander commented:

> From America we got a wonderful bunch of recruits. America was not yet in the war, so these scientists operated our gun-sites in the guise of civilians.

THE RADIATION LABORATORY

When Bowen and Cockcroft met with Loomis and members of his NDRC Microwave Committee at Tuxedo Park on September 13, 1940, they showed them the resonant-cavity magnetron and described its operation. They also talked about the potential of this device in guiding airborne interceptors, navigating bombers through dense cloud covers, detecting U-boat periscopes, and many other almost unbelievable applications. This was described by Committee member Dr. Edward Bowles: "I could scarcely believe what I was hearing. All we could do was sit in admiration and gasp."

In his true form, Loomis was ahead of everyone else. Based on these descriptions, he immediately proposed establishing a large, central microwave laboratory. Bowen and Cockcroft seconded the proposal, and it was agreed that this should be a civilian rather than military operation, staffed by scientists and engineers from universities and industries, and using as a model the Telecommunications Research Establishment (TRE) at Swanage and the Air Defense Experimental Establishment (ADEE) at Christchurch in Great Britain.

After the magnetron had been demonstrated at the BTL and orders placed for manufacturing 30 copies, Loomis pursued his idea of a central microwave laboratory with Vannevar Bush, Chairman of the NDRC. Bush immediately agreed with the concept and asked Dr. Ernest O. Lawrence, 1939 Nobel Laureate then leading the cyclotron development at Berkeley, to assist Loomis in its implementation.

Several preliminary planning meetings took place; then on October 12, 1940, in a meeting at Tuxedo Park, the Radiation Laboratory (Rad Lab) was formed. This name, the same as Lawrence's existing operation at Berkeley devoted to nuclear research, was selected to disguise the real purpose. MIT agreed to house the operation on its campus in Cambridge, Massachusetts. Lawrence would lead in enlisting an initial cadre of scientists. The NDRC would provide the initial operating funds. To start the activities, the Tuxedo Park microwave project led by Frank Lewis would be transfered to the Rad Lab.

Formation of the Rad Lab, Tuxedo Park, Oct. 12, 1940: L to R: Carroll Wilson, Frank Lewis, Edward Bowles, E. G. Bowen, Ernest Lawrence, and Alfred Loomis; John Cockcroft was present, but took the picture.

Initial projects to be undertaken were selected at a meeting the next day. In order of their priority, these were (1) a 10-cm airborne intercept radar code-named AI-10, (2) a 10-cm gun-laying system for anti-aircraft use, and (3) a long-range aircraft navigation system, at that time unspecified as to type. The first two were of great mutual interest in both nations, but the British had the major interest in the third. Bowen would outline the requirements for the AI project, and Cockcroft those for the GL project.

In his earlier briefings, Bowen had mentioned GEE, the hyperbolic navigation system being developed at TRE for use in blind bombing. Long-range navigation had been selected as one of the three first projects because of the importance to the RAF and the limitations of GEE. While the hyperbolic-navigation basis of GEE was a good concept, the short-wave signals were limited to line-of-sight distances, some 200 miles from a flight altitude.

It was agreed that a new system having a range of at least 1,000 miles and an accuracy of some 5 miles was needed. It was appropriate as one of the initial projects at the Rad Lab, but the technical concept was undefined. By the next morning, Loomis had come up with a brilliant concept: a system using long waves that could be reflected by the ionosphere, and with stations as much as a thousand miles apart. They would transmit synchronized pulses that could be precisely regulated by recently developed quartz timing devices. This became the general requirement for the third project.

A meeting of the Microwave Committee was called by Loomis for October 18, 1940. At this meeting, he announced that the NDRC would be sponsoring the development of a high-power radar using a multi-cavity magnetron. That same day, he also held a meeting with representatives from several select industries. After appropriate security formalities, he disclosed the British magnetron and the intended immediate development of a 10-cm airborne radar.

Loomis then asked for each representative to select those parts for which their company wished to tender bids to manufacture and deliver these within 30 days. When told that it might be difficult to get the bids in within a month, be replied: "No, you misunderstood me. I want you to submit your tender next week and deliver 30 days after that."

During this same meeting, industries for supplying initial major elements were selected: magnetron tubes from BTL, magnets from GE, modulated high-voltage supplies from Westinghouse and RCA, CRTs and IF amplifiers from RCA, and parabolic reflectors from Sperry.

MIT Bldg. 4 – Initial Rad Lab

MIT freed up 10,000 square feet of space in its Building 4; the NDRC provided a half-million dollars for funding the initial year; and the Rad Lab opened in November 1940 with 48 employees. Dr. Lee A. DuBridge, a dean at the University of Rochester, was named Director of the Rad Lab.

Lee Alvin DuBridge (1901-1994) was raised in Indiana and earned his B.S. in physics from Cornell College. After completing his Ph.D. from the University of Wisconsin in 1926, he received a fellowship to work with Professor Robert Millikan at California Institiute of Technology. In 1934, he was appointed Head of the Physics Department at Rochester

University, then later made Dean of the Arts and Sciences Faculty. During his career, DuBridge received 28 honorary degrees.

Dr. Isidor Isaac (I. I.) Rabi, a renowned physicist from Columbia University, was named the Deputy Director responsible for scientific matters. Somewhat later, Dr. F. Wheeler Loomis (no relation to Alfred Loomis) was hired as the Deputy Director for administration.

Similarly to what Dr. Ernest Rutherford had done in Great Britain in soliciting leading researchers to become active in RDF development, Dr. Ernest Lawrence became the primary recruiter for Rad Lab personnel.

Lee DuBridge

One of the best known and respected scientists of that day, Lawrence personally obtained the consent of over a dozen highly capable physicists and engineers to join an activity that he described as being "of vital importance to America," but unidentified for security reasons. The recruits even included one of the leading researchers at Lawrence's own laboratory, Dr. Luis W. Alvarez. In later years, seven of the early recruits became Nobel Laureates.

E. G. Bowen (seated), L. A. DuBridge, and I. I. Rabi

Before the end of 1940, the Rad Lab was firmly established in MIT Building 4, and the staff had enlarged to need a formal organization. The capabilities were divided into eight sections: Antennas, Cathode-Ray Tubes, Integration, Klystrons, Pulse Modulators, Receivers, Theory, and Transmitter Tubes (Magnetrons). With well-known leaders for each section, they would provide support as needed to the various design projects.

After briefly returning to England, E. G. Bowen was assigned to the Rad Lab as a consultant. Similarly, Frank Lewis was sent to Great Britain by the NDRC to provide liaison on microwave radar projects.

To the extent that could be done with a staff largely composed of university research personnel, Wheeler Loomis established procedures and work rules. Concerning these, one of the leaders later remarked:

We had no compunction whatsoever about bending the rules a little bit if necessary to get something done. . . . We were expected to hurry and we were not to be put aside for anything.

The "standard" work week was Monday through Friday at times needed to get the job done – usually many more then 8 hours – plus Saturday mornings. Somewhat like A. P. Rowe's "Sunday Soviet" at Bawdsey and TRE, DuBridge required all of the Rad Lab leaders to attend a Saturday afternoon meeting, where they openly discussed their work. Unlike Rowe's meetings, however, outsiders seldom participated.

In June 1941, the NDRC became part of the new Office of Scientific Research and Development (OSRD), also administered by Vannevar Bush. Created by an Executive Order and with Bush reporting directly to President Roosevelt, the OSRD was given almost unlimited access to funding and resources. It focused its efforts on weapons development, notably radar, as well as various forms of communication. In general, the OSRD negotiated contracts with research institutions, both universities and private firms, as well as with government agencies. The Manhattan Project to develop the atomic bomb started under OSRD auspices, but was subsequently transferred to the Army.

INITIAL MICROWAVE RADAR IN GREAT BRITAIN

After the demonstrations in August 1940 of the experimental set using the magnetron, microwave development at the TRE greatly increased. In the early fall, the activities were moved to Leeson House and many researchers were added to Dee's staff. In addition to resolving general problems concerning magnetron applications and providing support to emerging microwave developments of the Army and Navy, this group had the specific mission of developing a 10-cm (3 GHz) AI system for Blenheim night-fighter aircraft.

A nagging problem held up the total adoption of the magnetron and the dropping of further research on other microwave power tubes. With changes in operating conditions, the magnetron would occasionally lose stability and jump in frequency. This problem remained for some time, but was finally solved in August 1941 by Dr. James Sayers of Oliphant's laboratory. He simply inserted a copper strap between every other cavity; this was called a "strapped" magnetron and designated type CV56.

The addition of a strap not only stabilized the frequency but also increased the output power by a factor of five or more. This change was immediately made in Great Britain as well as in America; throughout the war, hundreds of thousands of strapped, multi-cavity magnetrons were produced.

A major problem in microwave adoption was the use of a common antenna for transmitting and receiving; this was especially important for airborne systems. Although the transmit-receive (T-R) switch had been developed for existing radars, and after the Tizard Mission exchange, the duplexer, invented for this function at the U.S. Naval Research Laboratory, was also known, these devices were not suitable for microwave frequencies. Even a small amount of the transmitted signal leaking back into the receiving channel would destroy the sensitive crystal-diode detector.

Dr. Arthur H. Cooke at the Clarendon Laboratories of Oxford University developed a microwave T-R switch in March 1941. Cooke's device used a small klystron without the electron source and containing a small amount of water vapor. With this, the low-level received signal could pass through to the diode, but the high-level transmitter signal immediately created a plasma in the tube that shorted the path. A special T-R tube (CV43) for this function was eventually available.

The development of the resonant-cavity magnetron at Oliphant's laboratory was under the joint sponsorship of the Air Ministry, the War Department, and the Admiralty. Thus, with the initial demonstration of an experimental microwave set by the TRE, an RDF Applications Committee was established under the Ministry of Defense. Under this, a Joint Ministries Investigation Group was formed, all leading to delays in starting specific projects within the three military ministries. In a few months, however, the TRE, the Army's Air Defense Experimental Establishment (ASEE) at Christchurch, and the Navy's Experimental Department of HMSS at Portsmouth had all initiated projects in microwave radar.

Microwave Radar for the Royal Navy

In the early 1940s, submarines of the *Kriegsmarine* were a major menace to merchant and Royal Navy vessels in the waters around the British Isle and the routes between Great Britain and the U.S. and Canada. Meter-wave RDFs from the Experimental Division of HMSS had some success in detecting surfaced U-boats, but offered no assistance in spotting the periscopes of submerged threats. Systems operating at UHF wavelengths were urgently needed. These would also be of great benefit in increasing the precision of rangefinders and air-warning RDFs.

In the pursuit of microwave technologies, the Experimental Division, as well as GEC under Admiralty contracts, initiated developments in the

500-cm (600-MHz) region using klystrons. They also experimented with an improved split-cavity magnetron from Oliphant's laboratory. With the development of the high-power, multi-cavity magnetron, this device appeared to be the solution to their needs.

The GEC Research Laboratory had improved the original magnetron, but was also under contract with the Admiralty to develop the 600-MHz system. Thus, while they likely recognized the potential value of changing to the new magnetron, they had to continue with their klystron-based system. In July 1940, one of the six-cavity Type 1198 magnetrons was received at HMSS Experimental Department. This facility, however, was heavily engaged in other RDF projects and, in addition, had little expertise or facilities for developing a microwave system.

In late October, Commander B. R. Willett and C. E. Horton, Head of the RDF Section, visited TRE for demonstrations of the experimental 10-cm set. From a 250-foot vantage point in front of Leeson House, the set detected ships in the Swanage Bay. Willett and Horton were not impressed; their deployed meter-wavelength systems could detect at greater distances while operating on masts at less heights above the sea. Their interest changed, however, when Skinner proposed fitting a 10-cm set on a destroyer for anti-submarine work.

In early November, more senior naval officials visited TRE to discuss the submarine problem. The equipment was demonstrated in detecting a submarine at 9 miles. A few months later than the other two services, the Royal Navy was finally convinced of the potential of microwave radar. Arrangements were made for a team from the Experimental Department to come to the TRE and, with assistance from Dee's group, build their own experimental set using a six-cavity E1198 magnetron.

With Dr. S. E. A. Landale as leader, C. A. Cockrance, J. R. Croney, and C. S. Owen from Portsmouth set up a laboratory in a trailer at TRE Lesson. On December 11, the completed experimental set was tested against the submarine HMS *Titlark*, and tracked it at a range of 13 miles. The set used two dipole antennas backed by parabola reflectors mounted on a swivel, producing a pencil beam.

Type 271 "Cheese" Antennas

The problem of maintaining the beam on a target while the ship rolled was immediately realized. To solve this, Lovell replaced the reflectors with 6-foot by 9-inch cylindrical parabolas that had been used on an earlier test system. This produced a fan-shaped beam, 20 degrees in the vertical plane

Microwaves Bridge the Atlantic

but still a narrow 1-degree azimuth. The results were excellent; so-called "cheese" antennas of this type subsequently became standard on microwave radars for smaller vessels.

The experimental set was taken to Portsmouth on December 19. There, the same team developed it into the Type 271 surface-warning set, the first operational microwave radar in Great Britain. The transmitter and receiver front ends were behind separate antenna reflectors. These were mounted on a pedestal that could be rotated from below, and the assembly was enclosed in a wooden and Perspex (plastic) lantern-like structure.

Type 271 Antenna Housing

Because of limitations imposed by using coax cabling, the enclosure was initially placed directly atop the radar room, where the antenna rotating wheel, power supply/modulator, and control/display station were installed. When waveguides for feeding became available, the enclosure structure was mounted on a mast. Also added was an electric drive for rotating the antenna.

Type 271 Installation

The system was first tested at sea on the corvette HMS *Orchis* in March 1941, detecting a ship at 7 miles, a surfaced submarine at 3 miles, and, at three-quarters of a mile, a periscope extended above a submerged submarine. Twelve sets were manufactured at Portsmouth, and a contract was let with Allen West & Company in Brighton for production.

The Type 271 became operational in August 1941, one year after the first demonstration of an experimental microwave set at the TRE. On November 16, a U-boat was sunk near Gibraltar, after first being detected by the Type 271 aboard the corvette HMS *Marigold*. By the end of 1941, 50 of these radars were in service on escort vessels.

Microwave Radar for the Army

The first priority for microwave development in the British War Department was for a 10-cm gun-laying (GL) system. While the existing

GL Mk 1 and Mk 2 systems were of some benefit during the Blitz raids, major improvements were needed in giving blind-firing capability for the anti-aircraft guns. This could only come through using the narrow beams of microwave radars.

At the formation of the Rad Lab in October 1940, John Cockcroft convinced the founding Microwave Committee that such a system should be in the first three projects pursued by this organization, and gave them the basic requirements for a microwave system as prepared in Great Britain by the ADEE. This ultimately led to three different microwave GL projects: one by the Rad Lab, one by the Canadians, and the originating one at the ADEE.

Improvements in the klystron at Oliphant's laboratory had been started with the GL application in mind. As early as March 1940, a 50-cm (600-GHz) klystron-based GL set was in test operation by the ADEE on a rooftop at Steamer Point, picking up echoes from as far as the coast of France. After the successful demonstration of the 10-cm experimental set at TRE, further development of the 50-cm set was essentially stopped.

Still in charge of developing GL systems at the ADEE, P. E. Pollard now turned to the high-power magnetron for this application. Recognizing that Dee's group at the TRE was well ahead in such developments, Pollard asked them for assistance. Rowe directed that major support be given to the ADEE on this project; Dr. Denis M. Robinson from the TRE was assigned to lead this support effort.

Unfortunately, the insertion of a Joint Ministries Investigation Group led to certain problems in responsibilities and associated delays. Upon his return to Portsmouth at the first of December, John Cockcroft was named Chief Superintendent of the ADEE and exercised his influence in putting the GL project on track.

Based on the prior models of the GL system and the experimental 10-cm radar, an experimental microwave GL set was completed at the TRE by Robinson's team before the end of the year. It was assembled in a GL Mk 1 van with a pair of 3-foot parabolic dish reflectors mounted on the top. The dipole antennas rotated, giving a split beam and allowing both azimuth and elevation measurements. Lovell assisted with the rotating dipole and the coupling from the high-power source. For this coupling, use was made of a capacity-sleeve arrangement that he was designing for the AI system.

Robinson's GL set was far from suitable for field use by the Army. Thus, in January 1941, Rowe had the set taken to the British Thompson-Houston (BTH) facilities in Rugby for further engineering. At the end of May, the completed BTH version was delivered to the ADEE at

Christchurch, just as this organization was renamed the Air Defense Research and Development Establishment (ADRDE).

The set was in two trailers, one for the transmitter and the other for the receiver and display. The transmitter trailer had two 4-foot parabolas on a searchlight turntable; the transmitter itself was on the mounting that extended as a drum inside the trailer. Under the responsibility of Dr. Edwin S. Shire of ADRDE, testing was performed during June and July. This indicated that the radar could follow typical planes out to about 10 miles, flying from very low altitudes to directly overhead. Bearing and elevation accuracies were of some 10 minutes of arc. These results clearly showed the advantages of the 10-cm GL equipment.

Donald H Tomlin and others at the ADRDE developed full requirements for the final design and production of a deployable system, now designated GL3. In addition to their use in Great Britain, these requirements were sent to the Canadian National Research Council and were also taken to the United States by Cockcroft when he participated in the Tizard Mission.

There remained differences between the Air Ministry and the War Department as to who had responsibilities for GL radars. Consequently, both the TRE and the ADRDE make plans for production. The existing version from BTH, called the A model, was believed by the TRE to be satisfactory. Their only significant change was to incorporate 6-foot parabolas. Twelve of these sets were ordered, but this model was later abandoned because of serviceability problems.

ADRDE GL3 Prototype

At the ADRDE, Shire led in improvements for rough use by the Army, resulting in the GL3B model. Based on this, BTH handled the final design for the transmitter, antennas, and trailers, and EMI finished the receiver and display units. Prototypes were delivered in January 1942, but with additional changes, the production contracts were not placed until August. BTH/EMI, Metro-Vick/Ferranti, and Standard Telephones each received orders for 500 systems, but full flow of production did not occur until August 1943.

By mid-1942, it was clear that the GL system needed to be fully automatic. As described elsewhere, when the American version, designated SCR-584, was received in September 1943, it included a full unit for automating the system. It also had better general performance characteristics than either the British or Canadian (GL3C) versions; thus, the SCR-584 was adopted. Two hundred SCR-584 systems were ordered; later in the war, these were credited with assisting anti-aircraft guns in downing many V-1s and in back-tracking V-2s to find their launch sites.

While the ADRDE had its major attention given to microwave gun-laying radars, development was also underway on microwave coastal-defense radars. A Type 271 radar set was borrowed from the HMSS Experimental Department and, with the assistance of Dr. A. E. Kempton, was installed in a rotating GL2 cabin with 6-foot reflectors on each side. In July 1941, this was set up at Dover on the 330-foot cliffs overlooking the English Channel. From this vantage point, it could detect ships at 45 miles, and track large vessels leaving Boulogne on the French coast.

With few other changes, this set was designated CD Mk IV. Before the end of the year, the shops at ADRDE built 12 of these; then 50 more were ordered from Metro-Vick. Some of these were also used by the RAF as Chain Home Extra Low (CHEL) or AMES Type 54, an adjunct to certain CH stations; these could successfully track aircraft to as low as 50 feet in altitude.

At the start of 1942, Harold Gibson led the development of a transportable version for overseas applications. Designated CD Mk V, this had a single 7-foot reflector and the cabin was modified to serve as a shipping container, commonly called a Gibson box. The Mk V quickly became a vital part of Mediterranean and other overseas port defenses.

Microwave Radar for the RAF

The 10-cm (3-GHz) AI radar, now called AIS with the S indicating the microwave band, was the development project of first priority at the TRE Leeson. Initially there were problems with GEC; this firm had a contract to develop a 25-cm (1.2 GHz) AI system using their greatly improved Micropup tubes. These tubes could produce several kilowatts of pulsed power, and GEC gave many reasons for continuing this effort. Two leading scientists from the GEC Research Laboratory at Wemberly visited TRE Leeson, and, after seeing the experimental set successfully track an aircraft, supported the change. GEC assigned one of these scientists, Dennis C. Espley, to lead their efforts.

A major advancement had been made when the cylindrical-parabola reflector was adopted, but separate transmitting and receiving antenna were required. To facilitate tracking of the two antennas, the TRE workshop built a swivel mount for holding the two units and simultaneously aiming their narrow beams. As the transmitted power was increased, the antennas had to be further separated, leading to difficulties in designing the mounting and tracking equipment.

It was assumed that the T-R switch problem would soon be solved, allowing a single antenna to be used in the nose of night-fighter aircraft. Lovell was responsible for developing a suitable antenna and finding appropriate materials for the covering dome. For the latter, he tested the microwave transmission characteristics of a number of materials and eventually selected a synthetic polymer of methyl methacrylate that was commercially available under the trade name Perspex. The RAE in Farnborough was tasked to make a nose dome for a Blenheim aircraft.

The cylindrical-parabola reflector could be fitted into the aircraft behind the dome, but the scanning of the beam was a problem. On the experimental set, Lowell tried moving the dipole across the face of the parabola and found that the beam could be shifted as much as 25 degrees without quality deterioration. Assisting Lowell was Alan L. Hodgkin, a physiology researcher from Cambridge, who also had a good knowledge of mechanical kinematics. Hodgkin suggested fixing the dipole and rotating the reflector in a spiral motion at a high rate; the spiral would be sinusoidal, moving it up and then down. He calculated that the resulting beam would scan a conical volume about 45 degrees each side of the center line.

The mechanical apparatus, driving a 28-inch diameter parabolic reflector, was built by the firm Nash & Thompson. The spin was 1,000 rotations per minute, with the spiral eccentricity covering 30 degrees in one second. It was so well-balanced dynamically that essentially no vibration resulted. Lovell noted that a glass of water would sit almost motionless on the frame.

Hodgkin also proposed using the spiral-scan mechanism to generate a signal for driving the time base of a cathode-ray tube. The result would be a two-dimensional display of the target information. (It might be noted that Hodgkin, based on his post-war research, shared the 1963 Nobel Prize in Physiology or Medicine.)

The original experimental set had used a type 1189 magnetron. The newer 1198, now designated CV38, with a much lighter magnet would be used in the prototype 10-cm AI transmitter. For this, Atkinson and Burcham, working with GEC, rebuilt the original modulator to deliver 1-µs pulses with a peak power of about 50 kW. As earlier described, the minimum range for an AI radar was set by the pulse width and shape. For the desired 500 feet (at which distance the pilot could identify the target visually), the 1-µs pulse with a sharp rise and fall was necessary. The pulse rate was eventually set at 2,500 per second.

The receiver for the prototype set was much the same as for the earlier experimental set. Sutton's 300-mW reflex klystron with external frequency tuning was now available and used in the local oscillator. Skinner's high-purity silicon crystal detector had been further improved, and, although not yet in production, was used as the detector. The now-standard 45-MHz IF stage was incorporated. The CRT display, designed by Hodgkin to show the sector covered by the transmitted beam, completed the system

By February 1941, the TRE had a prototype AIS radar complete except for a working T-R switch. Since multiple parabolic reflectors were not practical, GEC provided a temporary solution using a ring of quarter-wave coax lines and a dummy load. With this arrangement, three-quarters of the power was lost, and the crystal was still subject to occasional burn-out.

The RAF delivered a Blenheim to the Christchurch Airfield, and the prototype radar was installed. In early March, 1941, the first airborne tests started. These were conducted by Hodgkin and George C. Edwards, a leading engineer from GEC. Within a few days, the system detected targets at ranges two or three miles greater than the height above ground, a major shortcoming with the previous AI systems. In early July, Cooke's new T-R switch became available, and the target-detection range improved to 10 miles. The set was designated AI Mk VII, and 12 service test sets were ordered from GEC and Nash & Thompson; in September, sets were installed on four Bristol Beaufighters for service trials. During the last of 1941, the service tests were satisfactory, and the radar was ready to become operational.

At the same time that prototypes were being built, orders were placed with GEC and Nash & Thompson for 100 sets to equip a squadron of new Beaufighters. By March 1942, a number of the aircraft had operational AI Mk VII radars, and the first successful use in combat occurred in April.

Simultaneously with the AIS development, there was a relatively low-level project in applying the 10-cm technologies to the ASV. Led by Donald W. Fry, this was called the ASVS (with S designating the S-band). The AIS radar and scanner were fitted into a Blenheim aircraft and in March 1941 was tested as to its ability to detect surface ships. This was followed in April by tests in which it detected the conning tower of the submarine HMS *Sea Lion*.

Although the fundamental features of the ASVS were the same as the AIS, little progress was made. A contract was placed with Ferranti and Metro-Vick for a manufactured version, but this was not tested in a Wellington aircraft until the end of the year.

The Start of Microwave Radar Mapping

In 1937, when E. G. Bowen had experimented with the 200-MHz set that would become the AI and ASV RDFs, he had noted that ground targets such as wharves and airship sheds were "easily identified." In his report, he used these observations to predict the following:

> It is certain that towns and villages would show up from surrounding country, that hedges, trees, and possibly railway lines and power lines would also be in evidence.

At that time, this type of application was of low priority and little further experimentation was conducted.

As the RAF began large-scale bombing of Germany, the GEE and Oboe navigation systems were very important, but there was a pressing need for a system able to assist in navigating at extended ranges. In a Sunday Soviet meeting on October 26, 1941, Churchill's scientific advisor, F. A. Lindemann (now Lord Cherwell), insisted that the TRE must develop "self-sufficient RDF equipment in a bomber" for long-range navigation.

Dee was unaware of Bowen's earlier ground-imaging concepts, but he had seen experiments at the TRE using the first 10-cm radar that showed buildings in Swanage. He assigned Dr. Bernard J. O'Kane and Geoffrey S. Hensby to make tests using a 10-cm AIS radar in a Blenheim V6000 aircraft with the scanner modified so that the beam swept at a depression of 10 degrees. Modifications were quickly made and a camera was positioned to film the display on the CRT.

On November 1, the first flight test was made. Flying at 5,000 feet altitude, the display on the modified AIS showed town features that were clearly distinguishable from the ground clutter. The wet photos were rushed to Dee. When shown to Rowe, he exclaimed, "This is the turning point of the war!"

Scanner on Aircraft

Experimental H2S

Additional tests were run during the next two months. So great was the need for such a system that in late December EMI was given a contract for 50 sets; EMI assigned Alan D. Blumlein as the engineering lead. At the TRE, Bernard Lovell was assigned as the project director; other key development participants were O'Kane and Hensby. The project was given the code name H2S, standing for Home Sweet Home. Lord Cherwell gave the project this name "because it allowed the bomber to locate and home straight onto a target." This also indicated that the system was intended primarily for blind bombing, not general navigation.

The initial tests had been made using the scanner in the nose of a Blenheim aircraft. Since the radar would be used in larger, four-engine bombers, the Hadley-Page Halifax was selected for further testing. To provide 360-degree viewing, a Perspex cupola was added to the aircraft's underside, and a three-foot scanner placed inside. The same basic transmitter and receiver used in the microwave AI system were used in the preliminary system.

Halifax with H2S Cupola

At high-official levels, there were concerns over using a magnetron in equipment that might be potentially lost in flights over Germany. Therefore, a second H2S set was built, this one using a klystron with the highest available power. For comparative testing of the different H2S types, two of the modified Halifax aircraft were assigned to RAF Hurn.

The first test flight of the magnetron H2S was made in mid-April 1942. With the Halifax flying at 8,000 feet, crude images of ground features at 5-mile range were the best that could be obtained. This was disappointing to everyone.

INITIAL RADIATION LABORATORY PROJECTS

When the Rad Lab was being formed, the initial three projects were selected as a 10-cm airborne intercept radar (code-named AI-10), a 10-cm gun-laying system for anti-aircraft use, and a long-range aircraft navigation system. The navigation system would become the only significant non-microwave project of the laboratory.

By early December 1940, the Rad Lab staff had made a design and collected the special components for a 10-cm transmitter and receiver. In addition to one of the first magnetrons from the BTL, the components included waveguides, dipole antennas, parabolic reflectors, and a reflex klystron (from Great Britain). For the detector, a grounded-grid triode was used as a diode. The transmitter modulator, other portions of the receiver, and display unit were adapted from existing radars. On Bowen's recommendation, a rooftop room had been constructed atop Building 4 for hardware assembly and open-view testing.

On January 4, 1941, operating in the rooftop room, the breadboard set picked up echoes from buildings about a mile away in Boston. The set required separate transmitting and receiving antennas, and the beams of the two parabolic reflectors were very difficult to aim on a distant target. Because of the difficulties in aligning the narrow beams, some of the participants questioned if a microwave system could ever match existing radars in detecting and tracking aircraft.

The common-antenna problem would not be resolved until Cooke's T-R switch from Great Britain became available. A temporary, partially effective solution was made by inserting a klystron as a pre-amplifier to the diode detector. This isolated the diode from the high power, but burn-out still occurred in a short while. Then, through careful aiming of the single parabola, an aircraft at a distance of four miles was detected on February 7, 1941. This was reported with satisfaction to Loomis and the Microwave Committee meeting on the same day in Washington.

Project One – Airborne Microwave Radar

With the success of the breadboard microwave radar, the Rad Lab effort turned to Project One: developing an AI set suitable for a fighter aircraft. Although the AIS project directed toward the same type of radar was underway – and doing well – at the TRE in Great Britain, the Air Ministry had asked for this similar effort in America, hoping thereby to accelerate delivery of the final product.

Because it was the first microwave system project of the Rad Lab, essentially all of the existing key personnel took part. The leaders and their specialties included the following: Dr. Alexander J. Allen (scanning antenna), Dr. Robert F. Bacher (indicator), Dr. Kenneth T. Bainbridge, (modulator and power supply), Dr. I. I. Rabi (magnetron), and Dr. Louis A. Turner (receiver). Dr. Edwin M. McMillan was responsible for building and testing the engineering set.

All of these men were physicists on leave from universities. The only one with appreciable hands-on electronics experience was Bainbridge – he had earlier worked at General Electric and was also a former amateur radio operator (2WN).

The engineering set that was developed was primarily a packaging for the breadboard set. A 30-inch, parabolic reflector was used, with a scanning capability 180 degrees in azimuth and from –10 to +65 degrees in elevation. The radar operator had CRTs for azimuth versus range and for elevation versus range; a single, small CRT for range was available for the pilot.

The Army Air Corps would be the primary potential user of the AI radar; thus, for testing they provided a Douglas B-18 bomber. (At that time, they did not have in service a dual-engine fighter aircraft.) The B-18 was placed at the National Guard Hangar at Boston's Logan Airport, and by the end of February 1941, equipment installation began. Ground tests followed, and by mid-March echoes of airport objects were obtained.

The first flight test took place on March 27, 1941. Several researchers from the Rad Lab were aboard the B-18, including McMillan (in charge), Alvarez, and Bowen. The flights were made over the Cape Cod Bay using a small National Guard aircraft as the test target. On several runs, clear echoes were obtained at ranges up to three miles. The altitude of the B-18 was 10,000 feet, less than two miles; the maximum range was thus not limited by ground return, a primary advantage of microwave radar.

The B-18 was then flown to New London, Connecticut. There they found a surfaced submarine from the local Navy Base, and this large target was detected at ranges up to five miles.

The first day of flight testing had given excellent results. Bowen noted, "My colleagues from the Radiation Laboratory were wildly excited and could not contain their enthusiasm." The project was immediately expanded to include ship and submarine detection, and gained major interest from the Navy.

Chief Air Marshal Sir Hugh Dowding visited the Rad Lab in April. When taken for a test flight with the 10-cm AI, he asked to be shown the minimum range, a long-standing problem with the existing AI sets. The aircraft was flown to less than 500 feet from the target and the range indication continued to function, again showing the superiority of microwave operations. In June, upon Dowding's recommendation, the British Ministry of Aircraft Production placed an order with Western Electric for 200 of the Rad Lab AI radars.

Although the U.S. was not yet at war and had no immediate need for these systems, the requirement in Great Britain remained great. To compare microwave AI sets from the Rad Lab and similar sets from the TRE, a Boeing 247D on loan from the Royal Canadian Air Force (RCAF) was outfitted with the Rad Lab set and shipped to England at the end of June. Bowen accompanied the aircraft to England and remained for several months, giving demonstrations and assisting in comparative tests with the new AI Mk VII that were conducted at Christchurch and other RAF airfields. The results showed that the Rad Lab's transmitter gave superior performance, but the TRE's receiver was better.

After the incorporation of Cooke's T-R switch and certain other improvements from the TRE receiver, the Rad Lab set was designated

SCR-520, America's first microwave radar. Western Electric produced a number for testing by the just-formed Army Air Forces at Wright Field. At near 850 pounds, it was too heavy for the Lockheed P-38 Lightning, their first two-engine fighter. Through a weight reduction effort, this was decreased to some 600 pounds and by the end of 1942 found use on some P-38s and a few PBYs. As described later, the system was modified to become the SCR-518, an air-to-surface search set that saw limited service, and also upgraded to the SCR-720/AI Mk X, a system used extensively by both the U.S. and Great Britain throughout the war.

Project Two – Microwave Gun-Laying Radar

Incorporation of microwave technology into the GL radars started earlier in Great Britain. By August 1940, the ADEE had prepared basic requirements for a set designated as GL3. These requirements were used in Great Britain and also sent to Canada. Cockcroft had brought them to the United States, and they were used in initiating Project Two at the Rad Lab.

The same breadboard set that was the foundation of the microwave AI radar was used to start the gun-laying project. It was determined that the requirements could be met by basically the same transmitter and receiver but with a much larger antenna reflector. In addition, it was decided to incorporate automatic tracking.

The project was initially led by Dr. Louis N. Ridenour, but was soon taken over by Dr. Ivan A. Getting (1912-2003). A physics graduate of MIT, and a Rhodes Scholar with a D.Phil. from Oxford, Getting had a reserve commission in the Army Coastal Artillery Corps and thus had a personal knowledge of anti-aircraft guns and systems. The detailed technical development and testing was under Lee L. Davenport, a young physicist who also had engineering experience in industry.

Ivan Getting

At that time, MIT had an outstanding capability in servomechanism research and development; thus, the incorporation of automatic tracking was readily brought into the design. With Davenport leading, a machine-gun turret from a B-29 bomber was mounted on the roof and a 4-foot parabolic dish attached. In addition to dish rotation, conical scanning was needed; this was produced by rotating the antenna feed. To provide these motions, a servo control unit was designed and built. Upon detection of a target, the receiver output was used to put the servo control into a track-lock mode.

The control unit was completed, and a transmitter and receiver from the prototype GL set were ready to test. A camera was mounted on the turret to follow the beam direction. On May 31, 1941, tests using a small aircraft as the target were very successful.

A demonstration to Alfred Loomis resulted in a visit by Brigadier General Roger B. Colton, Chief of Research and Engineering at the Army's Signal Corps Laboratories (SCL). Colton had been the driving force behind the development of the Army's existing meter-wave GL radar, the SCR-268. Immediately recognizing the superiority of a microwave system, he promised Army support for the final development and recommended procuring a set for every AA battery. Having some reservations about the Rad Lab managing such a program, he also placed an order with the BTL for developing a similar system.

At the Rad Lab, work was started on the engineering model designated XT-1. The turret mechanism was redesigned, as was the servo control unit, adding helical scan for searching. Improvements were made on the modulator and the receiver, and a display unit was developed by Dr. Ernest C. Pollard. All of this was installed in four trucks, including the large power generator and a separate IFF unit.

Davenport and Getting on XT-1

In December 1941, preliminary testing was conducted by the SCL at Fort Monmouth, New Jersey. This resulted in the parabola being enlarged to 6 feet (to increase the range), and a PPI being added (to allow use of the system as an air-search radar). It was then sent to Fort Monroe, Virginia, for testing by the Antiaircraft Artillery Board using 90-mm guns and a BTL T-10 gun director. With the larger reflector, the range was now much greater than the 15,000 yards (8.5 miles) that the Army required. In March 1942, the Board concluded:

> The Radio Set XT-1 is superior to any radio direction finding equipment yet tested for the purpose of furnishing present position data to an antiaircraft director.

Contracts were negotiated with Westinghouse and General Electric for producing the system, now designated the SCR-584. For the antenna, drive mechanism, and vans (changed from original trucks), they turned to Chrysler Motors. Here the final design for production was placed under Frederic W. Slack in their Central Engineering Office.

The parabolic reflector was changed from aluminum to steel

SCR-584 Control Van

Assembly at Dodge Plant

so it could be formed using standard presses of automotive manufacturing. To keep the weight down, over 6,000 equally spaced holes were drilled through the face; these had no effect on forming the microwave beam. The entire drive mechanism was redesigned, resulting in improved performance, reduced weight, and allowing easier maintenance.

For the system to have automatic target tracking, BTL would provide the equipment "brain," an electronic analog computer containing 160 vacuum tubes. Called the M-9 Predictor-Corrector Unit, this was developed by David B. Parkinson and Dr. Clarence A. Lowell. With this computer, the system could automatically track targets to 18 miles. A single M-9 could direct four anti-aircraft guns.

General Electric and Westinghouse shared (60-40) the SCR-584 prime contract. A prototype was tested in May 1942, but because of bureaucratic delays deliveries did not start until early 1944. Eventually, about

M-9 Predictor-Corrector

1,500 of these systems were used in both the European and Pacific war theaters. Together with the proximity fuse, the SCR-584 is crediting with enabling anti-aircraft guns to destroy the majority of German V-1 "buzz bombs" attacking London following the Normandy invasion.

As noted earlier, General Colton had placed a separate order with the BTL for a similar 10-cm GL system. This was completed as the SCR-545, and had a 1.5-m array added to the dish to increase the field of view. Gun-laying 10-cm radars were also developed in Great Britain and Canada, designated as GL3B and GL3C, respectively. The performance of all of these systems, however, was greatly overshadowed by the SCR-584. Eventually, the British adopted this system, designating it GL3A. As later described, the U.S. Navy also adopted special versions of this radar; the SM for heavy carriers and the SP for smaller carriers.

Project Three – Long-Range Navigation

In selecting the initial projects for the Rad Lab, it had been agreed that a new navigation system with a range of at least 1,000 miles and an accuracy of about 5 miles was needed. The TRE in Great Britain had the GEE hyperbolic navigation system under development, but it operated in the short-wave region and was thus limited in range by line of sight. Loomis suggested a new concept: using long-wave transmitters, perhaps a thousand miles apart, emitting synchronized pulses precisely regulated by quartz timers.

The concept was turned over to the Rad Lab for investigation. This was clearly outside the other activities involving microwaves, so a new group of participants was needed. General Radio was a nearby Cambridge firm with a highly respected line of radio instrumentation. It was founded and run by Melville Eastman who, as the war approached, developed an absorbing interest in military electronics. Upon the formation of the Rad Lab, Eastman volunteered his services, and was soon asked to lead the long-range navigation project.

Scientists from the Rad Lab and the BTL visited the TRE and gathered some, but not all, additional information on GEE. Preliminary plans were developed for a transmitting system that, like GEE, had a master and two slave stations with pulse emissions. In the receiver, they turned to radar for a cathode-ray tube to display the received pulses.

The system was initially called LRN (Loomis Radio Navigation), but Loomis objected to using this name. The first laboratory model, built with the assistance of the BTL, was tested in the summer of 1941. Using a receiver in a station wagon and transmitters at several universities, comparative trials were made at different frequencies to evaluate ground-wave and sky-wave performance; this eventually led to the choice of 1.950 MHz as optimum. Initially, there was little success in synchronizing the different transmissions, but this changed when additional information was obtained on the GEE system, including details of an improved delayed and strobed timing technique

Robert Dippy, the inventor of GEE, came to America to assist in the project, and the Rad Lab team soon realized how far the British work had advanced. Project 3 was then abandoned, and the better points of GEE were incorporated in a new system. Coast Guard Captain Lawrence M. Harding, who represented the Navy at the Rad Lab and played an important role in the testing and implementation of land stations, suggested the acronym LORAN for Long-Range Navigation. Thereafter, the acronym was used for this technology and, unlike radar and sonar, was not converted to a name.

The LORAN Division, led by Donald G. Fink, was formed. Other leading participants included Dr. John A. Pierce, a Harvard professor with extensive experience in low-frequency radio propagation, and Dr. Julius A. Stratton, later President of MIT. Eastman continued as a consultant.

It was decided that LORAN would initially be for maritime navigation, highly important at that time for providing Atlantic routes to avoid known German submarines. The first LORAN pair (Montauk Point, New York, and Fenwick Island, Delaware), went on the air in June 1942. Additional stations were added along the Canadian east coast, and the system became operational in early 1943.

While transmitting stations were being built, the Rad Lab let contracts with a number of firms to provide receivers, initially for ships then later for aircraft. Dippy insisted that the LORAN and GEE receivers be made physically interchangeable so that any RAF or American aircraft fitted for one could use the other by swapping units.

In operation, the receiver would obtain pulses from two stations and display them on a CRT with a time base. The difference in arrival time would be read directly. They were then transposed to a LORAN lattice chart of hyperbolas and the position could be plotted.

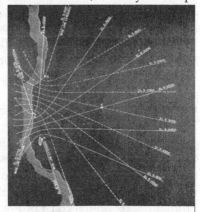

LORAN Chart of Hyperbolas

The master and slave stations could be separated by as much as 1,000 to 1,200 miles. They transmitted on four frequencies: 1.75, 1.85, 1.90, and 1.95 MHz. Since signals were pulsed, a relatively small 200-watt transmitter could deliver peak power in excess of 200 kW. The

LORAN Station Equipment

reliable operating distance was about 800 miles in daytime and 1,400 or more miles at night (the height of the ionosphere varies between day and night), with an accuracy of about 1% of the range.

While LORAN was important in the Atlantic and Eastern Europe, it was in the Pacific that the system made its greatest direct contribution to winning the war. As American forces moved westward, airfields were built on many of the small islands and atolls that dot the ocean beyond Hawaii. Most aircraft at that time had a range that required frequent refueling, and LORAN was ideal for finding these small airfields in the broad expanses of the Pacific Ocean. As bases were established close enough to Japan to allow a round trip, LORAN was invaluable for precision bombing of the mainland. The Japanese either failed or never tried to jam any of the LORAN transmissions.

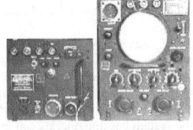

LORAN Aircraft Receiver

By the end of hostilities, navigational coverage of about 30% of the Earth's surface was provided by about 70 LORAN transmitter stations, and some 75,000 LORAN shipboard and aircraft receivers had been installed. Although it was neither radar nor involved microwaves, LORAN was certainly near the top of significant accomplishments of the Rad Lab.

CONTINUED DEVELOPMENTS IN GREAT BRITAIN

As the war continued, the *Luftwaffe* became a greater threat to the three radar development facilities on England's south coast. (It is noted that following the Tizard Mission to the U.S. in September 1940, the name RDF had been slowly replaced by the American name, radar.) From airfields on the coast of France, it was only minutes flying time to Portsmouth, Christchurch, and Swanage. Although research facilities, none of the three areas had been specific targets; it was assumed that this was mainly because the nature of work there was still unknown to the Germans.

Microwaves Bridge the Atlantic

During the Blitz, huge flights of bombers and protecting fighters often passed over the south coast heading for the large cities, and on return flights they sometimes emptied their munitions before leaving England. It might have been coincidental, but it seemed that machine-gun strafing of the main street of Swanage would increase on Monday mornings, just as the high-level attendees of Rowe's Sunday Soviet meetings were taking the train back to London.

The successful raid in February 1942 on Bruneval and capture of a *Würzburg* radar was a great boost for the British public's moral, but it also had its downside: it highlighted to the authorities how vulnerable installations close to the sea were to enemy commando raids. Prime Minister Churchill praised the three military services for their cooperating performance in the Bruneval raid, but reminded them that the Germans might do the same thing. Because of this, he gave orders that the research operations must be moved inland "before the next full moon."

Moving was not unexpected to the three operations. They realized their vulnerability and, in addition, all were out of space and were reluctant to add to existing facilities. The Army at ADRDE had already found a group of vacant buildings near the city of Great Malvern in Worceshire, only a few miles south of industrial center of Birmingham. The Air Ministry identified Malvern College, nearby to the new ADRDE facilities, as a suitable for relocating the TRE.

The Admiralty Signal Establishment

The Experimental Department of HMSS was the first to move, actually starting well before Churchill's instructions. Facilities at the Royal Barracks were inadequate, and some radar work was already occupying other space in Portsmouth. In late 1940, planning began for evacuating the Department to locations about midway between Portsmouth and Lomdon on the linking rail line. Two facilities were involved: Lythe Hill House, a 30-acre resort near Haslemere, and King Edward's School, a private institution for young men in Witley, both in Surrey and about 8 miles apart. Separately, the Valve Division was moved to the University of Bristol before the end of the year.

ASE Haslemere - Lythe Hill House

Renovations and new buildings at Lythe Hill were completed in August 1941, and a portion of the original Experimental Department moved at that time. This included the communications and infrared operations, as well as the administrative offices. The shops and production facilities moved to neighboring Whitwell Hatch.

ASE Witley - King Edward's School

King Edward's School was readily suited to the intended housing of radar development, but there had been a delay in its acquisition. Originally founded in London in 1553 as the Bridewell Palace, the School had moved to Witley and taken the new name in 1867. It had a Royal Charter, the Queen was its patron, and there was a powerful Board of Trustees; thus, notwithstanding the war, there was great resistance to the facility takeover. The Admiralty finally had to appeal directly to Prime Minister Churchill to get it commandeered.

Radar development activities finally moved from Portsmouth to Witley in September 1941. At that time, the combined operations became the Admiralty Signal Establishment (ASE), with Captain B. R. Willett assigned as the Commander. About 1,000 persons were involved in the overall move to the two locations, generating a severe strain on living accommodations in the area. King George VI visited ASE Haslemere in June 1942, and ASE Witley the following March.

The radar activities were totally reorganized at ASE Witley-Haslemere. Rather than being grouped on a "wavelength" basis as at Portsmouth, there were three equipment divisions supported by nine techniques divisions. The equipment divisions and the heads were Tactical and Anti-Sub under Dr. S. E. A. Landale; Gunnery under J. F. Coales; and IFF, Navigation, and Countermeasures under H. E. Hogben.

The techniques divisions were Aerials, Transmitters, Receivers, Displays, Test Equipment & Measurements, Servos, Mechanical Design, Research, and Theory. These represented the radar-related technologies being conducted. In addition, there was the Library, Publications, and Information Services, led by L. H. Bainbridge-Bell, an original member of Watson Watt's team, and who had transferred to Portsmouth following his investigation of the antennas on the scuttled German vessel, *Admiral Graf Spee*.

This first microwave radar, Type 271, had been intended for surface-warning on smaller vessels – primarily corvettes and frigates. Before the operations left Portsmouth, this radar was being modified for larger ships, and this work continued at ASE Witley. Landale and his team developed the Type 272 for destroyers and smaller cruisers, and the Type 273 with two 35-inch parabolic reflectors for major warships – cruisers and battleships. Some of these sets were in operation before the end of 1941. With larger antenna reflectors, the Type 273 also performed well as an air-warning radar against low-flying aircraft, giving a range of about 30 miles

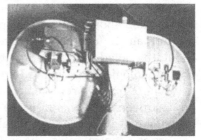

Type 273 Antennas

When the strapped magnetron (type CV76) became available, it was incorporated in a set designated Type 271Q; in 1943, it began to replace most of the earlier sets. This set had significantly higher power, and used electric drives for antenna rotation. On most vessels, the entire system, including the control cabin, was now mounted at a higher level.

An important addition was the Type 277 "nodding" 10-cm height-finder radar. It had a dual "cheese-style" antenna that was mounted vertically, creating a horizontally flattened beam. This was designed for both shipboard and coast defense, the latter using a trailer. This was the first British Naval radar with a PPI display. The trailer version was modified by the TRE and designated AMES Type 13.

Type 277 Installation and Antennas

A number of microwave gunnery-director radars were developed under Coales. These did not come into operation until 1944 and used newer, higher-power magnetrons. Type 274, a main gunnery director for ranging and shot-spotting, replaced the Type 284 on large vessels. This used a pair of "cheese" antennas, stabilized against pitch and roll. (It should be noted that Type numbers for Royal Navy radars were often not chronologically sequenced.)

Type 275 was a replacement for the Type 285 AA fire-control radar on destroyers and was also used as the secondary fire-control set on major ships. It had 4-foot parabolic reflectors mounted on opposite corners of a control cabin; the entire assembly rotated to direct the beam. Type 276 was an AA fire-control radar, replacing the Type 285 on selected destroyers. It used continuously rotating, 4-foot parabolic reflectors. All three of these gunnery radars incorporated PPI displays.

Type 262 on 40-mm Gun

Continued development of magnetrons by Dr. James Sayers at Oliphant's laboratory and at the GEC led to the E1487, producing 25 kW peak power at 3.2 cm (9.4 GHz). This became available as the CV208 and was used in the X-band Type 262 fire-control set for 40-mm Bofors weapons. Developed by GEC under the supervision of A. K. Solomon and A. V. Hemingway of the ASE, it had a dish antenna that spun off-center at a high speed. The system incorporated target lock and was fitted to a close-range fire-director mounted on the gun for blind firing. Although the radar was designed earlier, due to difficulties in the mounting it did not become operational until early 1945 on the aircraft carriers HMS *Ocean* and HMS *Bulwark*.

Another X-band radar was the Type 268, a high-resolution target-indication and navigation set for small vessels, and the Type 268W variant for submarines. These sets were actually designed and built in Canada under contract with the ASE. A prototype system was tested at Portsmouth in July 1943 and performed so well that 1,500 sets were placed on order.

The last major wartime project at the ASE was the development of a radar set to be used with an experimental guided missile under design for the Royal Navy. The radar would be used for target detection and tracking, with the signal also used for beam-riding on the missile. It would also need to carry out a search over a limited solid angle before locking on a particular target. P. E. Pollard was the project engineer.

To accomplish this, the antenna system was designed based on the principle of a reflecting optical telescope, using a 6-foot parabolic reflector with a small, slightly offset disc. The RF feed was first reflected off the disc, then formed into the beam with the large parabola. By simply rotating the disc, the beam could be nutated to rapidly sweep or track. Special strobing and filtering techniques were required to be developed, and coded information had to be added to the signal to accomplish beam-riding. Designated Type GMY4, the radar itself was successfully demonstrated in late1945, and the entire system was later placed into service as the *Sea Slug*, the Royal Navy's first surface-to-air missile.

The division under Hogben engaged in a variety of hardware and analytical activities in IFF, countermeasures, and navigation. Most IFF and countermeasures projects were under the TRE or joint efforts with America, but much analysis was required to interface their developments with the hardware systems.

The Germans had set up long-range guns on the coast near Pas-de-Calais and were able to accurately direct their fire at night on vessels in the Channel. Using a receiver, N. E. Davis determined that the guns were directed by a *Seetake* radar, and developed a countermeasure jamming transmitter. The Type 91 transmitter was tunable between 300 and 600 MHz, covering bands used by *Seeake*, as well as the *Würzburg* radars. A number of these sets were eventually set up in the Dover area, reducing the effectiveness of the German guns.

The ASE worked closely with the TRE in developing equipment to support the D-Day invasion. Hundreds of TRE-designed Mandrel wide-band jamming transmitters were built for the landing craft. The ASE also modified a number of H2S imaging radars, designating them as Type 970, to be carried on major invasion vessels; these provided CRT displays showing the coast, ensuring that they landed at the intended locations.

Representative of the independent analytical projects were those done by the Theory Division. Frederick Hoyle, a theoretical physicist from Cambridge who had previously done research in astronomy, joined the Experimental Department in 1941 to perform mathematical analysis of RDF signal propagation. When the Theory Division was formed at ASE Melville, Hoyle was named to lead a small group in this supporting activity. Two others with similar backgrounds who joined him were Thomas Gold and Hermann Bondi, both refugees from Austria, and who had initially been interned as enemy aliens at the start of the war.

Some of Hoyle's work at the ASE concerned estimating the height of target aircraft, anomalous propagation at centimeter wavelengths, effects of rain on radar signal propagation, and target detection in the presence of Window-type jamming. His height estimation technique was for meter-wavelength radars; it combined graphs, statistics, and operator bias in a process that was used with success for a number of years. In addition to the radar-related work, at night and other free time, Hoyle led Gold and Bondi in the preliminary formulation of astrophysical and cosmological theories for which they later received wide recognition.

The TRE at Great Malvern

At the end of May 1942, the laboratories and movable equipment of the TRE at Swanage were loaded into trucks and moved northward to Great Malvern, Worcestershire. More often simply called Malvern, this town is in the central part of England, about 125 miles north-northwest of London and 20 miles south of Birmingham. The area is called "the Switzerland of England," and Malvern is in a valley between mountains.

Malvern College, located in Great Malvern, was founded in 1865 as a public boarding school for boys. In October 1939, it had been temporarily taken over by the Admiralty for a special training facility. Then in April 1942, it was requisitioned to become the new facility of the TRE. (The College then operated in exile at Harrow School near London until after the war.)

Malvern College Main Building

In the following weeks, buildings and grounds of the College were almost totally transformed. The main building was converted to offices and limited laboratories. New buildings were constructed on available open spaces; included was the Preston Laboratory to house the expanding microwave activities. Most of the student houses were converted to test areas and workshops. Three electrical power stations were installed, and wiring throughout was upgraded or replaced. Three of the houses were retained as very limited living quarters, and a large canteen was shortly completed.

Upon arrival, most of the 2,000 people from TRE were on their own to find living accommodations. (About 500 people has remained at Swanage, but many later moved to Malvern.) The Army's ADRDE had just relocated from Christchurch to a large vacant facility at Pale Manor Farm near the College, and their personnel had filled much of the small amount of spare space in local homes. Bernard Lovell made note of the situation:

Microwaves Bridge the Atlantic

Our major problem was the hostility of the people of Malvern. The war had made little impact on the town and our arrival was a shock to the inhabitants who, for security reasons, could be given no idea of who we were or what we were doing.

In 1941, the RAF had built an airfield at Defford, about seven miles east of Malvern, and this was then assigned for use by the TRE. The Telecommunications Flying Unit (TFU) that operated flight trials for airborne equipment transferred its aircraft and personnel from Hurn to Defford. So hurried was the move that many of the RAF personnel had to be accommodated in tents for some time.

After a brief period, the operations of the TRE continued much as they had been at Swanage. A. P. Rowe remained as the Superintendent, and Dr. W. B. Lewis as the Associate Superintendent. The research and development personnel were in about 30 different groups, some based on technologies and others in large projects. The very important activities in centimeter radar under Dr. Philip Dee were assigned to the Preston Science building.

The first test flight of an experimental H2S blind-bombing radar occurred just before the relocation to Malvern. By the end of May 1942, Dr. Bernard Lovell and his team that included Dr. Bernard O'Kane and Geoffrey Hensby were on the top floor of the Preston Laboratory, and the two Halifax aircraft with their crews were at RAF Defford.

The RAF requirement for the H2S had been reduced from 30 to 15 miles in range while flying at an altitude of 15,000 feet, but the tests before leaving Swamage had given disappointing results; much more work was necessary.

As previously noted, many officials at high levels were opposed to using a magnetron in equipment that might be shot down over Germany; thus, a contract with EMI for 50 initial sets specified using a klystron rather than a magnetron. For comparisons, the set in one Halifax used a magnetron and a klystron set was in the other. Trials with the klystron set indicatedt that such a system would never meet even the reduced requirements. The debate, however, continued.

On June 7, 1942, Hensby and Alan Blumlein from EMI were in a Halifax demonstrating the magnetron set. Tragically, the aircraft crashed, killing all on board. Lovell and O'Kane went to the crash site and recovered the highly secret magnetron.

Without Hensby and Blumlein, progress at both TRE and EMI was greatly slowed. However, bombing runs deep into Germany had begun, and the need for H2S was great. In an early July meeting that included

Dee and Lovell, Prime Minister Churchilll gave orders that 200 sets must be built and installed on two sqaudrons of heavy bombers by October 15.

An emergency plan emerged whereby EMI would produce 50 sets, the Radar Production Shop (RPS) at the TRE would build the other 150, and Nash & Thompson would build all of the scanners, but the 200 finished sets could not be ready before the end of the year. The plan included the abandonment of a klystron set, and this received approval.

H2S Aircraft Installation

By the end of September 1942, a production prototype was installed on a Halifax and sent to the RAF Bomber Development Unit. There the navigators reported that the H2S would be "valuable to a high extent both as a navigational aid and as an aid to locating targets."

Continued development and testing went well, and EMI delivered their 50 sets on time; the RPS, however, failed completely in their delivery. By the close of 1942, H2S systems had been installed on 48 bombers. On January 30, 1943, the H2S was first used over Germany.

During the first half of 1942, the Battle of the Atlantic continued, with an average of about 400,000 tons per month being lost to U-boats. The Coastal Command increased the use of ASV Mk II radars, and the sinking of German submarines significantly increased and shipping losses decreased in areas covered by the aircraft. The German, however, soon found a countermeasure – using a special receiver that detected the radar transmissions – and the shipping losses resumed.

As a counter-countermeasure, a microwaves ASV was needed, with the belief that this would be more difficult for the U-boats to monitor. The ASVS, a 10-cm (S-band) version of the ASV, had been under development for a year, and this work was now accelerated. In parallel, the H2S was considered as an alternate. For this, Lovell and his team designed a new H2S scanner to fit into a Perspex "chin" under the nose of an aircraft.

By the autumn, preliminary tests indicated that the H2S would be suitable for this application. With the H2S already in operational use, the ASVS project was abandoned. The modified H2S system was designated ASV Mk III, and test flights began near the end of January 1943. On March 15, the Mk III was first used to detect an unspecting submarine.

Although a number of ASV Mk III sets were built by EMI and used on Coastal Command aircraft, this radar was not put into production. A similar radar developed by the Rad Lab and produced by Philco became available. This was adopted as the DMS-1000 for use on Liberator aircraft in patrolling the Northern Atlantic. By August, shipping losses dropping to about 100,000 tons per month, a decrease of some 75 percent

On the night of February 2, 1943, a Stirling bomber with an H2S radar was shot down over Holland. This is what opponents of the H2S had feared: Germany might now know about the magnetron. (It should be noted that German scientists had also developed multi-cavity magnetrons, but these were eventually abandoned because of their frequency instability, a defect corrected by "strapping" in Great Britain.)

Essentially nothing more than the magnetron itself survived the crash, so it was some time before the Germans realized that this was at the heart of microwave radar. It was six months before British Intelligence learned that 10-cm receivers were being used to give warning and subsequent jamming of microwave radars. Still shorter wavelengths were needed

Continued development of the magnetron had led to the CV208, with 25-kW peak power at 3.2 cm (9.4 GHz), now called X-band. The ASE incorporated this magnetron in designing the Type 261 fire-control set. In Dee's Division at TRE Malvern, a group had experimented with 3-cm sets, and Lowell had assigned I. J. Beeching to convert a H2S set to use these 3-cm components. Test flights with the 3-cm H2S were made, but the results gave little promise of improvement over the 10-cm set.

In late May, Lord Cherwell visited the Rad Lab, where he was given a demonstration of their 3-cm system then under development and was convinced of the superiority of this wavelength. Upon his return to London, it was decided that the TRE would convert the H2S to X-band, but the Americans would be expected to do the same for the ASV radars.

In a crash effort, the TRE improved the 3-cm performance and delivered six systems to the Pathfinder Force in November 1943. These gave excellent maps at an average range of about 24 miles when flying at 20,000 feet. On December 22, they were first used in a mission to "mark" Berlin, dropping incendiaries to show targets to successive navigators.

For the remainder of the war, 3-cm H2X radars were used with great success at night or under heavy cloud cover, with planes often flying as high as 30,000 feet. Joseph Goebbels, Nazi propaganda minister and

Lancaster with 3-cm H2X Chin

Hitler's successor, wrote in his diary: "The English aimed so accurately that one might think that spies had pointed their way."

The Countermeasure Groups under Dr. Robert Cockburn remained very active at TRE Malvern. As more information was obtained on German radars, Martin Ryle led several teams in developing ECM equipment for use against a variety of systems. These were both jammers and spoofers. The jammers included the broadband "Mandrel" that blinded *Freya*, *Würzburg*, *Wassermann*, and *Mammut* systems, and the "Shiver" for countering the *Würzburg*. (Both Cockburn and Ryle were knighted after the war, and Ryle was a Nobel Laureate in 1966.)

E. H. Cooke-Yarbrough on Ryle's team developed a swept-frequency receiver called "Bagfull" that marked the frequencies of transmission by *Würzburg* to allow future missions to use the correct jamming frequency. "Carpet" was an extension of Bagfull; it would quickly mark the *Würzburg* frequency then automatically set the jammer to this frequency. Another device from Cooke-Yarbrough was "Blonde." This recorded the pulse frequency and repetition rate of the radar signal and the mechanical characteristics of the antenna.

Spoofing was a more sophisticated approach, tricking the Germans into seeing "virtual" targets. Cockburn's team developed a device named "Moonshine" to spoof *Freya*. This was a "pulse repeater"; it listened for a pulse from a transmitter, and then fired back a spread-out pulse on the same frequency that fooled the receiver into "seeing" a reflection that appeared to be from a large aircraft formation.

After it was used in late 1942 with good results, the Germans caught on and Moonshine was discontinued. Ryle, however, returned to this technique in planning a major deception for the D-Day invasion. For this, six aircraft and four launches carried pulse-repeater equipment along the Straits of Dover, indicating that a massive invasion was converging on the French coast at Boulogne. This tied up a large segment of the *Luftwaffe* that could otherwise have defended the invasion at Normandy. During the actual invasion, 600 jammers (mainly Mandrels) were distributed in the assault vessels.

When RDF systems were initially being studied in Great Britain, Dr. R. V. Jones had suggested that a piece of metal foil falling through the air might create RDF echoes. In early 1942, Joan Strothers Curran, one of the few female physicists at the TRE, went back to Jones's suggestion and came up with a scheme for dumping packets of aluminum strips from aircraft to generate a cloud of false echoes. Generally called chaff, the strips needed to be about a half-wavelength long.

Because of practical length limitations, chaff was mainly useful against the microwave systems. It was code-named "Window" when cut to near 0.3 meters and "Rope" when 1.7 meters. This ECM was at first omitted from use, fearing that the Germans would then use it in turn during raids over England. However, when massive bombing of Germany began, Prime Minister Churchill personally ordered it used. As anticipated, the Germans then also started using this countermeasure.

Airborne countermeasures equipment developed at the TRE included units called "Boozer" and "Monica." Boozer was a receiver giving planes warning that they were being "painted" by German ground radars, and Monica was a small radar carried in the tail of planes to show when they were being approached by a night fighter. They functioned too well, giving out constant warnings, especially when used in large formations of aircraft, and were discontinued.

At TRE Malvern, Dr. B. V. Bowen led a continued development of IFF. The Combined Research Group (CRG), staffed by representatives of American, British, and Canadian military services, had been established at the U.S. Naval Research Laboratory (NRL) in late 1941 to resolve common problems. As British and American aircraft started joint flights over Europe, a common IFF was required. Representing the TRE, Dr. J. R. Whitehead was attached to the CRG for a long period, and was a leader in gaining acceptance of the British IFF Mk III as the common unit.

Programs such as GCI and CHL continued to have upgrades in transmitters and antenna systems, including changes to 50 cm and then 10 cm; this work was under Dr. William H. Penley. Development was completed on the first mobile GCI/CHL radar. Designated AMES Type 11, it was tunable between 50-60 cm to avoid jamming. Another innovation was a PPI that showed both the raider and the fighter.

There were three 10-cm sets: AMES Type 13 Nodding Height-Finder radar (converted from Navy Type 277), AMES Type 24 Long-Range Height-Finder radar, and AMES Type 14 High-Power Surveillance radar.

In January 1944, Penley was placed in charge of the Engineering Department, and at the end of the year he was named a Divisional Leader, responsible for fighter aircraft offensive and defensive radars, bomber defensive radars, and a number of other related systems.

The Circuit Division, under Dr. Frederick C. Williams, occupied what was called the Cricket Pavilion. He was renowned for his emphasis on circuit "designability" – the development of circuits whose performance could be accurately predicted before they were built. This operation attracted creative electronic types from throughout Great Britain. Among their developments were linear timing circuits that greatly improved the accuracy of radars, and a circuit that would later be called an operational amplifier – the building-block of analog computers.

Joint projects between the TRE and the Rad Lab required personnel from each organization to work closely with the other. The Rad Lab staffed a small laboratory operation at TRE Malvern. In the other direction, Dr. Denis M. Robinson was sent to the Rad Lab and represented both the TRE and ADRDE, and Dr. Edward G. Bowen became a long-time participant. For some time, Robert Hanbury Brown was attached to the U.S. Naval Research Laboratory.

There were numerous analytical and experimental studies of the propagation and reflection of radar signals conducted at the TRE, often in association with researchers at ADRDE, ASE, and the Rad Lab. One such study was led by Dr. Henry G. Booker and resulted in seminal information concerning environmental effects on electromagnetic signals. Booker also made studies that led to techniques for predicting the radar cross-section of aircraft (the "size" they appear to radar interrogation).

The last major hardware development at TRE Malvern before the end of the war was a complete mobile radar station and control center. The RAF placed a requirement with the TRE in mid-October 1944, and the total system, comprising 28 vehicles and designated AMES Type 70, was delivered in mid-January – 13 weeks from start to finish..

Called a Radar Convoy, the system was for the Second Tactical Air Force, which was supporting the Allied armies fighting their way through the Lowlands toward the Rhine. The system included a pair of AMES Type 14 high-power, 10-cm surveillance radars capable of detecting a single plane at 100 miles. These radars used a new lightweight antenna structure involving a huge parabolic reflector of mesh construction.

Two full radar sets were provided, one each for high- and low-elevations. Each transmitter, receiver, and antenna was mounted on a rotating cabin carried by a heavy truck. Two AMES Type 13 nodding height-finding 10-cm radars, modified from Naval Type 277, were included. There were also mobile diesel generators and a supply and service van.

The Group Control Center was composed of two controller's cabins, as well as cabins for planning and intelligence. Facilities for radio and telephone communications were included. When in operation, all of these cabins were arranged under a circus-type tent with large plotting displays along one side. Three weeks after delivery, the full system was operating in Holland.

The ADRDE/RRDE at Great Malvern

When it became necessary to evacuate from the southern coast, both the TRE and ADRDE decided on facilities at Great Malvern, Worcestershire, some 20 miles south of Birmingham. While the TRE had to await a major refurbishment of Malvern College, the facility selected by ADRDE was empty. and the move was made in mid-May 1942.

Located at Pale Manor Farm, about a mile from Malvern College, the ADRDE facility had initially been built for the possible evacuation from London of certain government departments. When such a move was considered unnecessary, it was taken over by the RAF as a Radio Training School, but this operation moved in early 1942.

The site included a number of existing buildings that were directly usable for laboratories, as well as a large building for living quarters and a full canteen for meals. Several special buildings called "giraffe houses" with high celings and entrances were constructed to accommodate the mobile units with antennas atop. A large and very complete shop was available. The site was surrounded by high fences, with guards drawn from civilian police forces.

Typical Giraffe House

When the ADRDE moved to Malvern, it had about 750 civilian and military personnel. John Cockcroft remained the Chief Superintendent, and in a short time organized 10 working groups, spread among various technologies and projects. There were about 50 scientists and engineers involved with radar; others were engaged in searchlight development and a variety of research areas such as acoustics, infrared devices, and electromagnetic propagation.

Since a large number of the personnel were housed within the main facility, there was not a major impact on the local residents. There was some resentment, however; the newcomers were "invading" and could not even hint as to their purpose in being at Malvern; throughout the war, their secret was well kept.

To maintain secrecy at the time of the move, few of those involved even knew the name of the destination city. Donald H. Tomlin, one of the radar engineers, has noted that few of the newcomers knew of the beautiful hills before arriving; it was almost a week before the heavy mists cleared to show their splendor. He noted that one morning a man emerging from the hostel remarked with surprise, "There's a mountain up there!"

Although the American SCR-584 system had been adopted by the British, activities continued at the ADRDE on gun-laying systems. BTH/EMI, Metro-Vick/Ferranti, and Standard Telephones each had orders for 500 of the British Army's version, the GL3-B, but the prototype had significant servo problems. Over the next year, a group at ADRDE led by J. M. Robson corrected these problems, and full production started in August 1943. The quantity ordered was eventually increased to about 2,000. These started to be available by mid-1944 and became the "work horse" of the British Army. Fifty sets were supplied to the USSR.

GL3-B Control & Power Trailer

By 1942, it had been decided that the GL systems needed an auto-follower, an electro-mechanical computer to control the radar after the target was identified. A cooperative development effort between ADRDE and BTH resulted in the Glaxo, an automatic-follower, for a 10-cm radar system designated GL3-B Mk 2. There was limited production of this system before the end of the war.

There was also continued work on the coastal defense radars. The 10-cm, high-power CD Mk IV was improved to become CD Mk V and then Mk VI; these were used by various Army installations along the eastern and southern coasts. The Mk VI was also adopted by the RAF as AMES Type 52, a low-altitude detection system.

GL3-B Presentation Unit

Before moving from Christchurch, Dr. A. E. Kempton and H. E. Rose started the development of a 10-cm high-power transmitter (HPT) for a radar that might be used in monitoring shipping out of Cherbourg on the coast of France. The HPT would be set at the Ventnor CH site that had an elevation of 700 feet. Even at this height, Cherbourg was beyond line-of-sight range, but propagation studies had indicated that extra bending of the microwave beam might give favorable results.

The equipment involved a 400-kW transmitter, and a 10-foot reflecting antenna to give additional beam forming and resulting gain. By the time ADRDE relocated to Malvern, the system was ready for testing. Initially, the system could see across the Channel about 25 percent of the time, but this increased as the operational procedures were

Microwaves Bridge the Atlantic

perfected. By August, some shipping was seen at 100-miles range, but the normal range was about 40 miles.

During the HPT project, much was learned about anomalous microwave propagation. Dr. Douglas R. Hartree had been assigned to ADRDE by the Ministry of Supply to examine the relationships between meteorological conditions and radio-wave propagation. For this, Hartree made use of the differential analyzer (an electro-mechanical computer) that he had earlier built at Manchester University.

When the ADRDE moved to Malvern, the Signals Experimental Establishment (SEE) relocated from London to occupy the facilities at Christchurch, and was renamed the Signals Research and Development Establishment (SRDE). This establishment was responsible for new communications technologies and had experimented with microwave transmissions.

W. A. S. Butement, then Assistant Director of Scientific Research with the Ministry of Supply, arranged for ADRDE to join with SRDE in the development of a new 10-cm communication system. Captain R. C. Cole led the efforts at SRDE, and Alan Oxford was the project scientist at ADRDE.

Using a 10-cm transmitter and receiver developed for radar, the Wireless Station No. 10 evolved. Called one of the electronic wonders of WWII, this was the first multi-channel, microwave communication system in Great Britain. It first went operational in July 1944, just after D-Day, and served as the central communications backbone for Field Marshal Bernard Montgomery's march across Europe

The contributions of Col. (Dr.) Basil F. J. Schonland must be mentioned. A native of South Africa, he compeled his bachelor's degree at age 16, and followed with a doctorate at Cambridge. Schonland led the development of radar in South Africa (see Chapter 8), then returned to Great Britain in mid-1941 to became Superintendent of the ADRDE Army Operational Research Group. His work centered on the applications of radar, as well as in intelligence analysis. He worked closely with R. V. Jones and, as earlier noted in Chapter 2, had a major role in planning the Bruneval raid.

Basil Schonland

There is an interesting side note concerning the Bruneval planning. To cover every aspect, Schonland set up a special team under Dr. D. Stanley Hey to analyze German jamming. For this, Hey used a mobile unit with a GL-radar receiver. In the days just before the raid, a number of AA stations with GL II radars reported a significant increase in static as seen on the CRT displays. On February 27, 1942 (the same day as the raid), Hey used his mobile unit to identify the sun as the source of this radio noise. A check with the Greenwich Observatory found that this coincided with the existence of a large sunspot at the center of the sun's disc. After the war and secrecy, Hey published a note in *Nature* (vol. 157, 1946) on this discovery, thus sharing with Germans Hans Plendl and Karl-Otto Kiepenheuer (see Chapter 5) initiation of the new science of solar radio astronomy.

Although seldom specifically noted, Schonland was personally involved in most of the analysis work at ADRDE. By the end of 1942, his activities had greatly expanded and was called the Army Operational Research Organization. It was comprised of 10 groups engaged in evaluating every aspect of the Army's operations, from weapons research to battle analysis. In 1944, Schonland became the Scientific Advisor to General Bernard Montgomery.

In early 1944, Cockcroft left the ADRDE to head the joint United Kingdom-Canadian Atomic Energy Project in Montreal. Charles W. Oatley (1904-1996) was then named Acting Chief Superintendent. At the same time, the operation was reorganized as the Radar Research and Development Establishment (RRDE). Oatley, a Cambridge graduate, had worked in E. V. Appleton's laboratory for 12 years before joining the RDF activities in 1939. He remained with the RRDE until the end of the war, then returned to research at Cambridge where he was instrumental in developing electron microscopy technology. P. E. Pollard succeeded Oatley as Chief Superintendent of the RRDE.

Charles Oatley

On the night following victory in Europe (V-E Day), the Searchlight Group at RRDE hauled two large units to a mountaintop near Malvern and lighted the sky with a giant, rotating "V".

CONTINUED DEVELOPMENTS IN AMERICA

With the establishing of the Rad Lab in late 1940, it was agreed that this operation would serve all of the military departments and have the major responsibility for developing microwave radars, but that the Navy and Army laboratories would continue to be responsible for the development and improvement of meter-wave radars. In addition, the military laboratories would have responsibilities in the performance and environmental testing of the microwave equipment, particularly in working with the engineering and manufacturing firms.

When the war came to America, the Rad Lab greatly expanded. Building 20 was built for the central offices and laboratories. Research Construction Corporation in Cambridge was started as a special manufacturing unit. The National Guard hangar at Logan Airport was taken over by the Rad Lab as its primary aircraft facility.

MIT Building 20 – the Rad Lab

The Radio Division of the Naval Research Laboratory (NRL) continued to have the basic responsibility for radar in the Navy. Robert M. Page was the technical leader of radar activities. Most of the research work was concentrated at the main NRL campus at Bellevue in Washington. D.C., but testing was performed at a number of locations.

During the war, the U.S. Army was divided into the Ground Forces, the Service Forces, and the Air Forces. The Signal Corps was a part of the Services Forces and officially served all other elements. The Signal Corps Laboratories (SCL) was centered at Fort Monmouth, New Jersey, and radar activities were in the Signal Corps Radar Laboratory (later Evans Signal Laboratory). The Air Forces had considerable independence and, although officially served by the SCL, had their own radar activities in the Aircraft Radio Laboratory at Wright Field, Dayton, Ohio.

Dr. Harold A. Zale was the technical leader for radar activities at the SCL, and the radar group at Wright Field turned to the SCL and Zale for advanced research. Both of the Army laboratories had developmental work on meter-wave radars and related technologies, but the Rad Lab and industries, particularly the Bell Telephone Laboratories (BTL), developed their microwave systems.

Meter-Wave Systems

As previously described, three British systems were sent to America as part of the Tizard Mission technology exchange: an ASV Mk II, an AI Mk IV, and an IFF Mk II. After being examined at the NRL, the ASV was sent to the nearby Anacostia Naval Air Station, where it was installed in a PBY flying boat. With Edward Bowen serving as the operator, it was first tested over the Atlantic on December 2, 1940. The tests were highly successful, including the detection of a capital ship at a range of about 60 miles.

ASE Yagis on PBY

With the addition of a duplexer, the 1.5-m (200–MHz) ASV Mk II was adopted by the Navy as the ASE for use on large patrol aircraft. Initially installed on PBY Catalinas, this became the first U.S. aircraft to carry radar in operational service. A Yagi antenna was fitted under each wing, skewed 7.5 degrees from the centerline; and lobe-switching was incorporated. Philco was given an order for 7,000 ASE sets.

The antenna on the ASE was too large for smaller aircraft; thus, a higher-frequency system with a correspondingly smaller antenna was needed. The NRL was already working on a 60-cm air-to-surface radar for the new fighter bomber, the TBF Avenger, that was under development by Grumman for the Navy. The receiver and display from the ASE were adopted for this radar. The transmitter used a new tube, the 15-E from the San Bruno Company; two of these provided 20 kW pulse power at the nominal operating frequency of 515 MHz (58.3 cm).

Like the ASE, it used a Yagi antenna fitted under each wing. First designated the XAT, then the ASB, this went into production by RCA just two days after Pearl Harbor. It was adopted by the newly formed Army Air Forces as the SCR-521. The maximum range was 15 miles on a surfaced submarine, 35 miles on a large vessel, and 70 miles on the coastline.

ASB Antennas on TBF

The last of the major non-magnetron radars, the ASB/SCR-521 was the most common airborne system of the Allies, with over 26,000 sets built. Installed almost universally on Navy carrier-based aircraft, it was known as the "Workhorse of Naval Aviation."

After being received by the NRL, the British 1.5-m (200-MHz) AI Mk IV was sent to Wright Field for examination. There it was installed and flown in a new fast-attack bomber, the Douglas A20 Havoc, and the testing went very well. Although the U.S. was not yet at war and thus the need for night interception was not urgent, an order was placed with

Western Electric for producing a small number under the designation SCR-540. The system, however, was obsolete before any large-scale production was needed.

A team from the U.S. Air Forces visited Great Britain in mid-1941, and was greatly impressed by the Ground Control Interception (GCI) system. They recommended that this system for the Air Corps, and the 1.5-m (200 MHz) GCI was adopted as the SCR-527. This was first tried in early 1943, and then found to be unsatisfactory in terrains other than the gently rolling English countryside for which it was originally developed.

Although not continuing with developing full radar systems, the NRL had major activities in improving radar technology throughout the war. One of the most significant efforts, directed by Robert Page, resulted in a technique that greatly increased the angular accuracy. Called simultaneous lobing or monopulse, multiple beams were simultaneously transmitted and the angular position determined for every individual pulse. First demonstrated in 1943, the technology was highly complex and was not immediately implemented. After the war, it was used in the FPS-16, likely the most popular tracking radar in history.

Combined Research Group – IFF

The IFF Mk II was the third system sent to the U.S. It was compared with a set, designated ABA by the Navy and SCR-515 by the Army, being developed by the NRL and General Electric. Unlike the British Mk II, this set was not dependent on the radar pulse. Primarily because it was already in use by the British, the IFF Mk II was adopted by the Navy as the ABD and ABE (different frequency ranges), and by the Army as the SCR-535. More than 10,000 of these sets were ordered from Philco and Raytheon.

Meanwhile, Dr. F. C. Williams at TRE Malvern developed the British IFF Mk III, and, somewhat later, Dr. C. E. Cleeton at the NRL developed a similar system. Both sets operated independent of the radar and used transponders with coded signals. With the prospects of combined U.S. and British flights over Europe, it was obvious that a common IFF system was necessary.

To resolve this problem, the Combined Research Group (CRG) was established in late 1941 at the NRL. Staffed by representatives of American, British, and Canadian military services, the CRG operated in a highly secure compound on the NRL campus, and was charged with selecting an Allied IFF system. Testing was conducted mainly at the nearby Camp Springs (later to become Andrews AFB).

Cleeton initially headed the CRG, and Dr. J. Rennie Whtehead from the TRE represented Great Britain. Robert Hanbury Brown was among other TRE personnel detached to the NRL during this period and participated at the CRG. After considerable analysis, testing, and compromising, the Mark III was selected for all of the Allies.

Starting in mid-1942, the Mark III went into American production by Hazeltine Research Corporation as the ABJ and ABK (Navy) and SCR-595 (Army). These were used on vessels and aircraft except fighters that were under GCI control. Later, the Mark IIIG came into existence, suitable for all types of aircraft; this was designated ABF and SCR-695.

IFF became Hazeltine's exclusive wartime project, producing thousands of the units. There were two general types of different electronic sets, and antennas were needed for essentially every type of aircraft and vessel – military and commercial – that would potentially be interrogated by radar. Antennas were very important; they had to be designed and mounted so as to be able to receive and reply to interrogations from any direction.

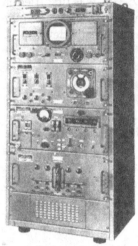

Typical Mark III Control Cabinet

Later, led by Commander Frank A. Escobar, the CRG eventually had a staff of about 200 military and civilian personnel. The NRL-developed system, ABA/SCR-515, was designated Mark IV and kept in "reserve" should the Mark III be compromised. Such a compromise did occur in early 1943, but rather than incur the great expense of converting to the Mark IV, the Mark III continued to be used with the code frequently changed on a pre-determined schedule. This led to many operating problems, particularly when the code required changing during a combat situation. There were also problems when large formations of aircraft were involved, resulting in what was called "IFF clutter."

Typical Mark III Aircraft Unit

Typical Shipboard Mark III Transponder

Microwaves Bridge the Atlanic

Even with its problems, the Mark III remained the primary unit. The CRG, however, continued to seek a more reliable and secure system. This was the Mark V, operating in the much higher 0.9-1.0-GHz frequency band and using pulse-coding; this set was eventually developed, but never put into production.

A parallel development was the United Nationals Beacons and Identification of Friend and Foe (UNBIFF). The UNBIFF was intended to be applied at all operational wavelengths, used for land, sea, and air, and give all Allies a universal system of transponder beacons. After the war, this became the Mark X and served as the genesis of a new air-traffic control system.

Radio Research Laboratory – Countermeasures

When the Rad Lab started, a separate laboratory directed by Dr. Frederick E. Terman was established at MIT to develop electronic countermeasures and counter-countermeasures (ECM and ECCM) to radar. Frederick Emmons Terman (1900-1982), earned his doctorate under Vannevar Bush at MIT in 1926, then joined the faculty at Stanford University and eventually became the Head of the Electrical Engineering Department. His book, *Radio Engineering* (McGraw-Hill 1932, 1938), was likely the best-known textbook on this subject in the world

With full funding from the OSRD and reporting directly to Dr. Bush, this operation soon became the Radio Research Laboratory (RRL), located at Harvard University, just a mile from MIT. Unlike the Rad Lab, the RRL never released significant details on its accomplishments; ECM and ECCM have always been closely guarded secrets by all nations.

Frederick Terman

As an example of early RRL's work on the hardware side, a commercial field-strength meter from General Radio covering 300 MHz to 3 GHz was modified by the RRL to become the Navy's ARC-7 and the Army's SCR-587 radar intercept receiver.

One of their jammers was "Tuba," a giant system generating two continuous 80-kW signals in the range of 300-600 MHz to jam *Lichienstein*

Tuba Antenna

radars. Tuba used Resatron tubes, developed by Drs. David H. Shone and Lauritsen C. Marshall and manufactured by Westinghouse. The Tuba used a horn antenna built of mesh wire 150 feet long and driven through 22- by 6- inch waveguides, possibly the largest ever built. Tuba was placed in operation in mid-1944 on the south coast of England. The radiated energy was such that it lighted fluorescent bulbs a mile away and jammed radars throughout Europe.

Using AN/SPT-3

Ultimately, many types of radio and radar countermeasures equipment were developed by the RRL. These were usually designated APT for Army units, APQ for airborne equipment, and SLQ and SPT for Navy shipboard radar jammers. Particular jammers for the *Wuerzberg* were the AN/APT-2 (airborne) and AN/SPT-2 (shipboard), and jammers for the *Freya* were AN/APT-3 and AN/SPT-3.

Signal receivers for countermeasures were developed by several organizations. The XARD from the NRL covering 50 MHz to 1 GHz was one of the first, going into service aboard submarines in 1942. The RRL also developed two excellent intelligence receivers: the ARC-1 (Navy) and SCR-587 (Army) covering 100 MHz to 1 GHz, and a similar unit, the AN/SPR-2 and AN/APR-2, that allowed 8-hours of recording on a electromagnetic recorder. The ARR was an airborne search receiver, tuning from 27.8 to 143 MHz in three bands; built by Hallicrafter, it was similar in performance to their S-27 and S-36 commercial receivers.

ARR Receiver

The origin and use of chaff – lightweight foil strips of about one-half a wavelength in length – have been previously described. The RRL conducted considerable analytical and experimental work on chaff. Dr. Fred L. Whipple, an astronomer, devised a formula giving radar cross-section at a given wavelength per kilogram of chaff. The U.S. produced about 20,000 tons of aluminum chaff; when it was authorized for use, it was very effective. The Germans were never able to fully counter this, although they offered a prize of 700,000 Reichsmarks to inventors.

The RRL made significant contributions to the basic understanding of methods, theories, and circuits at very-high and ultra-high frequencies for radio systems, particularly in signals intelligence gear and statistical communications techniques. At its high point, about 850 personnel were involved. After the RRL closed, this fundamental work was documented in *Very High-Frequency Techniques, Vol. I and II*, McGraw-Hill, 1947.

VT Proximity Fuse

In Great Britain, while he was testing the coastal defense radar in 1938, William A. S. Butement conceived the idea of a proximity fuse on the projectile to improve the kill probability. He made an excellent design, but was unable to interest defense officials in the development.

When the NDRC was formed, Dr. Merle A. Tuve of the Carnegie Institution was asked to head a section devoted to developing a proximity fuse. In August 1940, a laboratory devoted to this project was set up in the Department of Terrestrial Magnetism. The Tizard Mission brought Butement's design to the U.S., and his basic circuit was adapted.

Merle Antony Tuev (1901-1982), a native of South Dakota, was awarded the Ph.D. in physics from Johns Hopkins University for his work with Dr. Gregory Breit at Carnegie Institution in ionospheric research (see Chapter 3).

Merle Tuev

Concerning the fuse development, Tuev commented as follows:

> The proximity fuse was not invented by any one man; it was a composite of old inventions and re-inventions both here and in Britain. It was really a development, not an invention, and many individuals contributed to it.

VT Fuse

Designated the VT (Variable Time) Fuse, this was a substitute for the conventional fixed-time (set by the expected time of flight) or contact (detonation upon impact) fuses. The designation indicated that the time to detonation was determined by the variable flight-time to the target. This miniature, continuous-wave radar operated in the frequency range of 180-220 MHz. Using the body of the fuse as both a transmitting and receiving antenna, the Doppler-shifted RF return from the target beat with the original RF to form an audio signal, the amplitude of which was used to trigger the detonation. Screwed into the head of a projectile, it needed an operational range of only a few feet. By early 1941, the concept was proved through firings of un-miniaturized devices.

Dr. Richard Roberts led the efforts on electronic miniaturization and shock resistance. Central to this was the development of tiny vacuum tubes that could withstand the initial acceleration of 20,000 g and the high-rotation rate while on the trajectory. Starting with existing tubes for hearing aids, Raytheon and Hytron designed a suitable triode, pentode, and thyatron.

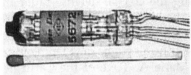

VT Fuse Tube
Compared With Matchstick

The circuit used a triode in the oscillator/detector circuit, a pentode and a second triode as a Doppler-signal (200-800 Hz) amplifier, and a thyatron output trigger.

A tiny battery with a long shelf life was another major problem. National Carbon came up with what was called a "reserve" battery – a small wet cell containing the needed electrolyte stored in a glass ampoule that broke upon launching. Premature detonation was a serious safety problem; this was solved by using a mechanical device that turned on the electronics in about a half-second after launch.

Many industrial and academic research organizations became involved in various parts of the project and, because of very tight security, most were not aware of each other or their accomplishments. Consequently, a number of organizations laid claim to developing the proximity fuse. The overall project was later transferred to the newly formed Applied Physics Laboratory of Johns Hopkins University.

The first device, the Mk 32 for the Navy's 5-inch antiaircraft weapon, was test fired in April 1942, and production by the Crosby Corporation started the next September. The first Japanese plane shot down using the Mk 32 was in January 1943. As the initial problems on the Mk 32 were solved, work began on the Mk 33, a smaller device for the British 4-inch antiaircraft gun.

The Army pressed for a suitable VT fuse for its field artillery, and the Mk 41 for a 3-inch projectile evolved. This was followed by a number of other types for a variety of Allied weapons. In most of these, as well as the improved Mk 32, the reserve battery was replaced by a small wind-driven electric generator.

Throughout the war, over 100 companies and research institutes were involved; 112 plants manufactured parts and assemblies for in excess of 22 million devices. In 1942, the Mk 32 cost $732 each; by the end of the war, the average device had an incremental cost of $18. The total

VT fuse program is estimated to have cost about $1 billion, one of the largest outlays of the war.

In Germany, several types of devices were designed and tested for anti-aircraft missiles being developed at Peenemunde, but none were put into production before the war ended. The American and British military units were initially prohibited from using VT fuses where there was a possibility of the enemy recovering an unexploded warhead.

This WWII radar-like device set the stage for post-war miniaturized electronics.

More American Magnetron-Based Radars

After the introduction of the resonant-cavity magnetron in late 1940, essentially all radar development in the United States was based on these devices. This section provides a brief discussion of the major systems. (There were many more.) The Rad Lab at MIT had the primary responsibility for most such systems, but there were also significant microwave developments, particularly for the Navy, at the Bell Telephone Laboratories (BTL) facility at Whippany, New Jersey.

The very rapid development of high-power magnetrons produced an astonishing number of these tubes. In addition to those made in Great Britain, it is estimated that by the end of the war well over a million magnetrons had been manufactured by at least six companies in the United States and one in Canada.

Magnetron development at the Rad Lab was under Dr. I. I. Rabi. In parallel with the 10-cm developments, Rabi's team started pushing the operating wavelength still shorter, first to 6 cm, then to 3 cm, and then 1 cm. As more frequencies were used, it became common to refer to radar operations in the following bands:

    P-Band – 30 - 100 cm (1 - 0.3 GHz)
    L-Band – 15 - 30 cm (2 - 1 GHz)
    S-Band – 8 - 15 cm (4 - 2 GHz)
    C-Band – 4 - 8 cm (8 - 4 GHz)
    X-Band – 2.5 - 4 cm (12 - 8 GHz)
    K-Band – Ka: 1.7 - 2.5 cm (18 - 12 GHz);
        Kb: 0.75 - 1.2 cm (40 - 27 GHz).

The K-Band was divided because of strong absorption by atmospheric water vapor.

The designation system for American radars might be mentioned here. The U.S. Navy used two types of ship-board radars: search (Bureau of Ships) and fire-control (Bureau of Ordnance), Initially, the former were type "S" and also carried a letter showing the model number; e.g., the SC was the third model of the search radar. (Not all models reached production, so certain letters are missing.)

The Bureau of Ordnance preferred the "Mark" designation, many with a modification shown; e.g., Mark 8, Mod 2 was the second modification of the type Mark 8 fire-control radar. Early fire-control radars were designated "F"; thus, the FH was also Mark 8. The "X" and "CX" designations were for experimental radars.

Early aircraft radars (Bureau of Aeronautics) used the same letter-type designation as those of the Bureau of Ships, but with an "A" suffix; e.g., ASE was the fifth model of an airborne search radar.

The Army, including the Army Air Forces, used the Signal Corps Radio (SCR) series for their radars. If a radar or other type of electronic set was approved – but not necessarily used – by both the Army and the Navy, the prefix "AN" or "A/N" was used.

P-Band Systems. The BTL's improvement of the original 10-cm resonant-cavity magnetron from Great Britain was previously noted. Dr. Mervin J. Kelly followed this by developing powerful magnetrons operating at higher and lower frequencies, as well as at higher power. By December 1940, Western Electric put into production the first American-designed resonant-cavity magnetron, actually a UHF device operating at 750-MHz (40-cm).

The BTL had also continued work on a successor to the FA (Mark 1), the Navy's first fire-control set. The new magnetrons were used in two new 40-kW fire-control systems: the FC (Mark 3), for use against surface targets, and the FD (Mark 4), for directing anti-aircraft weapons. For more accurate directions, the FC used horizontal lobe-switching antennas, and the FD antenna used both horizontal and vertical.

FC (Mark 3) Antenna

The FC was America's first magnetron-based radar to go into service; deliveries of both of these systems started in the fall of 1941. Ultimately, about 125 FCs and 375 FDs were produced. The Mark 3 was replaced by the very similar Mark 12 in

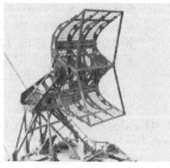

FD (Mark 4) Antenna

late 1943; the Mark 4, however, remained one of the most used fire-control radars throughout the war.

L-Band Systems. The most ambitious, long-term effort of the Rad Lab was Project Cadillac, the first air-borne early-warning radar system. Initiated by an Inter-Service Committee in mid-1942, it was intended to provide the Navy a surveillance capability unequaled by land- or sea-based systems. The need became much greater as the war in the Pacific progressed and the threat of *Kamikaze* attacks came into being. With Jerome B. Weisner as the project leader, eventually about 20 percent of the Rad Lab's staff would be involved.

The basic concept was to increase the range of surveillance radars by having them airborne. A 20-cm (1.5-GHz), 1-MW radar, the AN/APS-20, was developed. With an 8-foot radome and other equipment, it weighed 2,300 pounds. The ATM Avenger was the only carrier-based aircraft capable of such a load, and by mid-1944, tests began on an Avenger with the radome fitted under the aircraft.

Avenger with AN/APS-20

The resulting system could detect bomber-sized aircraft at ranges up to 100 miles. To make the radar findings more useful, the system included a television pickup of the radar's PPI display and a VHF communications link to convey the images up to 100 miles to the Combat Information Center (CIC) on the home carrier. The system was first flown in August 1944, but problems with signal interference delayed introduction into service until March 1945.

Shipboard Combat Information Center

As this activity was underway, the Navy was considering a more sophisticated method for airborne surveillance: placing the combat information operation aboard the aircraft. In mid-1944, the Rad Lab started Cadillac II with this objective. The PB-1 aircraft (the Navy's designation for the B-17) was selected. The AN/APS-20 would continue as the radar, but a CIC with three stations having 12-inch imaging tubes was built into the original bomb bay. In using the large PB-1's, the system would need to be land-based. While much progress was made, the war ended before the system could be demonstrated. However, Project Cadillac was the foundation from which the later AWACS concept evolved.

Here it might be noted that, as the war progressed, the CIC on aircraft carriers became the electronic nerve center of the vessel. With an operating staff of up to 50 persons, all of the radar information was displayed on large map-tables or hand-drawn on vertical transparent sheets.

S-Band Systems. The initial two projects at the Rad Lab were previously described. The first led to the SCR-520, America's first microwave radar. Intended as an air-interception set for the Lockheed P-38 Lightning, it weighed about 600 pounds and was too heavy for this aircraft. There was, however, an urgent need for radars to be used against the increasing U-boat threat, and the SCR-520 was modified by the BTL for this application. Designated SCR-517, by June 1941 a few went into coastal patrol operations on PBY Catalina flying boats and Liberator bombers. The SCR-517 had only forward scanning, and was thus of limited value. It was later upgraded to the SCR-717 and the ASC.

In early 1942, the Rad Lab and Western Electric, the contractor for the SCR-520, turned their efforts toward a lighter, 400-pound version. Designated SCR-720, this was also a 10-cm system. With a range of about five miles, it could also function as an airborne beacon/homing device, navigational aid, or in concert with interrogator-responder IFF units. It was tested later in the year at Wright Field and immediately adopted for the Northrop P-61 Black Widow night fighter.

SCR-720 on P-61

Voice communications were used by the radar operator in the two-place planes to steer the pilot to near the target; then the pilot used a smaller scope on his instrument panel to track and close for the kill.

The SCR-720 was also sent to Great Britain, where it was quickly adopted as the AI Mk X for the then-favored de Havilland Mosquito fighter aircraft. This set was eventually adopted as the standard RAF AI radar, and thousands were produced.

The second project resulted in the SCR-584, undoubtedly the best-known product of the Rad Lab. Described in detail earlier, it was primarily a gun-laying system for the Army, but also found many other uses throughout the war. Because of delays attributable to material priorities and security, it was a year longer in production than planned. In this time, however, the SCL did analyses and ran tests for such diverse uses as tracking trajectories of artillery shells, detecting ground vehicles, and correcting ballistic tables for the 90-mm gun. The M-9 Predictor-Corrector was a valuable adjunct.

SCR-584 with IFF Antenna

M-9 Trailer

As it was being designed, the Navy decided that a similar system would be very useful on aircraft carriers in directing fighter aircraft when close to landing. In cooperation with the NRL, two systems emerged: the SM for large carriers and later the SP for smaller carriers. The development primarily involved adapting the equipment for use on the unstable vessels, requiring a platform that eliminated pitch, roll, and yaw through the use of gyroscopes and servo-mechanisms. A second problem concerned a mechanically stronger antenna that could endure storms at sea.

The prototype, CXBL, was tested on the USS *Lexington* in March 1943, and two production SM sets went into service in October, actually before its parent, the SRC-584, was fielded. The ship-borne systems did not require the M-9 gun director or the separate power generator, but with the extra platform and heavier antenna the SM weighed near 20,000 pounds. This would have made the lighter vessels top-heavy, so the SP went into service in late 1944 without the stable platform; this lowered its weight by 7,000 pounds but greatly reduced its usefulness. The lighter-weight version was also adopted by the Army as the SCR-784.

In Great Britain, development of the Royal Navy's Type 271 had started after the experimental set demonstration at TRE Swanage, then, within a year, swept past other projects to become that nation's first microwave radar. In a similar manner, a radar for the U.S. Navy that was not even discussed in the Tizard Mission meetings started after the rooftop demonstration at the Rad Lab but then passed the first selected projects to become the America's first microwave radar.

At the request of the Navy, the project leading to the SG search radar and an airborne version, ASG, started as soon as the Rad Lab's 10-cm breadboard was demonstrated in January 1941. After a cooperative design effort involving the NRL and the BTL, project leader Dr. Ernest C. Pollard of the Rad Lab conducted the first sea trials of what would become the SG in late May on the USS *Semmes*, and production started in mid-year. The ASG followed shortly.

SG Antenna

The SG was the first Rad Lab radar to see combat (December 1943), and the SG/ASG were the first fielded radars using the NRL-developed Plan Position Indicator (PPI), a device that greatly improved their usefulness. Samuel Eliot Morison, official historian documenting WWII naval operations, called the SG the "greatest boon of scientist to sailorman since the chronometer."

The SG surface-search system produced 50-kW, 1.3-2.0-μs pulses and had a cut-parabolic antenna with a gyro-stabilized mount. It could detect large ships at 15 miles and a submarine periscope as far as 5 miles; it also allowed firing "blind" at targets in any weather conditions. It was also used for low-level air search. Manufactured by Raytheon, about 1,000 of these sets were installed on destroyers and larger ships.

The small, highly mobile Patrol Torpedo (PT) Boat was introduced in 1942. For this, Raytheon modified the SG to a more compact version, the SO. These sets gave the PT Boats a great advantage, particularly for operations at night.

PT Boat SO Radar (lower left)

Other direct descendants of the SG were the SF, a search set for destroyer escorts and lighter ships; the SH, a search-and-fire-control set used by the Navy's Armed Guard Service on large merchant vessels; and the SE and SL, search sets for small ships and similar to the SO. None of these were produced in significant quantities, but, with PPI displays, they were of great value in both defense and navigation applications.

The ASG, also designated AN/APS-2 and commonly called "George," was produced by Philco. This air-search radar was superior in performance to the ASB but was bulky and thus limited to large patrol aircraft and blimps. About 5,000 ASGs were built and used effectively against enemy submarines.

Blimp with ASG Under Gondola

The SCR-517 had only forward scanning. A few were modified by the Air Forces to provide full-circle scanning. Designated SCR-717 and incorporating a "B" scope display (almost a PPI), these were sometimes used on B-24 Liberator bombers for surveillance.

ASC Atop a PB2Y Coronado

The NRL made a more significant modification of the ASB, resulting in the ASC. This used a 24- to 30-inch parabolic reflecting dish that provided much better resolution. Positioned by a hand crank into a clear plastic blister above or below the fuselage, the antenna could revolve 360 degrees. A PPI was used for displaying this coverage. The range was limited to line of sight (30-35 miles at normal operational altitudes). It was used on PB2Y Coronado and other large patrol planes starting in late 1943.

Another early 10-cm radar development at the BTL was the SJ, a system primarily for supplementing the SD meter-wave radar on submarines. The antenna of this system could be used with the vessel at either surface-level or hull-submerged, sweeping the horizon for ships and aircraft. With the antenna at a very low level, the SJ's range was only about 6 miles, but with good accuracy. This system was first tried at sea in mid-1942. In 1945, this was upgraded to the SV with a larger antenna, increasing the range to 30 miles.

Radar Antennas on Submarines

The SCR-582 was an early 10-cm radar developed by the Rad Lab for the Army SCL. Primarily intended as a harbor-defense system, it had a 48-inch parabolic dish and was usually mounted atop a 100-foot tower. With a PPI display, it was ideally suited for guiding ships entering harbors and could also detect low-flying aircraft at 25 miles. It was the second Rad Lab radar (after the SG) to go into combat The SCR-682 was a transportable version.

SCR-582 Installation

The SCL was responsible for a number of other 10-cm radars used by the Army. Some of their air-transportable radars included the AN-CPS-1, an early-warning set from General Electric with a range up to 200 miles. The AN-CPS-4, nicknamed "Beaver Tail" from the shape of its beam, was a height-finder set from the Rad Lab; it was used with the SCR-270 and SCR-271. The BTL developed the AN-CPS-5, a GCI set that could track targets at more than 200 miles distance.

Representative SCL mobile-ground radars included the AN-GPN-2, a search set with a 60-mile range developed by Bendix, and the AN-GPN-6, a similar search set from the Laboratory for Electronics Inc. The AN-CPN-18 from Bendix was the surveillance radar portion of an air-traffic control system used by the Army Air Forces.

X-Band Systems. As the 3-cm magnetron came into being, Dr. Luis W. Alvarez (1911-1988), previously a physics professor at the University of California, Berkeley, used his expertise in physical optics to develop a new microwave antenna for shaping the radiation. Formed as either a "leaky" waveguide or a series of dipoles, this linear array produced a very narrow, fan-like beam that could be electronically swept.

The Alvarez antenna was incorporated in three new types of X-band systems. The first was an airborne radar in two configurations. For the Army Air Forces, the "Eagle" – later designated AN/APQ-7 – had a beam that electronically swept 30 degrees on each side of the aircraft heading. Intended for strategic bombing, the Eagle provided a map-like image of the ground about 170 miles along the forward path of the aircraft. With Dr. Emmeth A. Luebke as the project leader, a prototype was tested in mid-1943. Put into production by Western Electric, the first deliveries were in late 1944. About 1,600 Eagle sets were built. Although too late for the bombing of Germany, it was used on B-29s against Japan.

Microwaves Bridge the Atlanic

The same technology was also used by Norman F. Ramsey in developing the ASD, a search and homing radar for use by the Navy primarily in single-engine bombers. Eventually designated the AN/APS-3 and also known as "Dog," this system swept 150 degrees in the search mode, then switched to 30 degrees for target homing. With final design by the NRL and Sperry, it was put into production by Philco in late 1943. For carrier-based night fighters, ASD was converted to a lighter system, the ASH (AN/APS-4).

AN/APS-4 on P-38M

This system, however, required a second crew member – a radar operator crammed in behind the pilot – and thus saw limited use. It was eventually replaced by the AIA-1 (AN/APS-6), suitable for a single seat aircraft. With a display only two inches in diameter, it effectively made the AIA-1 a "radar gunsight," allowing the pilot to personally use this equipment.

GCA Operator Station

The second application of the Alvarez antenna was the Ground Control Approach (GCA) system for military airports. This blind-landing system used a 10-cm radar with a PPI for close-in tracking and two 3-cm sets for fine azimuth and elevation positioning. The X-band radars used linear arrays of dipoles. An electro-mechanical analog computer determined the aircraft's approach path, and final guidance was by a ground controller using voice radio.

Lawrence H. Johnson, a former student of Alvarez, was the GCA project leader and personally developed the associated analog computer. Much of the final system design was under contract with Gilfillan Brothers. Carried in several vans, it was first field-tested in 1942. Gilfillan also built about 15 units, designated AN/MPN-1, that were successfully used by the Army Air Forces in assisting bombers returning from Germany.

The Microwave Early Warning (MEW) was the third X-band system using the Alvarez antenna. The system had two back-to-back rotating antennas, both 25-foot wide parabolic "billboards," one 8-feet high for

AN/CPS-1 MEW Radar

low, long-range coverage and the other 5-feet high for high, near-in coverage. Each antenna was fed by a linear array of 106 dipoles. Dr. Morton H. Kanner led the project.

Designated the AN/CPS-1, the radar produced 700 kW. Five 12-inch CRTs were needed to display the coverage. Requiring eight transport trucks, the system weighed almost 70 tons. Certainly the physically largest equipment of the Rad Lab, a small number of the systems were built internally. The first AN/CPS-1 went into operation by the Army Air Forces in England in January 1944. To improve accuracy in the vertical plane, the system later used a separate height-finder radar, the AN/TPS-10 ("Li'l Abner").

The Rad Lab ran tests on the British 10-cm H2S and was not satisfied with the imagery. Believing that this could be improved with a 3-cm radar, in late 1942 they initiated development of such a system; Dr. George E. Valley led the project. The X-band Eagle project was already started, but there were problems with Alvarez's complex antennas. Valley gave up precision and incorporated a much simpler antenna for a system that would be used in area bombing – encompassing a

PBM-5A with AN/APS-15 Radar

target as much as a mile in diameter. In mid-1943, the 3-cm imaging set went into production, with Western Electric making the AN/APQ-13 for the Air Force and Philco producing the AN/APS-15 for the Navy. Both versions were commonly called "Mickey."

Mark 8 (FH) – America's First Phased-Array Radar

The BTL also had successes with X-band radars. They had experimented with phased-array technology before the war, and this was used for the first time in American radars on the Mark 8 (FH), a 3-cm fire-control system for the Navy. Developed by George E. Mueller, the antenna was an end-fired array of 42 pipe-like waveguides, called "polyrods," arranged in three

Microwaves Bridge the Atlanic

rows and phased such that the beam was steered. The system was produced by Western Electric starting in late 1942.

The original Mark 3 (FC) was deficient in determining the altitude of low-flying aircraft. In 1943, it was replaced by the Mark 12, also a 40-cm radar, but the 3-cm Mark 22 was developed by the Rad Lab as an adjunct. The Mark 22 had an "orange-slice" antenna, providing a very narrow, horizontal beam; this nodded to search the sky for targets. The land-based version of this radar used by the Army was AN/TPS-10, called "Li'l Abner" (named after Al Capp's comic strip character who liked to sit in a rocking chair).

Mark 22 (left) & Mark 12

The SS was another major X-band radar developed for the Navy by the BTL. This was a follow-on to the SJ and SV warning radars for submarines.

RADAR MAINTENANCE

When radar equipment started to become available to the armed services, it was widely recognized that existing maintenance capabilities were inadequate. In the United States, advanced training of Army personnel for this had traditionally been done at the Signal Corps School, affiliated with the Signal Corps Laboratories (SCL) at Ft. Monmouth, and for Navy personnel at the Radio Materiel School (RMS), on the campus of the Naval Research Laboratory. Both schools added topics in radar in 1940.

RMS at the NRL, 1941

As the war approached, the Navy began plans for greatly expanding the maintenance training. With the buildup of ships and aircraft for the expected war, it was realized that far more highly qualified technicians would be needed than the RMS could produce; further, the advent of microwave radar would require the maintenance personnel to have much higher knowledge and skills. In a crash effort in the weeks following the Pearl Harbor attack, a new training program was planned, approved, and actually started with

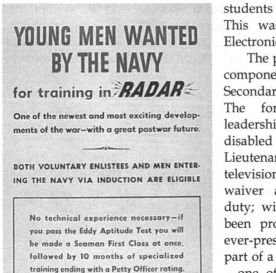

Typical Recruiting Advertisement

students by mid-January, 1942. This was commonly called the Electronics Training Program.

The program initially had two components: Primary and Secondary (Advanced) Schools. The former was under the leadership of William C. Eddy, a disabled (deafness), retired naval Lieutenant who was a well-known television expert. Eddy received a waiver and returned to active duty; within three years he had been promoted to Captain. His ever-present pipe was actually part of a miniature hearing device – one of his many inventions – passing the sound through clenched teeth to his ears

Eddy's first Primary School was in downtown Chicago, using space in experimental TV station W9XBK and with the station's engineering staff serving as instructors. The facilities and staff were provided without cost to the Navy by Balban & Katz Theater Corporation. Admission to this school was through an examination that came to be known as the Eddy Test. The school's lecture and laboratory instruction – given in 14-hour days over a 12-week period – covered the vital elements normally found in the first two years of a standard electrical engineering curriculum.

William Eddy, 1944

The Chicago school was the prototype of Primary Schools that were started the following March at six engineering colleges across the United States. Successful graduates then went to a 24-week Secondary (Advanced) School at the NRL in Washington or a duplicate school at Treasure Island, San Francisco; here, under tight secrecy, they were given the advanced theory and laboratory practice for all types of shipboard and shore-based electronic equipment.

A Secondary School for airborne equipment was established at Ward Island, near Corpus Christi, Texas. In addition to Navy personnel, there were trainees from the Marines and Coast Guard, as well as the Royal Air Force, the Canadian Royal Air Force, and a few men from Australia and Brazil.

Typical Advanced Laboratory

Later, with the demand for skilled technicians continuing to increase, several large Navy-operated Primary Schools were started, and a fourth Secondary School was formed at Navy Pier in Chicago. Even with the heavy filtering of the Eddy Test, failure rate in the program was unacceptably high, and several Pre-Radio Schools of four-weeks duration were opened in the Chicago area; these provided additional selection and preparation for the Primary Schools. During the war years, an estimated 500,000 persons took the Eddy Test, and about 30,000 men graduated from the overall program.

Details of the Navy's Electronics Training Program may be found in the author's book, *Solving the Naval Radar Crisis* (Trafford, 2007). It might be noted that the author graduated from this program, and then taught at the Advanced Airborne School at Ward Island.

In addition to the Electronics Training Program, the Navy developed radar schools for upgrading capabilities of technicians already serving in the fleet. The main schools were at the Naval Air Station, San Diego, and the Fleet Service Schools, Pearl Harbor and Norfolk.

The Army also had major needs in electronics maintenance, actually far greater than the Navy's in numbers of personnel. Earlier, as in the Navy, Army radio operators usually performed the maintenance, but the advent of radar changed this. In 1941, the Signal Corps School greatly expanded and included advanced training in electronics maintenance.

With the start of the war, the capacity at Monmouth was exceeded, and new schools were started at nearby Camp Edison and Camp Charles Wood; the overall operation was renamed the Eastern Signal Corps Training Center. This was also soon filled, and new centers were opened at Camp Crowder, Missouri; at Camp Kohler, California; and on the Davis campus of the University of California. All of the schools gave both basic recruit training and as elementary signal training. By the end of 1942, approximately 50,000 technicians entered the Army Signal Corps through the Enlisted Reserve Corps and its training program.

Camp Murphy, located near Hobe Sound in Southeast Florida, was opened in mid-1942, as the Signal Corps's secret facility for radar operation and maintenance training. Later renamed Southern Signal Corps School, this encompassed some 1,000 buildings and had a complement that reached a maximum of about 5,800 personnel. Through 1944, it graduated thousands of enlisted men and officers from the Army, as well as the Marines and certain Allied countries.

The airport at Boca Raton, Florida, was converted into the Army Air Forces's only radar training station during the war.

Unlike the Navy electronics schools, those for Army enlisted personnel were along the lines of traditional military training. While the Army sought very capable students, there was nothing equivalent to the Eddy Test or the Electronics Training Program. Instruction in the Army schools was generally short on theory and long on standard routines. The advanced portions gave specialization in particular equipment. Nevertheless, the programs were very adequate for the Army's needs, and their radar systems provided excellent service throughout the war.

As the radar hardware evolved, the need for engineering specialists in this technology was apparent. In June 1941, MIT established a Radar School for select military and civilian personnel. The course involved two months of instruction in the basic principles of radar, followed by separate Army and Navy hardware programs of three-months length.

For students needing preparatory studies, mainly officers who did not have engineering degrees – the Navy set up such a school at Harvard University. The Harvard Pre-Radar School was five months in duration and covered much of the same instruction as was given in the Pre-Radio and Primary Schools of the Electronics Training Program. Graduates of Pre-Radar went directly into the Navy's 12-week hardware program at MIT. Later, a second Pre-Radar School was set up at Bowdoin College, located in Brunswick, Maine.

The MIT Radar School trained some 8,800 persons in this technology. The instructional notes were compiled as *Principles of Radar*, the first comprehensive book on this subject. This was first printed as a secret document in 1944, and then openly published by McGraw-Hill in 1946.

There were significant programs for training radar technicians in the British Commonwealth countries. A major training facility in Great Britain was at the Radio School Yatesbury in Wiltshire, where ground radars such as CH and CHL were taught. The RAF operated Radio School Prestwick in Scotland, providing maintenance instruction for airborne systems such as AI, ASV, and IFF equipment.

Nearby the TRE facility in Swanage, the private Forres School was taken over in 1941 for offering a variety of radar training programs.

Included in the first instructors was a group of 18 WAAF science graduates who were personally selected by Watson Watt. Among these was the later Lady Renie Adams who commented:

> I soon found out the calibre of these people [the students] coming from all walks of civilian life. . . . They were splendid and spent the time at the school working very hard. . . . Some evenings we had to lock them out of the classrooms to make them take a break from work.

His Majesty's Signal School in Portsmouth, later the Admiralty Signal Establishment, provided maintenance training for naval radars. The Army's Air Defense Research and Development Establishment at Christchurch included training for radar mechanics. A number of other schools existed during wartime years, but the secrecy surrounding them led to a loss of their identification.

The critical need for radio mechanics, as they were called in the Commonwealth, was emphasized by the High Commissioner for the United Kingdom when he stated in 1941, "Radio technicians have the highest priority of all requests for manpower from the United Kingdom." In June 1940, Canada agreed to provide RDF (radar) technicians for worldwide assignments, as well as to give training for other Commonwealth nations. The Royal Canadian Coastal Services offered the first RDF course in November 1940 at Halifax, Nova Scotia.

After a group of signal officers were trained in Great Britain, an RDF school that included maintenance was established by the Air Council of Canada in May 1941, at Clinton, Ontario. In addition to students from Canada and other Commonwealth nations, initial classes included U.S. Army, Navy, and Marine students. For Canadians and others from the Commonwealth, initial training was in 10- to 12-week preparatory courses provided by Universities and Technical Schools across Canada. Those who passed a tough exam completed their training at the highly secret #31 RAF school in Clinton; eight weeks were devoted to ground radars and six to airborne radars.

Concerning the secrecy, U.S. Navy Radioman Eugene H. Fellers, who was sent to Clinton for maintenance training, told an interesting story:

> The girls in town were friendly and would tease us a bit, asking, 'What are you doing here?' We would reply, 'We're studying radio.' They would laugh and say, 'You mean radar, don't you?' We thought that no one outside the school was supposed to know that word! They would then say, 'Don't worry; we know the name radar and what it does.' And that was that.

During the war, about 6,500 Canadians and 2,300 Americans went through courses at Clinton. At its high point in early 1945, the #31 School had a staff of 478. The airborne radar training program set up by the U.S. Navy at Annapolis in late 1941 was patterned on this school.

In 1942, the Canadian Army set up RDF maintenance schools at Debert, Nova Scotia, and at Camp Barriefield, near Kingston, Ontario. The Barriefield school was enlarged in 1944 to become the Royal Canadian Electrical and Mechanical Engineers (RCEME) training center.

The Royal Canadian Navy's Signal School at St. Hyacinthe, Quebec, was the central location for training Canada's shipboard and shore-base communications personnel during World War II. Moved from an original location in Halifax in the summer of 1941, the St. Hyacinthe school had 73 buildings spread over a 25-acre site. At a given time, there were about 3,200 officers, enlisted men, and Wrens (Women's Royal Canadian Naval Service), involved in all phases of training in communications: signalmen, telegraphers, communications coders, RDF (radar) operators, and radio/radar artificers (mechanics or technicians).

The course at St. Hyacinthe for Radio Artificers was 42 weeks in length. The first 20 weeks were devoted to basic training, with studies of electricity theory, mathematics, and shop skills. During the second half of the course, they applied this basic knowledge to wireless and RDF equipment. On graduation, they were qualified as Radio Artificers Fifth Class, just below the rating of a Petty Officer. Officers taking RDF courses at St. Hyacinthe were all university graduates in electrical engineering, physics, or similar fields. Their course, lasting several months, entitled them to use the letter R after their rank on graduation

In Australia, even before their first indigenous radar was built, training programs were begun in September 1941. These provided 6-month courses for signal officers at Sidney University and for maintenance mechanics at Melbourne Technical College. As previously noted, personnel from Australia were also trained at the #31 RAF radar school in Canada. Although the radar activities in South Africa were relatively small, the Electrical Engineering Department at the University of Natal provided special training courses in radar operation and maintenance.

CLOSURE

The last major application of wartime radar must be mentioned. On August 6, 1945, when the atomic bomb ("Little Boy") was dropped on Hiroshima, and three days later another ("Fat Man") on Nagasaki, radars strapped to the bombs were used to initiate their detonation at about

Microwaves Bridge the Atlanic

1,900 feet – and they worked exactly as planned. These trigger radars were built from older 415-MHz tail-warning systems, AN/APS-13, with the Yagi antennas directed downward. On the Nagasaki mission, there was heavy cloud cover, and the B-29 was guided to the target by a AN/APQ-13 blind-bombing radar.

The Rad Lab was one of America's largest wartime projects, at peak involving about 4,000 people and developing more than 100 different radar designs. The operation was officially closed on December 31, 1945, but continued on a transitional basis until July 1, 1946. At that time it was incorporated into MIT's new Research Laboratory for Electronics.

Rad Lab accomplishments were documented in the MIT Radiation Laboratory Series of 28 volumes published by McGraw-Hill over the next several years. The knowledge generated by this Laboratory strongly influenced many areas of post-war science and technology. Nine former Rad Lab employees or consultants were recipients of the Nobel Prize – but for other accomplishments, not for their radar work. These included I. I. Rabi, who received the 1944 Nobel Prize in Physics, and Luis Alveraz, who received the same prize in 1968.

In 1945, the BTL started developing the Nike anti-aircraft missile system for the Army. The associated radars eventually evolved to the highly accurate systems used in modern ballistic missile defense. The BTL, however, ceased doing defense work in 1970.

Throughout the war, the NRL remained responsible for the Navy's non-microwave radar improvements, and cooperated with the Rad Lab and the BTL in their microwave activities. With some 900 projects ranging from radar and sonar to lubricants and uranium isotope separation processes, the NRL grew to a peak of 4,400 personnel in 1946.

Following the war, the NRL returned to its primary mission of performing basic and applied research. Robert Page was named the NRL's first Director of Research in 1949. When he retired from the NRL in 1966, Page had 65 patents in his name. Hoyt Taylor and Leo Young also remained with the NRL until they retired; Robert Guthrie succeeded Taylor as Superintendent of the Radar Division.

The Army's SCL, like the NRL, greatly expanded during the war, eventually reaching near 14,000 total personnel. The specific part devoted to radar was the Evans Signal Laboratory, located just south of Fort Monmouth. After the war, the Evans Signal Laboratory remained responsible for radar research and development in the Army Signal Corps. In 1946, this Laboratory gained world-wide attention with Project Diana, the first radar signals bounced off the moon.

The Aircraft Radio Laboratory, located at Wright Field, became a part of the Air Materiel Command when the United States Air Force was established in 1946. The Wright Air Development Center was formed in 1951, taking the responsibility of all research and development of the Air Force, including radar activities.

In Great Britain, the Telecommunications Establishment (TRE) at Malvern increased to about 3,500 personnel. At the close of the war, a staff of some 1,500 persons continued radar work, but moved from Malvern College to nearby facilities formerly used by the Royal Navy's Duke Training Establishment.

The Radar Research and Development Establishment (RRDE) remained at Malvern, but with a greatly reduced staff. In 1953, the TRE and RRDE were combined to form the Radar Research Establishment, later called the Royal Research Establishment.

The Admiralty Signal Establishment (ASE) consolidated the operations of Watley (King Edward's School) with those at Haslemer (Lythe Hill House) and remained there while facilities at Portsmouth were being arranged. In 1948, all of the Royal Navy's radio and radar R&D activities were combined to form the Admiralty Signal and Radar Establishment, located at Portsmouth.

Robert Watson-Watt established a private consulting firm and, for a while, operated in Canada. Arnold Wilkins returned to the Radio Research Station at Slough and eventually became the Superintendent. Edward G. "Taffy" Bowen remained at the Rad Lab until 1944, then migrated to Australia, where he joined the CSIR as Chief of the Radiophysics Division.

A. P. Rowe and William Alan S. Butement also migrated to Australia; Rowe established the Scientific Advisory Board for the Department of Defense, and Butement was named the first Chief Scientist for Defense. John Cockcroft went to Canada in 1944 to head their Atomic Program, then returned to England in 1945 to serve as Director of the Atomic Energy Research Establishment, Harwell; he received the 1951 Nobel Prize in Physics.

A large part of the senior scientific staffs involved in wartime radar for the RAF, Army, and Admiralty returned to academic positions. Five of these individuals became Nobel Laureates, all for work independent of their radar accomplishments.

Bibliography and References for Chapter 4

Baxter, James Phinney, III; *Scientists Against Time*, MIT Press, 1968

Bowen, E. G.; *Radar Days*, Inst. of Physics Pub., 1987

Brown, Louis; *A Radar History of World War II – Technical and Military Imperatives*, Inst. of Physics Pub., 1999

Bryant, J. H.; "The First Century of Microwaves – 1886 to 1986," *IEEE Trans. on Microwave Theory and Tech.*, vol. 36, p. 830, 1988

Buderi, Robert; *The Invention That Changed the World*, Simon & Schuster, 1996

Callick, E. B.; *Meters to Microwaves: British Development of Active Components for Radar Systems 1937-1944*, Peter Peregrinus Ltd., 1959

Churchill, Winston S.; *The Hinge of Fate*, vol. IV of *The Second World War*; Cassell, 1951

Coales, J. F., and J. D. S. Rawlinson; "The Development of Naval Radar 1935-1945," *J. Naval Science*, vol. 13, nos. 2-3, 1987

Cockcroft, J. D.; "Memories of radar research," *IEE Proceedings*, vol. 132, Oct, p. 327, 1984

Comb, Cyril; "Fred Hoyle and Naval Radar 1941-45," *J. Astrophy. & Space Sci.*, vol. 285, July, p. 293, 2003

Conant, Jennet; *Tuxedo Park: A Wall Street Tycoon and the Secret Palace of Science That Changed the Course of World War II*, Simon & Schuster, 2002

Drury, Alfred T.; "War History of The Naval Research Laboratory," in the series, *U.S. Naval Administrative Histories of World War II*, Department of the Navy, 1946

Fisher, David E.; *A Race on the Edge of Time*, McGraw-Hill, 1988

Gebhard, Louis A.; "Evolution of Naval Radio-Electronics and Contributions of the Naval Research Laboratory," Report 8300, Naval Research Laboratory, 1979

Guerlac, Henry E.; *Radar in World War II*, vol. 8 of *The History of Modern Physics 1800-1950*, Amer. Inst. of Physics, 1987

Hanbury Brown, Robert; *Boffin: A Personal Story of the Early Days of Radar, Radio Astronomy and Quantum Optics*, Taylor & Francis, 1991

Hay, J. S.; *The Evolution Of Radio Astronomy*, Elek Scientific Books, 1973

Heil, Oskar E., and Agnesa Arsenjewa-Heil; "On a New Method for Producing Short, Undamped Electromagnetic Waves of High Intensity," *Zeitschrift fur Physik*, vol. 95, p. 752, 1935 (in German)

Hinsley, F. H., and Alan Stripp, (Eds.); *Codebreakers The Inside Story of Bletchley Park*, Oxford University Press, 1993

Howse, Derek; *Radar at Sea: the Royal Navy in World War II*, Naval Institute Press, 1993

IEEE Center for the History of Electrical Engineering; *RAD Lab: Oral Histories Documenting World War II Activities at the MIT Radiation Laboratory*, IEEE Press, 1993

Jones, R. V.; *Most Secret War – British Scientific Intelligence 1939-1945*, Hamish Hamilton Ltd., 1978

Johnson, Bryan; *The Secret War*, British Broadcasting Corp., 1977

Jurgen, Ronald K.; "Captain Eddy: the man who launched a thousand EEs," *IEEE Spectrum*, Nov. 1975

Latham, Colin, and Anne Stobbs, (Eds.); *Radar – A Wartime Miracle*, Sutton Publishing, 1996

Lovell, Sir Bernard; *Echoes of War: The Story of H2S Radar*, Adam Hilger, 1991

Megaw, Eric C. S.; "The High-Power Magnetron: A Review of Early Developments," *J. of the IEE*, vol. 93, p. 928, 1946

Mawer, Bertie; *King Edward's School: Bridewell to Witley 1553-2005*, Old Witleiens' Association, 2005

Merchant, P. A., and K. M. Heron; "Post Office Equipment for Radar," *Post Office Electrical Engineering Journal*, vol. 38, Jan. 1946

Nissen, Jack; *Winning the Radar War*, St. Martin's Press, 1987

Page, Robert Morris; *The Origin of Radar*, Anchor Books, 1962

Penley, W. H.; "Penley Radar Archives—the Early Development of Radar in the UK," 2006; http://www.penleyradararchives.org.uk

Radar R&D Sub-Committee; "Operational Characteristics of [U.S.] Radar Classified By Tactical Applications," FTP 217, Joint Committee on New Weapons and Equipment, 1 August 1943; http://www.history.navy.mil/library/online/radar.htm#contents

Redhead, Paul A.; "The Invention of the Cavity Magnetron and Its Introduction into Canada and the U.S.A.," *Physics in Canada*, Nov./Dec., p. 321, 2001

Randall, J. T. and H. A. H. Root, "Historical Notes on the Cavity Magnetron," *IEEE Trans. on Electron Devices*, vol. 39, p. 363, 1976

"Radar: A Report on Science at War," Office of Scientific Research and Development, distributed by Office of War Information, 15 Aug. 1945; http://www.ibiblio.org/hyperwar/USN/ref/Radar-OSRD/index.html

Rowe, A. P.; *One Story of Radar*, Cambridge University Press, 1948

Rowlinson, Frank; *Contributions to Victory*; Metropolitan-Vickers Electric Co., 1947

Skolnik, Merrill I.; "Fifty Years of Radar," *Proc. of the IEEE, Special Issue on Radar*, vol. 73, p. 182, 1985

Skolnik, Merrill I.; *Introduction to Radar Systems*, 3rd ed., McGraw-Hill, 2001

Southworth, George C.; "Survey and History of the Progress of the Microwave Arts," *Proc. of the IRE*, vol. 50, p. 1199, 1962

Stewart, Irvin; *Organizing Scientific Research for War; Administrative History of the OSRD*, Little, Brown, 1948

Stout, Wesley W.; *The Great Detective*, Chrysler Corp., 1946

Swords, S. S.; *Technical History of the Beginnings of Radar*, Peter Peregrinus Ltd., 1986

Tomlin, Donald H.; "From Searchlights to Radar – The Story of Anti-Aircraft and Coastal Defense Development 1917-1953," IEE Conference on the History of Electrical Engineering, p. 110, July 1988

Tomlin, Donald H.; "Pale Manor Hosted Vital War Work," *Worcester News*, Aug. 5, 2004;
http://archive.worcesternews.co.uk/2004/8/6/73958.html

Tomlin, D. H.; "Fifty years of the cavity magnetron," Symposium Lecture at the University of Birmingham, February 1990

Vaeth, J. Gordon; *Blimps and U-Boats: U.S. Navy Airships in the Battle of the Atlantic*, Naval Institute Press, 1992

Varian, R. H., and S. F. Varian, "A High Frequency Oscillator and Amplifier," *J. Appl. Phys.*, vol. 10, p. 321, 1939

Watson, Raymond C., Jr.; *Solving the Naval Radar Crisis:: The Eddy Test – Admission to the Most Challenging Training Program of World War II*, Trafford Publishing, 2007

Watson-Watt, Robert; *Three Steps to Victory*, Odhams Press, 1957

Watson-Watt, Sir Robert; *The Pulse of Radar*, Dial Press, 1959

Whitehead, J. (James) Rennie; "Memories of a Baffin,"
http://www3.sympatico.ca/drrennie/memoirs.html

Willoughy, Malcolm Francis; *The Story of LORAN in the U.S. Coast Guard in World War II*, Arno Pro, 1980

Winterbotham, F.W.; *The Ultra Secret*, Harper and Row, 1974

Zimmerman, David; *Top Secret Exchange; The Tizard Mission and the Scientific War*, McGill-Queens University Press, 1996

Chapter 5

FUNKMESSGERÄT DEVELOPMENT IN GERMANY

What eventually became known as World War I began in August 1914, and ended with the Treaty of Versailles signed in June 1919. Germany, led by Kaiser Wilhelm II and seeking to gain control of the western mainlands of Europe, invaded Belgium with the intent of easily passing through to France and Paris. The French counterattacked, and, backed by the British, established a battle line of trench warfare from the North Sea through eastern France to the border of Switzerland. While this remained the central area for land combat, there was also extensive naval warfare, as well as fighting in several other countries. The United States entered on the side of France and Great Britain in April 1917, forming the leadership of the Allied Powers.

In France, the Germans made a desperate attempt to break through to Paris in June 1918, but were repelled and by early November were in full retreat. The German top generals realized that they now had no hope of winning the war and approached the opposing countries for negotiations and peace talks. This was refused unless Germany was placed under a Democratic government. In early November, the *Reichstag* (national assembly) met and proclaimed a new government. The Kaiser then abdicated his throne, and a Parliamentary Republic was formed.

On November 11, 1918, representatives of the German and Allied armies signed an armistices agreement. After a series of exchanges and negotiations, the Treaty of Versailles was signed the following June.28. Some historians have called the terms of the Treaty the most severe in history, taking away much land, requiring massive reparation payments, and greatly reducing the *Heer* (German Army) and *Reichsmarine* (Realm Navy), and eliminating the Army's Air Corps. It did not, however, adequately restrict overall military rebuilding, and this started in a very short while.

Most of the population of Germany never understood World War I, and did not support the fragile democratic government. Millions of the people were in such a state that they were easy prey to Adolph Hitler and his Nazi Party, who promised better times while asserting Aryan superiority. When the Great Depression struck in 1929, it led to some 43 percent of the workforce being unemployed, and the Nazi party came to dominate the Parliament. In early 1933, Hitler was named Chancellor;

shortly thereafter he was given dictatorial powers and took the title *der Führer* (the Leader). Major anti-Semitic laws were passed in 1935.

Adolph Hitler, der Führer

The Nazis initiated extensive building programs for housing, public facilities, and highways. As Hitler had promised, the economy was booming. The *Heer*, supposedly limited to 100,000 soldiers by the Treaty of Versailles, was greatly increased, using the excuse that it was needed to defend the nation against the Communists (the USSR).

In 1935, the *Kriegsmarine* (War Navy) was formed from the *Reichsmarine*; this and the *Heer* were placed under the authority of the *Oberkommando der Wehrmacht* (OKW, High Command of the Armed Forces), reporting directly to Hitler. Also, in 1933, Hermann Göring had been ordered to establish the *Luftwaffe* (Air Force), and this now became the third arm of the OKW. All of these changes were in violation of the restrictions of the Treaty of Versaille.

The *Schutzastaffel* (SS, Protective Squadron) and the *Geheime Staatspolizei*: (Gestapo, Secret State Police), both commanded by Heinrich Himmler, were totally devoted to Hitler and were responsible for the vast majority of war crimes perpetrated under the Nazi regime..

Hitler's plan for expanding Germany began to be exercised in 1936. The Rhineland, an ethnic-German area, was annexed. The same year, an alliance was made with Fascist dictator Benito Mussolini of Italy, forming the Berlin-Rome Axis. In a short time, Japan also joined the Axis. In March 1938, the Germans annexed Austria, and Czechoslovakia agreed to become a German protectorate. Great Britain, under Prime Minister Neville Chamberlain, adopted a policy of appeasement; France and other European countries would not act without Britain's backing.

On September 1, 1939, the German forces invaded Poland, marking the start of World War II.

TECHNOLOGIES DEVELOPMENT

In 1923, only five years after the defeat of Germany in World War I, the *Reichsmarine* begin rebuilding its weapons research and development capabilities. One of the facilities set up was the *Nachricjtenmittel-Versuchsanslalt* (NVA – Experimental Institute of Communication Systems) in Kiev. The NVA had been tasked by another Navy weapons facility, *Torpedoversuchsanstalt* (TVA, Torpedo Research Establishment), to devise improved methods of using sound for determining the direction and range of targets.

The NVA established a laboratory devoted to acoustical research, and this made considerable progress in developing depth sounders and equipment use by ships and submarines for detecting surface targets – the technology that would eventually be called sonar.

In 1928, Dr. Rudolph Kühnhold joined the NVA's acoustic laboratory, and soon made significant contributions in the research supporting TVA. By 1933, however, he concluded that while acoustics were very suitable for underwater detection equipment, this technology could never meet the desired precision for open-air applications, both surface-to-surface and surface-to-air. He then turned to techniques using radio waves.

The technology using electromagnetic waves for detection was actually quite old in Germany, going back to 1888 and the experiments of Dr. Heinrich Rudolf Hertz (1857-1894). After earning his doctoral degree at the University of Berlin under Gustav Kirchhoff and Hermann von Helmholtz, two of the foremost physicists of that time, Hertz became a university lecturer and researcher. In a corner of his physics classroom at the Karlsruhe Polytechnic University in Berlin, Hertz generated electromagnetic waves using a circuit containing a spark gap, and detected them in a similar circuit a short distance away. In this, he not only proved the existence of these waves but also showed that, like light waves, they were reflected by metal surfaces. Thus, his waves "detected" the presence of the metal. Unfortunately, he felt that this discovery would have no practical applications and only pursued it from an intellectual standpoint.

Hertz and his Experiment

It was completely different with Christian Hülsmeyer (1881-1957). After losing a friend in a ship collision, 22-year old Hülsmeyer formed a company in Dusseldorf to apply Hertz's reflected waves in a *Funkmessgerät* (radio measuring device) for detecting the presence of ships. In April 1904, he registered German and foreign patents for an apparatus, the *Telemobilskop* (Telemobiloscope). An article, "The Telemobiloscope," was published in a 1904 issue of *The Electrical Magazine*. This was described as an anti-collision system using a 50-cm wavelength spark-gap transmitter and a coherer detector. The radiated

Christian Hülsmeyer

signal was beamed by a funnel-shaped reflector and tube that could be aimed. The receiver used a separate vertical antenna with a semi-cylindrical, movable reflecting screen.

While developing his sensing apparatus, Hülsmeyer realized that in a river full of ships all using his device at once, the ships could receive false warnings. In January 1906, he received a patent for a time-limiting electromechanical mechanism for making the receiver respond only to signals "from the proper transmitter." He performed several successful demonstrations and the press and public opinion were very favorable. However, the *Kaiserliche Marine* (Imperial Navy) was unimpressed, and none of the European shipping lines he approached wanted to use his invention.

At that time, Marconi's Wireless Company tightly controlled the navel communication industry and would not tolerate any competition. Hülsmeyer soon gave up his endeavor, and little more was done on radio-based detection in Germany for almost three decades.

Carl Braun

During the first third of the 20th century, there were many advances in radio technology in Germany. Two developments having significance in radar will be noted. Dr. Karl Ferdinand Braun (1850-1918) Professor at the University of Strasbourg, made many early contributions, including the crystal-diode detector and patents for tuning circuits. His greatest and most lasting development was the cathode-ray tube, still called the Braun tube in many nations. In 1909 he shared the Nobel Prize with Marconi, citing Braun's "contribution to the development of wireless telegraphy."

With the advent of de Forest's triode, it was assumed that this device would soon be the used in generating high-frequency oscillations. It was found, however, the spacing between electrodes in triode gave an upper limit to frequencies that it could handle. A technique called velocity modulation was theorized to overcome this. In 1920, Drs. Heinrich Barkhausen and Karl Kurz at the Dresden Technische Hochnische used this theory in developing the retarded-field triode that could operate at ultrahigh frequencies. Although severely limited in output power, the

Barkhausen-Kurz tube was quickly adopted world-wide for UHF research.

Many types of *Funkmessgerät* (radio measuring device) systems were developed in Germany just before and throughout the war. The major firms producing these were GEMA, Telefunken, Lorentz, and Siemens & Halske; these will be described, along with a few of the many radar systems. Radars and related equipment adopted by the German military were officially designated with a FuG (*Funk-Gerät*, Radio Device) number; the *Luftwaffe* used FuMG (*Funkmessgerät*), followed by a model number and often a letter designating the manufacturer. The code names – what they were commonly called – are used herein.

GEMA

Upon completing his doctorate in physics from the University of Gottingen in 1928, Rudolf Kühnhold (1903-1992) became a scientist at the NVA. Here he worked on the task from TVA in improving technologies and devices for acoustical detection of targets. In this, he made a number of contributions to signal processing and obtained a patent for an underwater distance-measuring device. In 1931, he was promoted to the position of Scientific Director of the NVA.

Rudolf Kühnhold

The accuracy of acoustical measurement of direction and distance to targets was poor, and in 1933 Kühnhold turned his attention to radio techniques for open-air applications. His preliminary analysis indicated radio signals with a wavelength of 10 to 20 cm could be focused into sufficiently narrow beams and possibly reflected from targets of interest.

The firm of Julius Pintisch in Berlin had developed a transmitting and receiving set operating at 13.5 cm (2.22 GHz) for a communications link. The security of this set came from its narrow beam. Both units of the set used Barkhausen-Kurz tubes functioning beyond their normal frequency limit, and the transmitter produced only 0.1 watt. To experimentally test his proposed method for target detection, Kühnhold purchased one of these sets and placed it atop his laboratory building, with the transmitting and receiving antennas behind separate 81-cm parabolic reflectors. To identify any returned signal, the transmitter was modulated with a 1000-Hz tone. Initial attempts to detect reflections from a building about 2-km distance were totally unsuccessful.

The failure was attributed to a lack of transmitter power, and Pintisch began an effort to increase the output. In late 1933, Kühnhold also approached Dr. Wilhelm Runge, Acting Director of the Radio Receiver Laboratory at Telefunken, about developing a radio-based detection system for the NVA, but was told that tubes with significant power at these wavelengths were years off. (Actually, Runge was already having some success with such tubes and possibly listened to Kühnhold's concepts to gain information for his own use.)

A small firm, Tonographie, had supported NVA in developing recording equipment for underwater sound. Its young owners, Paul-Gunther Erbsloh and Hans-Karl von Willisen, were amateur radio operators and had started a project at Tonographie in a narrow-beam, VHF system for secure communications. Kühnhold asked for their assistance in his radio-detection project and they enthusiastically joined in the effort.

In January 1934, Erbsloh and von Willisen formed a new company, *Gesellschaft für Electroakustische und Mechanische Apparate* (GEMA). The name not only reflected their well-known prior work in acoustics, but also served as a cover for the real military activities. From the start, the firm was always called simply GEMA

Work on a *Funkmessgerät für Untersuchung* (radio measuring device for reconnaissance) began in earnest at GEMA. Drs. Hans E. Hollmann and Jakob Theodor J. Schultes, both affiliated with the prestigious Heinrich Hertz Institute in Berlin, were added as consultants in high-frequency devices. Schultes soon became an employee of GEMA and led the Microwave Department for many years; Hollmann was the chief technical consultant in the period 1934-37.

Following its invention by Dr. Albert W. Hull at General Electric, the magnetron had been improved by researchers in several countries. In Japan, Dr. Kinjiro Okabe developed a split-anode magnetron, providing details in the April 1929 issue of the *Proceedings of the IRE*. At the Philips Research Laboratory in The Netherlands, Klass Posthumus had designed a split-anode device that generated up to 70 W at 50 cm (600 MHz). This was advertised for sale by Philips, and one was ordered by von Willisen for testing at Tonographie.

Frequency instability of the Philips magnetron prohibited its use in Tonographie's secure communication systems, but it was taken by GEMA and used with some success in their detection experiments. To work with the magnetron, Hollmann built a regenerative receiver using a Barkhausen-Kurz tube, and Schultes developed Yagi-array antennas. The magnetron was used to feed one Yagi array and, to prevent saturation, the receiver and second Yagi array were set up some distance away. In June 1934, large vessels passing through the Kiev Harbor were detected by Doppler interference at a distance of about 2 km.

The 13.5-cm transmitter from Pintisch was improved to produce 0.3 watt, and then also detected (again with Doppler interference) the moving vessels at about double the range of the 50-cm system. This led GEMA to conduct extensive studies of the relative influences of frequency, antennas, power, and reflecting surfaces. Like in the experiments performed earlier at the NRL in the U.S., the detection by both systems was in the form of a beat effect, caused by the Doppler-shift in returns from the moving vessels.

Kühnhold worked closely with GEMA and led their attempts to improve the continuous-wave system, but also retained his position at the NVA. In October 1934, strong reflections were observed from an aircraft that happened to fly through the beam; this opened consideration of targets other than ships and brought funding from NVA.

At that time, the success of a number of researchers in using pulsed-transmission for measuring the height of the ionosphere was well known. Also, underwater acoustical detection used pulsed transmission. Thus, the GEMA team turned its attention to developing a pulsed radio system for combined detection and range determination. Pulsing would also allow the receiver to be time-isolated from the transmitter.

The pulsed system used a new Philips magnetron operating at 50 cm (600 MHz) and with better frequency stability. It was modulated with 2-μs pulses at a PRF of 2000 Hz. The transmitting antenna was an array of 10 pairs of dipoles with a reflecting mesh. The wide-band regenerative receiver used Acorn tubes from RCA, and the receiving antenna had three pairs of dipoles and incorporated lobe-switching. A blocking device shut the receiver input when the transmitter pulsed. For displaying the range, it had a Braun tube (a CRT), improved in the late 1920s by Manfred von Ardenne.

The equipment was placed atop a tower at a NVA test facility beside the Lubecker Bay near Pelzerhaken. This pulse-modulated system first detected returns from woods across the bay at a range of 15 km in May 1935, a few months behind similar accomplishments by American and British researchers. The GEMA system, however, had limited success detecting a research ship, *Welle*, only a short distance out on the bay.

The receiver was rebuilt, becoming a super-regenerative set with two IF stages at different frequencies. With this improved receiver, the system readily tracked vessels at up to 8-km range. In September 1935, a demonstration was given to Admiral Erich Raeder, Commander-in-Chief of the *Kriegsmarine*. The system performance was excellent; the range to the cruiser *Koenigsberg* was read off the screen of the Braun's tube with a tolerance of 50 meters (less than 1 percent variance), and lobe-switching had given a directional accuracy of 0.1 degree.

To demonstrate that the equipment could operate in the harsh shipboard environment, it was then mounted on the *Welle*. There it detected other ships at up to 7 km and the coastline at 20 km. Historically, this marked the first naval vessel equipped with radar.

Kühnhold was convinced (correctly) that future systems would mainly operate in the microwave portion of the spectrum. In continued work on the 13.5-cm set, the Barkhausen-Kurz transmitter tube was replaced by a newer magnetron, but it was still unstable and the set was not put into production.

Kühnhold remained with the NVA and also consulted for GEMA; he is often credited in Germany as being the inventor of radar. In a rare cooperative activity between the services, he worked with Dr. Hans Plendl in 1936-37, then with the *Luftwaffe's* Laboratory for Aviation, on *Knickebein* (Bent Leg) and other radio navigation system.

Just before the beginning of the war and for a while thereafter, some research on microwaves was continued by Kühnhold and Dr. Anton Röhrl at NVA. Aside from this, little further work on microwave radars was done in Germany until after early 1943 when a British multi-cavity magnetron was found in a downed RAF bomber.

Seetakt Radar

The *Kriegsmarine* funded GEMA for further development and production of two systems in what was called the *Seetakt* (from *Seetaktisch*, navy tactical) series. A 50-cm (600-MHz) system was codenamed *Dezimeter Telegraphie* or simply *DeTe*, the name used in all communications concerning this general detection technology. This used a "mattress" antenna very similar to the "bedspring" on the American CXAM. In a major intelligence blunder, a 1938 book from a German publisher included a picture of a warship clearly showing the mattress antenna; the naval authorities that approved the picture apparently had no idea that they were looking at secret equipment.

The first modification to this system was the *DeTe-I*, a surface-search radar. In a parallel project to increase power and range, the wavelength was increased to 2.4 m (125 MHz), and it became the *DeTe-II*, an air-warning radar. For these, GEMA developed a high-frequency triode, the TS1; this was very similar to the Doorknob tube made by the BTL and provided 8 kW in a push-pull configuration. A PRF of 500 Hz was used.

Funkmessgerät Development in Germany

The *DeTe-I* was for both ship-board and shore installations, but the *DeTe-II* was only for shore. Inadequacy of range eventually drove the *DeTe-I* wavelength first to 60 cm, then to 80 cm. A 60-cm version was aboard the pocket battleship *Admiral Graf Spee* that, after an epic battle with a British cruiser pack in January 1939, was scuttled off the coast of Uruguay. A British expert examined the exposed antenna, confirming that the Germans had radar.

TS1s Push-Pull Module

Throughout the war, GEMA provided a wide variety of *Seetakt* shipboard sets including several types for U-boats; in fact, almost every installation was different – virtually custom modified. Most had a very sophisticated range-measuring module called *Messkette* (measuring chain) that provided range accuracy within a few meters regardless of the total range.

Flakleit G Radar

The 80-cm *DeTe-1* became the *Flakleit*, capable of directing fire on surface or air targets within an 80-km range. It had an antenna configuration very similar to the U.S. SCR-268. The fixed-position version, the *Flakleit-G*, included a height-finder.

Although the *DeTe-II* was an excellent air-warning system, no one in the *Luftwaffe* was even aware of its existence until July 1938, when Hermann Göring, Commander of the *Luftwaffe*, saw it demonstrated for Hitler at the TVA.

Freya Radar

Although the *Kregsmarine* attempted to keep the GEMA from working with the other services, the *DeTe-II* was shown to Maj. Gen. Wolfgang Martini, Chief of the *Luftnachrichtentruppe* (Air Signal Corps). Martini, a very capable technical officer, was highly impressed. In the fall of 1938, he ordered development of the *Luftwaffe*'s own version of the *DeTe-II* and gave it the name *Freya*. This was a ground-based radar operating around 2.4 m (125 MHz) with 15-kW peak power giving a range of some 130 km. The electronics of this set became common for essentially all German air-warning radars.

Air Transportable Freya

As an additional inducement, GEMA offered Martini an adjunct to the radar that would allow the identification of German aircraft (an IFF system). Called *Erstling* (First Born), this would add an interrogation signal to the radar, and a responder on the aircraft would transmit an identifying signal on a different frequency. In early 1939, Martini ordered 3,000 *Erstling* sets.

The *Freya* radar system used three antennas. In the most common configuration, these were stacked on a tower with the transmitting antenna on the lower level, receiving antenna at the middle, and the *Erstling* identifying antenna on top. The transmitting and receiving antennas each had 12 vertical, full-wave dipoles. The basic *Freya* radar was continuously improved, with over 1,000 systems built throughout the war.

In 1940, Lt. Hermann Diehl of Experimental Air Signals introduced *Freya* improvements to allow better vectoring of fighters to attack positions. He worked with GEMA in developing a lift to raise the antenna 20 m on a track to alter the pattern and give better target-height estimates. He also introduced lobe-switching on the receiving antenna to improve the horizontal accuracy.

Two enlarged versions were introduced in 1941; one was the *Mammut* (Mammoth) using 16 *Freyas* linked into a giant 30- by 10-meter (100- by 33-foot) antenna with phased-array beam-directing, a technique that would eventually become standard in radars. It had a range up to 300 km and covered some 100 degrees in width with an accuracy of near 0.5 degree. About 30 sets were built, some with back-to-back faces for bi-directional coverage.

Mommut Radar

The second version, the *Wassermann* (Waterman), had eight *Freyas* also with phased-array antennas, stacked on a steerable, 56-meter (190-

foot) tower and giving a range up to 240 km. A variant, *Wassermann-S*, had the radars mounted on a tall cylinder. About 150 of all types were built starting in 1942. Soon after they were introduced, then-Lt. Gen. Martini applied both the *Mammut* and *Wassermann* in fighter ground-control systems.

TELEFUNKEN

Telefunken GmbH was formed in 1903 as a joint venture of Siemens & Halske and AEG (*Allgemeine Elektrizitäts-Gesellachaft*, General Electricity Company). It had the specific purpose of developing radio equipment, and soon became the largest supplier of consumer radio products in Germany.

In the early 1930s, Aryanization and Nationalism took a strong hold in Telefunken, resulting in the dismissal of Dr. Emil Mayer, its highly capable Jewish Director General, and the addition of research in military-related equipment. This research achieved considerable success, particularly in decimeter FM radios for line-of-sight relay stations. By the start of WWII in 1939, Telefunken had over 23,000 employees. Siemens shares in the company were bought by AEG in 1941, and Telefunken then operated as an independent subsidiary of AEG.

Wilhelm Tolmé Runge (1895-1987) was the son of the famous mathematician, Dr. Carl Runge. After service as a signals instructor during WWI, he gained a strong interest in radio and earned his Dr.Eng. degree in electrical engineering from Darmstadt Technische Hochschule. In 1923, he joined Telefunken's Radio Research Laboratory. Having done his dissertation in high-frequency technology, he focused his work during the next several years on this area. When Kühnhold visited him in late 1933, Runge was already experimenting with high-frequency transmitters and had the Telefunken tube department working on cm-wavelength devices.

Wilhelm Runge

Runge's interest in radio-based detection increased over the next year. Now the Laboratory Director, in the summer of 1935 he initiated an internally funded project in this activity. Using Barkhausen-Kurz tubes, a 50-cm (600-MHz) receiver and 0.5-W transmitter were built. With the antennas placed flat on the ground some distance apart, Runge arranged for a Ju-52 aircraft to fly overhead and found that the receiver gave a strong Doppler-effect signal. This was similar to the experiments of Hyland and Young in 1930, and the Daventry experiment in early 1935, starting points of radar development in America and Great Britain, respectively.

An article in the September 1935 issue of the magazine *Electronics* was headlined as follows: "Telefunken firm in Berlin reveals details of a 'mystery ray' system capable of locating position of aircraft through fog, smoke and clouds."

Experimental Darmstadt

Although little interest was shown in this detection technology by Telefunken officials at the time, Runge, with Hans Hollmann as a consultant, continued in developing a 1.8-m (170-kHz) system using pulse-modulation. Engineer Wilhelm Stepp developed transmit-receive device for allowing a common antenna. Stepp code-named the system *Darmstadt* after his home town, starting the practice in Telefunken of giving the systems names of cities. The system, with only a few watts transmitter power, was first tested in February 1936, detecting an aircraft at about 5-km distance.

Würzburg Radar

By this time, the *Luftwaffe* had leaked information to Telefunken that their competitor, GEMA, was developing equipment of a similar type. With Runge's latest success, this was sufficient to generate interest by Telefunken management, and authorization was given for developing a full gun-laying system, including a new triode, the LS-180 capable of about 10-kW pulse-power at 50 cm (600 kHz). The system was code-named *Würzburg*.

Telefunken received a contract from the *Luftwaffe* in late 1938 to build the *Würzburg*. The transmitter had a 2-µs pulse width and a PRF of 3,750; the antenna used a 3-m (10-ft) parabolic reflector built by the Zeppelin Company. With the increased power, the range was up to

Würzburg Radar

Funkmessgerät Development in Germany

40 km for aircraft. The *Würzburg* was demonstrated to Hitler in July 1939. With Stepp now serving as the project leader, several versions of the system evolved. On one version, a rotating, slightly off-set dipole was inserted at the parabola focus, giving a conical scan and resulting in very accurate directional determination.

Requiring only one operator, the *Würzburg* came to be the primary mobile, gun-laying system used by the *Luftwaffe* and *Heer* during the war. About 4,000 of the various versions of the basic system were eventually produced.

The accuracy of the *Würzburg* was inadequate for guiding night-interception fighter aircraft. The solution came from Leo Brandt, who had taken over further development of this system. The *Würzburg-Riese* (*Giant Würzburg*) had a 7.5-meter (25-foot) dish and was mounted on a railway carriage; this well-engineered structure was another product from Zeppelin. The system also had an increased transmitter power; combined with the enlarged reflector, this resulted in a range of up to 70 km, as well as greatly increased accuracy. The PRF was decreased to 1,875 Hz to accommodate the longer range. About 1,500 of this model were built.

Würzburg-Riese Radar

In 1938, the *Luftwaffe* had inherited the *Flak* batteries from the *Heer*. After making comparative tests with similar systems from Telefunken, GEMA, and Lorenz, then-Maj. Gen. Martini made the decision that the *Würzburg* should be used for directing *Flak* fire. The first bomber brought down using a *Würzburg*-directed *Flak* battery occurred in September 1940, ensuring the acceptance of this system.

As British bombings increased and *Flak* became more critical, Telefunken started changes to the *Würzburg*, optimizing it for this function. This resulted in the *Mannheim*, still a 50-cm (600-MHz) system but with increased power and considerably improved electronics, including a phased-array antenna. The system had a 30-km range with 10-m accuracy and an azimuth accuracy of 0.15 degree, three times the accuracy of the basic *Würzburg*. Some of the later models included electronics that provided automatic tracking after the target was acquired.

The *Mannheim* went into production at GEMA in 1942; about 400 sets were built for *Flak* batteries. Since the systems looked the same, the

Mannheim was more often called a *Würzburg*. A few *Mannheim-Riese* sets were built for research purposes, but it was never put into production.

Lichtenstein Night-Fighter Radar

In early 1941, Air Defense recognized the need for radar on their night-fighter aircraft. The requirements were given to Runge at Telefunken, and by the summer a prototype system was tested. Hans Muth was assigned as the project technical leader. Code-named *Lichtenstein*, this was a 62-cm (485-MHz), 1.5-kW system, generally based on the technology now well established by Telefunken for the *Würzburg*.

Lichtenstein Antennas

The primary developmental problems were reduction in weight, provision of a good minimum range (very important for air-to-air combat), and an appropriate antenna design. An excellent minimum range of 200 m was achieved by carefully shaping the pulse. The antenna array had four dipoles with reflectors, giving a wide searching field and a typical 4-km maximum range (limited by ground clutter and dependent on altitude).

Rather than lobe-switching for angular accuracy, a rotating phase-shifter was inserted in the transmission lines to produce a twirling beam. The elevation and azimuth of a target relative to the fighter were shown by corresponding positions on a CRT display.

The first production sets (*Lichtenstein B/C*) became available in February 1942, but were not accepted into combat until September. In May 1943, a B/C-equipped Ju-88 landed in Scotland; it had flown the wrong way against an *X-Leitstrahlbake* directional beacon – or perhaps the pilot had defected. This was not realized by the crew until too late to destroy the radar or the *Erstling* identification (IFF) set. The British immediately recognized that they already had an excellent countermeasure in Window (the chaff used against the *Würzburg*); in a short time the B/C was greatly reduced in usefulness.

Lichtenstein Display

Funkmessgerät Development in Germany

From its introduction, there had been dissatisfaction with the range of the B/C. To correct this, development of a version with higher wavelength – and thus longer range – was underway. When the chaff problem was realized, it was decided to make this wavelength variable, allowing the operator to tune away from chaff returns. In mid-1943, *Lichtenstein SN2* was released, operating with a wavelength changeable between 3.7 to 4.1 m (81 to 73 MHz).

The *SN2* sets were primarily used on Me-110 and Ju-88 night-fighter aircraft. An additional benefit was that the increase in wavelength moved the *Lichtenstein* operation closer to the *Freya* band, making it more difficult to identify. This also resulted in a wider beam, making it easier to locate targets. On the negative side, the longer wavelength required larger dipole antennas, resulting in greater drag on the aircraft and reduced speed – greatly to the dissatisfaction of the pilots.

LORENZ

The firm of C. Lorenz AG dated from 1870 when Carl Lorenz started manufacturing electrical lighting products. It later took this name and was an early entrant in the wireless hardware field; in 1907 it supplied the first naval radiotelephone systems in the world. In 1930, it was bought by *Standard Elektrizitätsgesellschaft*, a branch of the American firm International Telephone and Telegraph, but Lorenz continued to operate independently.

Since before WWI, Lorenz had led in modular-manufactured communication equipment for the German military and was the main rival of Telefunken. In late 1935, when Lorenz found that Runge at Telefunken was doing research in radio-based detection equipment, they started a similar activity under Dr. Gottfried Müller. A pulse-modulated set called *Einheit für Abfragung* (EFA, Device for Detection) was built. Using a type DS-310 tube (similar to the Acorn) operating at 70 cm (430 MHz), it had identical transmitting and receiving antennas made with four rows of two half-wavelength dipoles backed by a reflecting screen, all mounted on a rotatable mast.

In early 1936, initial experiments gave reflections from large structures such as the Berlin Cathedral at a distant of about 7 km, but a transmitter with greater power was needed. Two type DS-320 triode tubes in push-pull were used to give up to 2-kW pulse power. In mid-1936 the equipment was set up on cliffs near Keil, and good detections of ships at 7 km and aircraft at 4 km were attained.

The success of this experimental set was reported to the *Kriegsmarine*, but they showed no interest; they were already fully engaged with

GEMA for similar equipment. Also, because of extensive agreements between Lorenz and many foreign countries in their LEF air-navigation system, the naval authorities had reservations concerning the company handling highly classified work.

Land-Based Radars

After their naval radar was rejected by the *Kriegsmarine*, Müller's research team at Lorenz turned to developing a land-based detection system for supporting *Flgzeugabwehrkanone* (*Flak*, anti-aircraft guns). In operating over land, ground returns (clutter) were a problem, and research in more directional antennas was started. The wavelength was decreased to 62 cm (385 MHz), the limit of the tubes, and the antennas were changed to dipoles backed by 2.4-m parabolic reflectors. The transmitter and receiver were attached directly behind the parabolas, and the synchronized aiming was controlled by an electromechanical drive. Lorenz developed a special Braun tube (CRT) that allowed the range to be shown in a circular display.

Kurfürst Radar

This system was demonstrated to the Ordnance Office of the *Heer* in early 1938, giving good returns from aircraft at up to 13-km range. A contract was received to develop a prototype *Flak*-aiming set code-named *Kurfürst* (Cure prince). Dr. Hermann Berger joined Lorenz and developed a series of triodes delivering a new level of VHF power; the first of these, the RD-12Tf, was much better than tubes from either GEMA or Telefunken. A mobile gun-mount was used to support the equipment, with a tubular mast holding the reflectors substituted for the gun barrel. Tests now gave a range of about 30 km.

Responsibility for German Air Defense was transferred from the *Heer* to the *Luftwaffe*. In late 1938, Martini had tests run at the Air-Defense Training School, comparing the prototype *Furfürst* with prototypes of the *Freya* from GEMA and the *Würzburg* from Telefunken. These tests showed that the *Furfürst* range and accuracy of 100 m would be satisfactory for *Flak* applications; the bearing angle would be determined by a separate infrared detection system.

At Martini's request, two improved versions were developed. The *Kurpfulz* had a mobile, enclosed shelter for the equipment, and antennas were pivot-mounted in a forked frame. The *Kurmark* was similar, but was transportable rather than mobile. Twenty each of these improved sets were ordered for further evaluation. The *Würzburg*, however, was eventually determined to better meet the requirements for directing *Flak*.

Later in the war, the basic *Kurpfulz /Kurmark* was modified to form two new systems. The *Tiefentwiel* was a mobile system built to complement the *Freya* against low-flying aircraft. The *Jadgwagen* was a 54-cm (560-MHz) mobile unit used for air surveillance; this system had a plan-position indicator. Both of these systems were produced by Lorenz in small numbers starting in 1944.

Kurpfulz Radar

Airborne Reconnaissance Radar

In June 1941, an RAF Hudson bomber, equipped with an ASV (Air-to-Surface Vessel) Mk II radar, made an emergency landing in Brest, France. Although the crew had attempted to destroy the set, the remains were sufficient for the German Laboratory for Aviation to discern the operation and its function. This was the Germans' first indication of anti-surface vessel equipment. Tests by Dr. Paul von Handel on the repaired set – called *Eule* (Owl) by the Germans – indicated the merits of such a radar. Martini, then a Lieutenant General, also saw the value and tasked Lorenz to develop a similar system.

Having just lost the *Flak* work to Telefunken, Lorenz received this news at an opportune time. With backgrounds in the *Leitstrahl* aircraft navigation equipment and the *Kurpfulz/Kurmark* aiming systems, the company had excellent capabilities for this project. Dr. Fritz Trenkle was the Research Director, and he assigned Dr. Karl Christ to lead the effort.

Before the end of the year, they had built a set based on the general *Kurpfulz/Kurmark* design using their RD-12Tf tube, but greatly reduced in size and weight and with improved electronics. Called *Hohentwiel*, it producd 50-kW pulse power at 55 cm (545 MHz) and had a very low PRF of 50 Hz. The set used two separate antenna arrangements,

Hohentwiel Antennas on FW 200

providing searching either forward or side-looking, with the input to the receiver changed by a remotely controlled switch.

The *Hohentwiel* was demonstrated in detecting a large ship at 80 km, surfaced submarine at 40 km, submarine periscope at 6 km, aircraft at 10 to 20 km, and land features at 120 to 150 km. The range accuracy was between 50 m (short distances) and 1 km (long distances). A bearing accuracy of about 1 degree was obtained by rapidly switching between two receiver antennas aimed 30 degrees on each side of the transmitter antenna direction.

Put into production by Lorenz in 1942, the *Hohentwiel* was highly successful. It was first used on large reconnaissance aircraft such as the Focke-Wulf 200 Condor. In 1943, the *Hohentwiel-U*, an adaptation for use on submarines, provided a range of 7 km for surface vessels and 20 km for aircraft. Altogether, some 150 sets per month were delivered.

Hohentwiel-U Monitor

Guidance Systems

Early in the development of radio, Otto Scheiler invented a system of four antennas set in the corners of a large square that generated an array of overlapping, very narrow beams. In 1932, Dr. Ernst Kramer of Lorenz used the Scheiler antenna in developing a system called *Ultrakurzwellen-Landefunkfeuer* (LEF) or simply *Leitstrahl* (Guiding Beam). Transmitters drove the antenna to radiate a dot-dash tone to one side of the beam and a dash-dot on the other; when on path, the tone would be continuous. Commonly known as the Lorenz System, it was sold worldwide for aircraft guidance in bad weather and at night.

At the *Deutsche Versuchsanstalt für Luftfahrt* (DVL, German Laboratory for Aviation) in Rechlin, Dr. Hans Plendl had experimented with changes in the LEF commercial system to allow more direct guidance for aircraft of the newly formed *Luftwaffe*. His basic concept was to have two narrow beams originating at widely separated

Funkmessgerät Development in Germany

transmitters and use their crossing point to give a relatively concise location to the aircraft. If on a bombing mission, this might be the location to release the bombs. The complex Scheiler antenna was greatly simplified by Dr. Rudolph Kühnhold at the NVA.

Code named *X-Leitstrahlbake* (Directional Beacon), this was tested and accepted by the *Luftwaffe* in 1937. Lorenz received a contract for supplying the ground stations, and the aircraft receivers were the same as used in the LEF. By 1939, stations radiating into other countries were installed; these did not raise suspicions since the signals were on the same frequency (30-MHz) and with the same tones as the standard Lorenz System. Because of the shape of the transmitting antennas, they were called *Knickebein* (Bent Leg); this also became the nickname of the system.

Knickebein Antenna

When Germany started night-time bombing of Great Britain in 1940, the *Knickebein* was used to guide the lead aircraft, as well as to give the bomb release point. The British quickly code-named the German beams "Headaches," and developed countermeasure transmitters called "Aspirins." The *Knickebein* system was soon abandoned, and the Battle of the Beams followed.

Late in 1940, a more sophisticated modification of the Lorenz System – the *X-Gerät* (Device) using several beams – was placed into service. Countermeasure transmitters (called "Bromides") again neutralized the system, and it was changed to the *Y-Gerät*. This involved the retransmission from the aircraft of the main beam, and, when received at the main transmitter, the phase comparison gave a concise distance and allowed direct signaling for bomb release.

In 1939, Lorenz developed an aircraft beacon system called *Elektra* that, when used to augment the LEF, allowed a highly accurate determination of flight quadrant and position. The beacon operated at 300 kHz (1-km wavelength) and used three fixed antennas in a row, spaced one wavelength apart. If an aircraft was on a path roughly perpendicular to the antennas, the navigator could take two fixes at different positions and use triangulation to find the aircraft's location. The system was improved and given the name *Sonne* (Sun), and soon became known as "Console" to the Allies; it was so useful that it remained in service well after the war.

SIEMENS & HALSKE

The firm of *Telegraphen-Bau-Anstalt von Siemens & Halske* was formed in 1847. It became Siemens & Halske AG in 1897, at which time about half of its 5,500 employees worked in plants outside of Germany. Between the World Wars they were involved in the secret rearmament of Germany and participated in the "Nazification" of the economy. By 1939, Siemens & Halske had become the world's largest electrical company.

As one of the original founders and co-owners of Telefunken, Siemens & Halske did not have radio electronics as a major product area, but as Germany entered WWII, they were drawn into this activity. In a supporting role, they contributed in the production of the *Wassermann* and *Mammut* systems, as well as in the development and manufacturing of *Neptun*. For the *Heer's* rocket research and development program at Pennemünde, Siemens & Halske modified the basic system to become the *Neptun V2*, which served as a precision tracking instrument.

Panorama Tower

A system with great range was needed to track the British and American bomber formations as they crossed Germany. For this function, Dr. Theodor Schultes at GEMA, with the assistance of consultant Dr. Hans Hollmann who had originally suggested such a system, designed an experimental 2.4-m (125-MHz), 30-kW radar called *Panorama*. Built by Siemens & Halske in 1941, it was placed atop a concrete tower at Tremmen, a few kilometers south of Berlin. The antenna had 18 dipoles on a long horizontal support and produced a narrow vertical beam; this rotated at 6 rpm to sweep out 360-degrees of coverage to about 110 km.

Based on the operation of *Panorama*, Siemens & Halske improved this system, and renamed it *Jagdschloss* (Hunting Lodge). They added a second switchable operation to 150 kW at 1.2 m (250 MHz), increasing the range to near 200 km. The information from the receivers was sent via co-axial cable or a 50-cm link

Jagdschloss Antenna

from the tower to a central command center, where it was used to direct fighter aircraft. Hollmann's polar-coordinate (PPI) CRT was used in the display, the first German system with this device; it was also added to the *Panorama*.

The *Jagdschloss* could make small changes in frequency to accommodate the *Erstling* identification (IFF) interrogation, resulting in the enhanced display of fighters. The *Jagdschloss* entered service in late 1943, and about 80 systems were eventually built.

Jagdschloss PPI Display

The *Jagdwagen* was a mobile, single-frequency version; operating at 54 cm (560 MHz), it had a correspondingly smaller antenna system.

OTHER ORGANIZATIONS AND CONTRIBUTORS

While all of Germany's main wartime radar systems were built by GEMA, Telefunken, Lorenz, and Siemens & Halske, many other electronics firms, both public and private, were involved; some of these will be noted here. Also, several individuals who made significant contributions to radar development will be noted.

Luftwaffe Radar Laboratories

The *Flugfunk-Forschungsinstitut, Oberpfaffenhofen* (FFO, Air Radio Research Institute, at Oberpfaffenhofen), was founded by the *Luftwaffe* in 1937 for theoretical and experimental studies in *Funkmesstechnik* (radio-measurement techniques). In addition to their central operation in Oberpfaffenhofen (near Munich), they had research laboratories in Gaulting, Seeshaupt, and other locations.

The FFO had a close companion, the *Deutsche Versuchsanstalt für Luftfahrt* (DVL, German Laboratory for Aviation). DVL was a hardware-oriented facility; among other activities, during the war they examined captured or salvaged radar and related equipment and built jamming devices.

The FFO had a large, very competent staff working in high-frequency technologies, including signatures of aircraft, jamming of enemy signals, antenna development, and fusing devices. They designed and built a number of klystrons and magnetrons, and made significant advancements in crystal detectors for microwaves. A proximity fuse for *Flak* projectiles and "smart" bombs was developed (but not put into production). Their research on countermeasures is described in another section. As the war drew to a conclusion, all of the documentation and developmental products for advanced research were destroyed by the FFO; thus, few details are known of their accomplishments.

In the FFO's only contribution to full systems, they designed a radar that was code-named *Neptun* and intended for smaller night-fighters. This radar operated at around 1.7 m (175 MHz) with a pulse-power of 2 kW and a range of 100 m minimum and 5 km typical maximum (limited by ground clutter and dependent on altitude). About 150 of these systems were built by Siemens & Halske, with some installed on the World's first jet fighter, the Me-262.

Neptun Antennas on ME 262

Reich Postal Ministry

From the earliest days of radio, the *Reichspostzentralamt* (Realm Post Office Central Office) had been the German agency with oversight of all communication operations. This included the licensing of amateur radio operators. The technical work came under its Department VIII. At the start of 1937, this Department was expanded to become the *Forschungsanstalt der Deutschen Reichspost*, or, in short, the *Reichspostministerium* (RPM, Reich Postal Ministry).

Headed by Wilhelm Ohnesorge, the RPM was responsible for research and development in high-frequency technology, television engineering, wide-band cable transmission, and related metrology. In a 500,000-square meter facility at Miersdorf near Berlin, Director of Research Dr. Friedrich Wilhelm Banneitz had hundreds of engineers and scientists pursuing civil and military projects.

The RPM had a major expertise in radio propagation, and understood well the nature of signal interference; thus, this organization was given the primary responsibility for developing jamming techniques and hardware. Led by Dr. Friedrich Vilbig, a high-frequency specialist and deputy to Banneitz, a wide variety of systems evolved; these are specifically described in the section on Countermeasures.

There was an active involvement of the RPM in at least one full radar system. As the V-2 rockets were prepared to become operational in 1943, the Germans had no means of checking the actual accuracy of their

Funkmessgerät Development in Germany

inertial guidance. The *Neptun V2* supplied by Siemens & Halske gave information at launch, but there was no radar to track the descent as it reentered the atmosphere. RPM, assisted by Telefunken, hurriedly designed and built the *Elefant-Rüssel*, a radar with an intended capability of giving some level of long-distance tracking.

Patterned after the British Chain Home system, the transmitter – *Elefant* (Elephant) – had a floodlight emission over a 120-degree arc on the 10- to 15-m (30- to 20-MHz) band. The receiver – *Rüssel* (Elephant's Trunk) – was located about one km away and had a steerable antenna for direction finding. It had a range up to 800 km, but very poor accuracy. Several sets were built and installed along the Western coast, but, because of the lack of accuracy, they provided little useful information.

Private Firms

One of the largest private firms, AEG, is often included in the list of primary radar firms and will be particularly noted. Some of the smaller firms will also be mentioned.

AEG, officially *Allgemeine Elektrizitäts-Gesellachaft* (General Electricity Company), was founded in 1883 for building lighting products under a license from Thomas Edison. Concentrating on electrical equipment and supplies, it soon became one of the largest companies in the world. Supporting Germany's military efforts, between 1910 and 1918, it was a supplier of airplanes, building some 28 different models of tactical bombers and canon-armed fighters. After WWI, AEG entered the electronics field; among its products was the first practical magnetic tape recorder, the K1 Magnetophon, released in 1935.

As one of the original founders of Telefunken, AEG was a major supplier to this firm as well as Siemens & Halske (the other Telefunken founder) as they built radars and related electronics in the late 1930s and the following years. During WWII, the AEG factory in Nuremberg used hundreds of female inmates of the Keiserwald Concentration Camp as slave laborers assembling electronic and electrical goods.

Metox Société (Corporation), a manufacturer of high-quality radio receivers and vacuum tubes, was founded by Metox Gardin and operated in Paris (Occupied France).

Rohde & Schwartz was founded in 1933 by Drs. Lothar Rohde and Hermann Schwartz. Located in Munich, it was the premier manufacturer of radio instruments in Germany.

Blaupunkt, GmbH, goes back to 1923, when it was named Ideal and built radios. It used a blue dot as a symbol of quality and customers came to ask for *Blaupunkt* (Point of Blue) products. The Berlin firm adopted this name in 1938. During the war, it built transmitters and receivers.

Special Development Leaders

Many highly capable men made significant contributions to the development of radar in Germany. Brief bios of some of these were given in the preceding sections. Three others who deserve special attention are given here noted here.

Hans Erich Hollmann (1899-1960). The contributions of Hans Hollmann to radar in Germany were spread across many organizations. After receiving his doctorate in electrical engineering from Darmstadt Technische Hochschule in 1928, he joined the Heinrich Hertz Institute in Berlin and developed instruments for measuring the Heaviside-Kennelly layer (the ionosphere). He designed and built the first microwave communications link in Germany, and assisted Telefunken in setting up a microwave department. As earlier noted, he was closely involved with GEMA in their early radar developments. In 1933, he also became a lecturer at the Technical University in Berlin.

Hans Hollmann

During 1933-36, Hollmann wrote the first comprehensive treatise on microwaves, *Physik und Technik der ultrakurzen Wellen (Physics and Technique of Ultrashort Waves,* Springer 1938). Included in this two-volume set was the application of microwaves in measuring devices. For these, he asserted that crystal diodes would make the best detectors.

In this period Hollmann also worked on multi-cavity magnetrons (receiving a U.S. patent on his device in July 1938). He patented a number of types of CRTs, including the first "polar coordinate oscillograph" (a PPI); this was publicly disclosed in 1936, but not incorporated in German systems until several years later. In his career, Hollmann received about 300 patents, of which 75 were filed in the U.S.

When writing his book on microwaves, Hollmann was consulting through GEMA on projects for the *Kregsmarine*, and they censured portions of the text. Because of this, he would no longer work on *Kregsmarine* projects.

Hollmann later formed the *Labor für Hochfrequenztechniken und Electromedicine* (Laboratory for High-Frequency Techniques and Electromedicine). This was adjacent to Manfred von Ardenne's laboratory in Lichterfelden, Berlin, and the two of them collaborated in many endeavors. In 1939-40, as his last major radar project, he worked with GEMA and Siemens & Halske in designing the *Panorama*, an experimental, wide-coverage surveillance system at Tremmen near Berlin. Eventually, this was the first German radar using a PPI.

In 1942, his laboratory in Berlin was bombed, and he relocated to Thuringian. During the next years, Hollmann did fundamental research in microwave technologies and medical instrumentation. He was able to gain oversight of certain research laboratories in occupied countries and, through this, prevented the deportation of a number of scientists.

Wolfgang Martini

Wolfgang Martini (1891-1963). While attending the Gymnasium (high school) in his hometown of Lissa, Province of Posen, Wolfgang Martini had been a radio enthusiast. Upon graduating in 1910, he joined the Army as a Cadet, and his talents were such that he soon became a Lieutenant and made Company Officer of a Telegraph Battalion. During WWI, he had a number of leadership positions in radio operations, being promoted to First Lieutenant and then Captain. At the end of the war, he was a Radio Specialist in the Grand Headquarters and Commander of the Army Signals School at Namur.

After the Treaty of Versailles was signed in 1919, Martini was one of the few officers allowed to remain in the Army. For the next five years, he served as a signals instructor at several Army schools, then from 1924 to 1928, he was the Signals Staff Officer with a District Command. Between 1928 and 1933, he was promoted to Major and served as an Equipment Advisor with the Reichs Defense Ministry.

Upon the formation of the *Luftwaffe* in 1933, Martini transferred to this new service and shortly became the Chief for the Board of Radio Affairs. Between 1934 and 1937, he was a Lieutenant Colonel, then Colonel (1936), serving as Leader of the Department of Communication

Affairs. In 1938, he was promoted to Major General and named Chief of Communication Affairs of the *Luftwaffe*.

When Martini was first shown the *DeTe* (radar) in July 1938, he immediately recognized the great value to the military of this technology. As reflected in the references to him in the earlier sections, for the remainder of his military career he was the primary promoter of radar to the German High Command. Although not university educated, his grasp of this technology was instinctive and his involvement was perhaps the greatest impetus to the ultimate development of wartime radar in Germany.

With the start of the war in 1939, Martini continued to lead radar development, and in 1940 was promoted to Lieutenant General. In 1941, he was elevated to *General der Luftnachrichtentruppe* (General of the Air Signal Corps) and remained in this position until the end of the war in May 1945.

Johannes (Hans) E. Plendl (1900-1992). After earning his Dr.Ing. degree in electrical engineering from the Muenchen Technische Hochschule (Munich Technical University) in 1924, Plendl joined Telefunken in Berlin as a radio research engineer. His work included investigations of the Heaviside-Kennelly layer (the ionosphere) for determining limits of high-frequency communications; this led to an interest in sun-spots and other solar activities and their effect on the layer.

In the early 1930s, Plendl was responsible for developing the radio communications to be used in flights by planes and the *Hindenburg* Zeppelin between Germany and Rio de Janeiro, Brazil. For this, improvements of the Lorenz radio-navigation system were needed. While making the changes, he saw how the commercial system might be refined to provide special guidance for military aircraft.

Hans Plendl

Plendl joined the *Deutsche Versuchsanstalt für Luftfahrt* (German Laboratory for Aviation) in 1934 to develop his concept. In a few years he demonstrated a radio-navigational system called *Knickebein* that he developed with the cooperation of Dr. Rudolf Kühnhold of the *Kriegsmarine*. This was followed by more sophisticated navigational systems, the *X-Gerät* (Device) and then the *Y-Gerät*.

Funkmessgerät Development in Germany

In 1938, Plendl returned to Telefunken to participate in its new contract for developing the *Würzburg* radar. Concerned that the system could not distinguish between friendly and enemy aircraft, he developed an identification set (IFF) called *Stichling*, demonstrating it in late 1939.

Plendl's capabilities were noted by Field Marshal Hermann, and in late 1940 Göring brought him into the *Luftwaffe* Command to serve as the *Bevollmächtigter fuer Hochfrequenzforschung* (Plenipotentiary of High Frequency Research). In this capacity, he had oversight of all radar development for the *Luftwaffe*. Although still a civilian, Plendl functioned as a *Luftwaffe* officer, operating from a research laboratory at Rechlin. He also served as a technical advisor to Dr. Wernher von Braun, the Director of the Pennemünde Rocket Research Center.

Near the end of 1942, Plendl also received the honorary appointment of *Staatsrat* (State Councillor). Upon his request to the *Wehrmacht*, some 1,500 highly qualified persons were brought in from the battlefields and added to various research efforts.

Plendl's interest in sun spots and their effect on radio communcations had continued from his earlier studies. With *Luftwaffe* funding, he and Dr. Karl-Otto Kiepenheuer established the *Zentrales Institut für Solarforschung* (Central Institute for Solar Research), and in 1942 started a network of solar observatories all over Europe. They also used instruments on V-2 test rockets to measure the sun's spectrum from outside the atmpsphere. From this, Plendl and Kiepenheuer are often called the fathers of the science of space weather. (Dr. Stanley Hey in England had made some similar observations – see Chapter 4.)

As the war went on, maintenance of electronics became a significant problem since the equipment, although well built, started failing due to age. Plendl initiated an extensive training program for technicians, and also set up a repair facility in the Dachau Concentration Camp using scientists and engineers (mainly Jews) who were confined there. As a result of this latter effort, he was relieved of his positions in 1944.

OTHER SIGNIFICANT ACTIVITIES

A number of other wartime activities were significant to Germany's radar developments. Some of these are briefly described.

Air Defense System

At the start of the war, the *Luftwaffe* fighter organization was concerned mainly with defending targets in Northern France and the Lowlands. The bulk of aerial combats were taking place in the relatively small area over those countries and over the English Channel; a warning system, consisting of a coastal belt of *Freya* radars and visual observers, was adequate.

Josef Kammhuber

Colonel (later General) Josef Kammhuber was assigned to organize an enlarged Air Defense System in July 1940. He first set up *Helle Nachtjagd* (Illuminated Night Fighting), in which a *Freya* found a target at a distance and passed the azimuth to a *Würzburg*. The *Würzburg* then directed searchlights that illuminated the targets for the defending night-fighters. British pilots were startled by how quickly the lights found them.

The Air Defense System was continually upgraded. To improve the range, Telefunken developed the *Würzburg-Riese* and GEMA had increased the *Freya* dipoles to make the *Wassermann* and *Mammut*. Eventually, this was set up with a large number of cells, each containing a *Freya* or successor, two *Würzburg*, and a fighter, but no *Flak* weapon or searchlight (these had been ordered back to Germany to defend the cities). The cell was code-named *Himmelbett* (four-poster bed), and covered an area some 45 km wide and 30 km deep. These were set 50 to 75 km behind the early-warning radars on the coast. There was also a radar warning ship, the former freighter *Togo*, which operated in the Baltic Sea; this was equipped with a *Freya* for long-range detection and a *Würzburg-Riese* for tracking.

The Air Defense System also had *Y-Dienst* (Y-Service), receivers with directional antennas that detected emissions from approaching radars, IFFs, jammers, and radio communications. Through triangulation, a set of *Y-Dienst* stations could give the emitter's location. The *Y-Dienst* operation had the distinction of being un-jammable. The early-warning vessel *Togo* had a pair of these receivers.

An Observer Corps network with strategically located posts also supplied aircraft warning information, while in some instances patrolling aircraft shadowed the attacking aircraft. All of these systems were connected to three well-fortified Fighter Control Centers, called Kammhuber *Kinos* (cinemas), where filtered information was projected on a large wall-map for use by squadron-control officers; one such center was located at the Berlin Zoo.

Hans Hollmann had offered his map-like display (the PPI) in 1940; however, Hermann Göring personally dismissed it saying, "Such a comprehensive development is no longer worth while; the war is already as good as won." The *Jadgschloss* and *Panorama* radars with PPI displays finally came in late 1943.

A highly ingenious passive system for early-warning of bomber formations leaving England was put into operation in August 1943. Called the *Kleine Heidelberg Parasit* (KHP, Small Heidelberg Parasite), it used two adjacent *Y-Dienst* receivers set to a British Chain Home signal, one with the antenna aimed directly at the CH station and the other with a rotatable antenna to receive the weaker reflection from the bomber formation. By comparing the CRT display of pulses from the two receptions, the time delay, and thus the additional path difference, was found. From geometry, this described an ellipse along which the longer path must lie, and the focal points were the known transmitting and receiving positions. A line from the receiving location and at the azimuth angle crossed the ellipse at the actual target position.

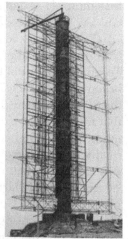

Wassermann and KHP Antennas

Five stations were built along the coast, with the receiving antennas either as back-to-back adjuncts to tall *Wassermann-S* antennas or simply installed on a *Wassermann-S* mast. This technique could give the range at up to 400 km with an accuracy of some 2 km and a bearing error as small as 1 degree, both depending on the geometry. There was, however, a major obstacle: as a general countermeasure, the CH radars randomly changed their frequency and pulse characteristics, each time requiring a restart of the monitoring system.

Since the *Wassermann* antenna and mount were used, it is likely that the special hardware for the KHP was built by the GEMA; the concept, however, possibly came from FFI analysts.

Identification Systems (IFF)

The first German identification system was the *Erstling* (First Born), developed by GEMA to work with the 2.4-m (125 MHz) *Freya*. In late 1939, Maj. Gen. Martini had ordered 3,000 sets for *Luftwaffe* aircraft. In this same time period, Dr. Hans Plendl, then with Telefunken, developed *Stichling* (a fish), an identification system that responded to the 50-cm (600-MHz) interrogation signal from the *Würzburg*. Upon learning of this, Martini demanded that a system responding to multiple wavelengths be developed; this, however, was not done at that time.

It might be noted that it was not until this identification problem came to light in late 1939 that the three military services were fully aware that multiple radar systems were being developed by the three firms: GEMA, Telefunken, and Lorenz.

The primary concern was identifying aircraft targeted by *Flak* batteries using the *Würzburg*. To meet this need, in early 1941 the Technical Bureau of the *Reichsluftfahrtmisisterium* (National Air Ministry) brought out the *Zwilling* (Twin) identification system, a modification of the *Stichling*. About 10,000 of these units were built before it was realized that there were major operational flaws. It was not until July 1942 that a modified *Erstling*, compatible with both the *Würzburg* and *Freya*, was finally available from GEMA. The unusable *Zwilling* sets were mainly dismantled to obtain parts for modifying *Ersting* sets.

In spite of Martini's continuing efforts, identification remained a major problem in Germany throughout the war. It was previously noted that a Ju-88 night-fighter carrying an *Ersting* showed up in Scotland in May 1943, providing the Allies full information on this set. A good replacement, the *Neuling* (Newcomer), was designed but was never deployed.

Alhough GEMA started deliveries in 1939 of *Seetakt* shipboard radars for the *Kresgmarine*, it was not until 1942 that an associated identification set was developed. Called *Wespe* (Wasp), this was initially only to allow the 80-cm (370-MHz) *Seetakt* on large vessels to recognize similar vessels (mainly battleships). The *Wespe* was gradually expanded, however, for other *Seetakt* models and for use on smaller ships as well as submarines.

Microwave Radar

Although early research in Germany on *Funkmessgerät* had begun in the microwave portion of the spectrum, no practical systems had been developed. The only available RF power sources were the Barkhausen-Kruz and klystron tubes and elementary magnetrons; the first two could not produce power at sufficient levels and magnetron, while continuing to improve in power level, was too unstable in frequency. In addition, there was a general belief (without experimental verification) in the scientific community that microwaves impinging on large surfaces would result in "specular" reflection – like from a mirror – and only a very small fraction would be returned to the source. Consequently, by the start of the war further research for the military at these wavelengths had essentially been discontinued.

Funkmessgerät Development in Germany

On the night of February 2, 1943, a British Stirling bomber was shot down near Rotterdam, Holland. In the wreckage, the Germans found an H2S radar, a ground-mapping system that had just been placed into use for precision bombing. Although the equipment was badly damaged, it contained an intact 10-cm (3-GHz) resonant-cavity magnetron, the most secret item then in the British arsenal. (The cavity actually resonated at 9.4 cm, but this was commonly referred to as 10 cm.)

The operation of the *Rotterdam-Gerät* (Device) – the name given to the magnetron – was readily discerned, and, in typical German fashion, a large commission was formed to assess its applications. Not included in the commission were Drs. Hollmann and Kühnhold, the two scientists who likely knew more about magnetrons and their potential applications than anyone else in Germany. (It might be noted that six years earlier, well before the British invention, the Japanese had developed a similar magnetron, but never shared this with the Germans.)

An additional surprise from the bomber wreckage was some co-axial cable. Like earlier high-frequency cables in America and Great Britain, the Germans had used a plastic insulator (polyisobutylene) but resorted to a hollow, almost inflexible cable with ceramic spacers for the highest frequencies. The chemical firm IG Farben identified the plastic (polyethylene) in the newly found cable to be the same as their Lupolen H, a previously overlooked material with excellent high-frequency characteristics. Co-axial cables using the Farben material were immediately put into production.

Telefunken was given the mission to reverse-engineer the damaged H2S equipment. Sanitas GmbH, a firm that built medical X-ray equipment, copied the magnetron and the vital T-R switch; the 10-cm magnetron was designated LMS-10. The crystal diode used as the detector was unlike any used in German radio equipment, and the *Physikalisch-technische Reichsanstalt* (Physical-technical Institute for Realm – similar to the U.S. Bureau of Standards) – took on this project. An equivalent diode evolved, but, as was earlier found in the U.S. and Great Britain, it was very difficult to manufacture reliable devices. A 10-cm radar, called *Rotterdam*, was demonstrated by Telefunken in mid-1943. Eventually, six sets were built; however, these were essentially useless since the *Luftwaffe* had discontinued their large-scale bombing missions.

Dr. Adolf-Echard Hoffmann-Heyden of Telefunken was placed in charge of developing a gun-laying radar for *Flak* applications using the LMS-10 magnetron. Near the end of 1943, an experimental set with a 30-cm parabolic reflector detected an aircraft at 8 km. Changing to a 3-m *Mannheim* reflector increased the detection range to 30 km. A 10-cm gun-laying set code-named *Marbach* was produced in limited quantities and used in *Flak* batteries around a number of large industrial cities. The most important characteristic of this microwave radar was its relative immunity to Window – the chaff used by the British as a countermeasure against *Würzburg* and similar 50-cm radars; thus, microwave radars were then recommended for extensive use.

Attempts were made to develop several other 10-cm systems, but none made it into mass production. One was *Jagdschloss Z*, a panorama-type experimental set with 100-kW pulse-power built by Siemens & Halske. *Klumbach* was a similar set but with only 15-kW pulse-power and using a cylindrical parabolic reflector to produce a very narrow beam; when used with *Marbach*, the combined fire-control system was called *Egerland*.

Klumbach 10-cm Radar

Although Telefunken had not been involved with previous radars of any type for fighter aircraft, in 1944 they started the conversion of a *Marbach* 10-cm set for this application. Downed American and British planes were scavenged for radar components; of special interest were the swiveling mechanisms used to scan the beam over the search area. An airborne set code-named *Berlin* was completed in January 1945, and about 40 sets were built and placed on night-fighter aircraft in March. A few sets, code named *Berlin-S*, were also built for shipboard surveillance.

Berlin 10-cm Radar

Near the end of 1943, the Germans also salvaged an American AN/APS-15 and British H2X, both blind-bombing radars containing 3-cm magnetrons. The American set was duplicated as the *Meddo* (the town in Holland where it was found) and the British H2X as the *Rotterdam-X* (the letter possibly indicating that the Germans were aware of the "X-band" designation used by the Allies), but they were never produced. An important application of both the 10-cm and 3-cm equipment was in the German development of countermeasures, particularly warning receivers.

Radar Countermeasures

From the start of serious development of *Funkmessgerät* in Germany, it was assumed that other countries would do likewise; thus,

Funkmessgerät Development in Germany

simultaneous work on countermeasures was performed. Identification techniques (IFF) were closely tied to the radars themselves and have been basically described previously. Other radar countermeasures involved active devices such as jammers and deceptive emitters, passive measures included chaff, camouflage, and warning devices. The German engineers and scientists were highly ingenious in developing countermeasures and their accomplishments were many; only a few, however, will be noted in this section.

It should be mentioned that countermeasures were considered by the top German military leaders as defensive techniques. Since Hitler's emphasis was on offensive operations, development of countermeasures was not, at least initially, given very high priority.

Before the war, aircraft flights between Berlin and London passed over Orfordness, a peninsular along the North Sea coast of Great Britain. In early summer of 1935, reports were made in Germany of large radio masts being built on Orfordness. Subsequent checks with radio receivers indicated pulsed signals at 7.5-15 m (20-40 MHz) in the HF band; since frequencies in this band were known to be used in Great Britain for monitoring the Heaviside-Kennelly layer, the radiations were dismissed as scientific research. In fact, this was the early development of what would become known as the Chain Home, Great Britain's first RDF (radar) system.

As time went by, similar masts were reported at a number of other locations along the North Sea coast. Wolfgang Martini, now a Major General and Chief of Communication Affairs of the *Luftwaffe*, became suspicious that these might be used in a British development of *Funkmesssgerät*. In 1939, he arranged for the *Graf Zeppelin II* (LZ-130), then operated by the *Luftwaffe*, to carry him and a small team of radio specialists with a variety of receivers along the east coast of England. On May 9, when the flight was made, a dense fog covered the region and the airship became lost; no suspicious signals were received.

On August 2, just before the start of the war, a much better prepared *Spionagefahrt* (espionage trip) started. This time Martini had with him 28 *Luftwaffe* radio specialists headed by Dr. Ernst Breuning, and a broad assortment of radio receivers and wire recorders. Under the pretense of testing radio direction-finding equipment, the airship remained aloft for 48 hours. Flying along England's east coast, the motors were often cut, allowing the airship to drift while those on board checked radio signals and photographed the masts. However, the monitoring emphasized the VHF band, then used by German radars; since the Chain Home operated in the HF band, no signals of interest were found.

Breuning's equipment did include HF receivers, but during much of this time an impulsive tranamitter radiating broadly in this band was also being used aboard the airship for researching "radio-weather" – the best wavelengths for communications – a continuous concern of Martini. These impulses severely disturbed the highly sensitive HF receivers in the 10- to 12-m band, the very waveband in which the Home Chain systems were operating at that time. As a result of this, Martini concluded that the British did not yet have a operational radar; thus, immediate attention was not given to countermeasures such as jammers and warning receivers.

Jamming. At the start of the war in September 1939, the British CH system began continuous operation. With the fall of France in June 1940, the Lutwaff's *Signalisiert Nachrichtendienst* (SND, Signals Intelligence Service) set up radio monitoring posts along the coast, with a central station at Calais. The DVL (Aviation Research Institute) was tasked to interpret the signals; these included pulse transmissions at 1.5 m (200 MHz), coming from Chain Home Low (CHL) sets, recemtly added at some stations to detect low-flying aircraft.

In a short time, a team led by Dr. Paul von Handel determined the general characteristics of CH operations, including the map locations of all existing stations. It was concluded that the different CH stations were connected with each other through their pulse modulation. With the possibility that interfering with one station might affect all of the others, the DVL built a station to produce barrage jamming centered at 25 MHz (12 m).

Barrage jamming was used with brief success in July 1940, but the British soon used the counter-countermeasure of changing the frequency (each station had four switchable frequencies between 20 and 55 MHz). In September, the jamming was changed to a technique of interfering on a pulse-by-pulse basis; the British then applied another counter-countermeasure of changing the pulse width and PRF.

To overcome the Chain Home counter-countermeasure of switching frequencies, five 1-kW jamming stations were set up along the coast to sweep the 22- to 28-MHz range. One of the stations had another transmitter operating in a higher band. Called *Breslan*, this used a motor-driven frequency sweep and was modulated to synchronize with the CH pulses. This system was used with modest success for some time.

As a countermeasure for the second signal (the CHL), GEMA developed *Olga I* operating between 167-200 MHz (1.8-1.5 m) with a power of 300 W. It was soon supplemented by a better set, *Olga II*, built

Funkmessgerät Development in Germany

by Blaupunkt GmbH; this set covered 167-215 MHz (1.8-1.4 m with 450 W. Both type sets had a selectable pulse frequency of 500, 700, or 900 Hz.

As previously noted, the Reich Postal Ministry (RPM) had the primary responsibility for jamming equipment. Under efforts led by Dr. Friedrich Vilbig, a wide variety of systems evolved; a few of these will be described.

During the mass bombing by the *Luftwaffe* of cities in Great Britain, the RAF fighter planes were guided to the bombers using voice communications. The *Caruso* was two 30-kW jamming transmitters set up on the coast and sweeping 100-110 and 110-120 MHz, the VHF bands used for ground-to-air communications. *Caruso*, however, did not go into full operation until after the bombing raids stopped.

When the British and Americans started large-scale bombing on Germany, *Karl* jamming transmitters were built to disrupt Allied air-to-air communications. These sets operated in four ranges of the 90- to 250-MHz band and used a saw-tooth modulation of 150 kHz. Installed in pairs, *Karl I* had a power of 0.4 kW and *Karl II* had 2 kW. About 150 of these were eventually put into widespread operation.

Karl Jammers

GEE was a hyperbolic navigation system used by British bombers. With transmitters on the English coast and operating in the 20-86 MHz range, it covered Holland and the Rhur Valley industrial area of Germany. Starting in March 1942, GEE was used to guide the massive 1,000-plane RAF bombing attacks. An aircraft with a GEE set was shot down and the receiver sent to the DVL where the GEE function was discerned; the FFO then determined a countermeasure function.

In late 1942, the RPM set up the *Heinrich* network as a GEE countermeasure. This had 1.5-kW, tunable jamming transmitters, modulated with 200-kHz signals and a 100-Hz hum. *Heinrich I* transmitters operated at 20-80 MHz, and *Heinrich II* at 50-75 MHz. Two other jammers sometimes used against GEE were *Feuerstein* (Flint) at 20-200 MHz supplied by Telefunken and *Feuerhilfe* (Fire Helper) at 50-90 MHz from Köthen. As a counter-countermeasure, the British adopted the practice of switching GEE frequencies on a pre-determined schedule during an attack, requiring the jammers to be retuned. By the end of the war, there were about 280 stations for jamming GEE.

To better use the jamming resources, in late summer 1944 several large groups of transmitters and warning receivers were set up in

Störderfer (jamming villages). These were particularly used to countermeasure the GEE navigation system. For these villages, Siemens & Halske developed the *Feuerzange* (Fire Pliers) jammer with the huge power of 1,000 kW at frequencies between 42-52 MHz. The first unit was installed at Feldberg village in November 1944, but it was destroyed in an air raid three months later.

The *Kobold* shipboard jamming transmitter was designed to counter the British 200-MHz ASV Mk II airborne radar. The jammer operated between 165 and 250 MHz, modulated with a 500-Hz signal. Testing indicated that the *Kobold* was only effective when the ASV was at a considerable distance from the vessel; thus, the development was terminated.

After equipment (the H2S) containing the *Rotterdam-Gerät* (the British 10-cm magnetron) was found in February 1943, there was an urgent need to have jamming for this and subsequent microwave radars. Siemens was tasked for this development, and within a few months came up with the *Roderich*. This was a highly unsatisfactory device with a klystron producing only 4-W power. Used for limited jamming over a small area, it was aimed at the aircraft using a *Korfu* receiver system (described later). It was of no value in jamming the new 10-cm H2S/ASV radar. Little more was done on microwave jamming for over a year.

After Germany built high-power magnetrons, Siemens used one producing 30 kW in the *Roland* jammer. This was modulated at 100 kHz with a pulse of 300 W; using a large horn antenna with a 30-degree beam, it could jam the 10-cm systems at up to 30-km range. The *Roland* went into service in February 1945.

Another microwave jammer, and the last such countermeasure device built during the war, was the *Feuerball* (Fire Ball). This used a newly developed klystron tunable over 8.6-9.4 cm. Having a power of about 80 W and with noise modulation, the device had a 70-degree funnel that provided jamming to 30-km range. It was guided by a *Korfu* receiver system and was first used was in March 1945.

Warning Receivers. A major function of the *Luftwaffe* SND (Signals Intelligence Service) was to find new sources of emission and, through this, identify new threats. This led to special receivers being built to warn of the presence of these threats. Because they received the signal directly (rather than via a reflection), their detection range was at least twice that of the sending radar. Many types of these warning receivers were built; some of those most commonly used are briefly described.

In 1940, the British Navy started testing 85-MHz (3.5-m) shipboard radars (eventually designated Types 280 and 281, the latter becoming

Funkmessgerät Development in Germany

their most-used naval radar of the war). As a countermeasure, the *Kriegsmarine's Kommunikationen Forschungsinstitut* (Communications Research Institute) conceptually designed a warning receiver covering 60-160 MHz (1.88-5.00 m). Designated *Metox R600*, the sets were developed and about 1,000 manufactured by the firm Metox in Paris. Installations on German vessels started in early 1941. A number of different directional antennas were used, including a phased array for battleships.

Metox R600

In mid-1962, the British started using a very effective operation involving an airborne searchlight (the Leigh Light) combined with a 200-MHz (1.5-m) detection radar (ASV Mk II) carried on the same plane. Again, a warning receiver built by Metox was used as a countermeasure. The *Metox R203* swept through 80-330 MHz (3.8-0.9 m) and had a Magic Eye tuning indicator. These sets were installed on U-boats and used when they were surfaced.

Rohde & Schwarz, a highly regarded radio instrumentation firm in Munich, developed in 1943 the *Samos RS1/5*, the most advanced receiver of its time. Covering 90-470 MHz (3.3-0.66 m) in four bands and with high sensitivity, this was intended as a replacement for the *Metox R203* warning receivers in submarines. Near the end of 1943, Lorenz built these receivers for the *Luftwaffe*.

At about the same time, Rohde & Schwarz also developed the *Fanö*, the first receiver having coverage in the microwave region. This was in two models, one covering 38-75 cm (790-400 MHz) and the other 19-38 cm (1.6-0.79 GHz).

The Allied Forces began the invasion of Europe in July 1942 at Sicily. For the next 18 months as they moved into southern Italy and upward toward Rome, German reconnaissance aircraft carrying *Metox, Samos, Fanö*, and other receivers flew around the region to identify types and locations of American and British radars.

Upon the discovery in February 1943 that the Allies had microwave radar (the H2S/ASV Mk III) operating at 10 cm (actually 9.4 cm), the requirement for warning receivers at this wavelength was established. Obtaining a reliable detector was the most difficult problem, but this was finally solved by Telefunken with a crystal diode based on one in equipment on the downed aircraft at Rotterdam. *Naxos*, a warning receiver for 8-12 cm, was developed before the end of March. When mounted on the *Würzburg* dish, this was called the *Naxbugh*.

Because of interservice secrecy and rivalry, information concerning the British 10-cm radar and the *Naxos* receiver was not made available to the *Kriegsmarine* until later in the year.

During the second half of 1943, use of *Metox* and the new *Samos* super-heterodyne receivers on submarines had been suspended; there were suspicions that radiations from local oscillators on these sets were being detected by Allied aircraft. (A captured British pilot had planted this false story.) Later in the year, checks using a *Naxos* showed that the aircraft were using new 10-cm radars (the ASV Mk III). In December, some *Naxos* sets with parabolic-reflector antennas were installed on submarines; the reflectors had to be taken down when the submarine submerged. Although the range was less than 8 km, it was a welcome addition.

A better microwave warning receiver, the *Korfu 912*, was developed by Blaupunkt. The *Korfu 912* was a super-heterodyne set using a magnetron as the local oscillator and covering 9-12 cm. In late 1943, the Germans also began instaling these on submarines.

Korfu Warning Station

For the *Luftwaffe*, a network of 15 warning stations using *Korfu* receivers was set up in the fall of 1943; these fed information to the Fighter Control Center in Berlin. With a rotating horn antenna, the stations could locate H2S and other 10-cm transmitters at a range up to 100 km and a bearing accuracy of 4 degrees.

In 1944, other *Korfu* receivers were developed, covering the bands of new shorter-wavelength magnetrons being used on U.S. radars. The new warning receivers included the *Tunis*, covering the 3-cm band; this entered service in May. The *Athos* was another 3-cm set, and the final warning receiver developed during the war. This had very advanced electronics and included a PPI CRT display.

Other Countermeasures. Conditioning or desensitizing has an important role in countermeasures. Martini was a strong advocate of this and used it in the breakout of three German warships that were trapped in the harbor of Brest, France. As an expert in radio communications, he was very aware of atmospheric disturbance patterns. At the same time over several days, he had jamming transmitters covering the frequencies of British naval radars slowly brought up, leading the radar operators to

Funkmessgerät Development in Germany

become accustomed to this as a periodic natural disturbance. On February 12, 1942, while the radars were unsuspectedly being jammed, the ships escaped without being noticed.

All of the main types of radars in Germany were initially designed to operate on a single frequency, optimized for their function. On February 27, 1942, the British conducted an assault on a *Würzburg-Riese* installation on the coast near Bruneval, France, capturing the major parts of this system. The Germans then realized that the British would learn of the single-frequency limit and use this to more easily jam the radars.

As a counter-countermeasure, the major radars were modified to have quickly changeable frequencies (frequency diversity). The single original and modified ranges were as follows: *Würzburg*, 560 MHz (492-555 MHz); *Seetakt*, 368 MHz (250-430 MHz); *Freya*, 125 MHz (120-130 MHz). In an urgently implemented effort, the modifications were essentially all made by September.

Window was light-weight aluminum strips to be dropped in a cloud from British aircraft as a countermeasure to *Würzburg* radars. Telefunken also tested this type of countermeasure, calling it *Düppel* (after the town where testing took place). The *Luftwaffe* found it so effective that its use was prohibited for fear that it would be copied by the Allies. The British first used Window in July 1943, blinding the *Flak* radars defending Hamburg during a night raid. At the same time, the *Freya* radars were disabled by jammers aboard the bombers. The effectiveness of Window is indicated by the loss of only 3 of the 800 bombers.

As a counter-countermeasure to Window, a device called *Nürzburg* was added to the *Würzburg*; it used the Doppler effect to allow the radar to distinguish between the slow-falling chaff and the high-speed aircraft. The *Nürzburg* became available in September 1943. A better Window counter-countermeasure was the *K-Laus* signal processor provided by Telefunken in the summer of 1944. With this, slowly changing signals cancelled themselves, leaving the desired aircraft targets.

A mention must be made of the highly advanced radar countermeasures research that was conducted at the FFO, the Air Radio Research Institute at Oberpfaffenhofen. Two of these projects were in what would later be called stealth technology. One involved sensing the frequency of impinging radar signals, amplifying them and inverting the phase, and then reradiating it to cancel out the normally reflected return. Some work using klystrons was apparently done in this to countermeasure microwave radar.

A second FFO project involved improving the conductivity of exhaust gases from jet aircraft. This increase in their negative charge was expected to induce a positive charge on the front edges, leading to an electromagnetic bending-path around areas that normally have a high radar reflectivity. There is no indication that either of these research activities led to actual stealth techniques.

RADAR MAINTENANCE

Historically, German-built equipment has been of the highest quality. This was true of their radar hardware; the electronic and mechanical designs included very high reliability. Little field maintenance was required; the sets were built in modular form, and any failed module was usually replaced and the faulty item returned to the factory. Consequently, there was essentially no training of maintenance technicians in the German military. Also, due to the high level of security, technical manuals were not allowed to be kept with the equipment.

As the war continued and equipment aged, maintenance began to be a problem. This was particularly true as factory personnel were pressed into military service and sent to the front with a gun to defend the country. One partial solution involved inducting teen-aged boys, mainly radio amateurs, into *Flak* battalions where they divided the day between regular studies and assisting in equipment maintenance; these were called *Flakhelfer* (AA Gun Aid).

Dr. Hans Plendl, then Director of High Frequency Research in the *Luftwaffe*, addressed the maintenance problem and in 1943 enacted two other partial solutions. The first was in establishing a year-long school giving exceptionally capable young men the foundations of radio and radar theory and maintenance; passing a very difficult examination was required for entrance. (This appears to have been very similar to the Electronics Training Program given by the U.S. Navy; see Chapter 3.) Plendl's son himself was in the *Flakhelfer* and also went through the training program in Germany.

Plendel's second solution involved using the many highly qualified engineers and scientists who had been arrested by the Gestapo and sent to concentration camps. A special repair laboratory was set up at the Dachau camp with a number of inmates, mainly Jews, assigned. It was led by inmate Dr. Hans Mayer, formerly the Director of Research at Siemens & Halske who had been imprisoned for "incautious talk." This ultimately came to the attention of Hitler; in March 1944, Plendl was dismissed from his position because he had "passed secret materials to non-Germans." (Race Laws passed in 1935 deprived German Jews of their rights of citizenship, with the status of "subjects" in Hitler's Reich.)

CLOSURE

Near the end of the war, GEMA led the German radar work, growing to over 6,000 employees. The central facilities were largely destroyed by Allies' bombings. Following the war, they were taken over by the Russians (they were in the sector of Berlin occupied by the USSR), and soon GEMA ceased to exist. A firm started in East Germany in about 1947 with this name had no connection with the original GEMA.

GEMA Central Facility - Berlin

After the war, Lorenz lost its operations in Eastern Germany to the Russians. Prohibited by the U.S. from building their primary products of radio and aircraft navigational systems, Lorenz almost immediately started manufacturing parts and components in their factories in Berlin and Western Germany, obtaining permission through their American ownership. In a short time, they turned to consumer products and railway control and safety equipment.

Dr. Hans Hollmann's laboratory in Thuringian was in the Soviet-occupied region. Although prohibited from continuing his microwave work, during 1946 he was a consultant to the local Siemens factory and also taught at the University of Jena. In 1947, he came to the United States under Project Paperclip to serve as a scientist for the Naval Air Missile Test Center, Point Mugu, California. Later, he was Director of Research at Dresser Industries. As noted in a tribute on his 60th birthday, he was remembered not only for his technical contributions but also for his "humor and calmness, unswerving sense of justice and right, and wartime assistance in taking care of colleagues."

General Wolfgang Martini, like other officers at his level, was held by the U.S., then the British until 1947; however, no charges were placed against him for his service to the *Luftwaffe*. After his release, he returned to high positions with the new German Air Force and then NATO. In 1962, he was awarded an Honorary Doctorate in Engineering. In these years, he also established relationships with radar pioneers in other countries. One of these, Sir Robert Watson-Watt, included the following in his 1959 autobiography, *The Pulse of Radar*:

I have a very dear postwar friend in General Wolfgang Martini, a shy, modest, charming, and very perfect gentleman . . . His many claims on my affectionate respect include his failure to endear himself to Göring, from whom the qualities I have just tried to summarize may have concealed General Martini's very high technical competence, wisdom, and resource.

Dr. Hans Plendl was also invited to come to the United States under Project Paperclip. He dedicated himself to basic research in solid-state physics and infrared spectroscopy at the U. S. Air Force Cambridge Research Laboratory, then returned to Germany upon his retirement in 1970. R. V. Jones, the British scientist who had worked on the other side of the channel to analyze and jam Plendl's navigation beams, became a good friend. Jones's book, *Most Secret War*, provides details of Plendl's wartime work from another perspective.

At the close of the war, Russian search teams operated in Germany to identify and "requisition" equipment, materiel, intellectual property, and personnel useful to the Soviet missile, radar, atomic bomb, and other projects. A number of German scientists willingly went to the Soviet Union and assisted in their post-war weapon programs. Included were Manfred von Ardenne; Dr. Gustav Hertz, Nobel laureate and Director of Siemens Research Laboratory; Dr. Peter Thiessen, Director of the Kaiser Wilhelm Institute; and Dr. Max Volmer, Director of the Physical Chemistry Institute at the Berlin Technische Hochschule.

Bibliography and References for Chapter 5

"Air Scientific Intelligence Interim Report, Heidelberg" IIE/79/22, Nov. 11, 1944

Bauer, Arthur O.; "Some Aspects of German Airborne Radar Technology, 1942 to 1945," DEHS Autumn Symposium, Sheivenham, Oct. 2006; http://www.cdcandt.org/airborne_radar.htp

Brown, Louis; *A Radar History of World War II: Technical and Military Imperative,* Inst. of Physics Pub., 1999

Goebel, Gregory V.; "The Wizard War: WW2 & The Origins of Radar," a book-length document; http://www.vectorsite.net/ttwiz.html

Guerlac, Henry E; *Radar in World War II*, vol. 8 in the series *The History of Modern Physics 1800-1950*, American Inst. of Physics, 1987

Heil, Oskar E., and Agnesa Arsenjewa-Heil; "On a New Method for Producing Short, Undamped Electromagnetic Waves of High Intensity," *Zeitschrift fur Physik*, vol. 95, p. 752, 1935 (in German)

Hentschel, Klaus (editor), and Ann M. Hentschel (translator); *Physics and National Socialism: An Anthology of Primary Sources*, Birkhäuser, 1996

Hepcke, Gerhard; "The Radar War, 1930-1945," translated by Hannah Liebmann; M. Holliman; http://www.radarwar.org/radarwar.pdf

Hoffman-Heyden, A. E.; "German WWII anti-jamming techniques," in *Radar Development to 1945*, edited by Russell Burns, Peter Peregrinus Ltd, 1988

Hoffmann-Heyden, A-E.; *Die Funkmessgeräte der Deutschen Flakartillerie* (*The Radars of the German anti-aircraft artillery*), in German, Verkehrs und Wirtschafts-Verlig, 1957

Hollmann, Hans E., *Physik und Technik der ultrakurzen Wellen, Band 1 und 2, [Physics and Technique of Ultrashort Waves, Book 1 and 2]*, Springer, 1938 (in German)

Hollmann, Martin; *Hans Erich Hollmann, Pioneer and Father of Microwave Technology*, Aircraft Designs, Inc., 2003
http://www.radarworld.org/flightnav.pdf

Hollmann, Martin; Son of Hans E. Hollmann, personal communications

Hülsmeyer, Christian; "Hertzian-Wave projecting and receiving apparatus adapted to indicate or give warning of the presence of a metallic body, such as ship or train, in the line of projection of such a wave," British Patent issued Sept. 2, 1904

Jones, R. V.; *Most Secret War, British Scientific Intelligence 1939-1945*, Hamish Hamilton, 1979

Kroge, Harry von; *GEMA: Birthplace of German Radar and Sonar*, translated by Louis Brown, Inst. of Physics Publishing, 2000

Kummritz, H. "German radar development up to 1945," in *Radar Development to 1945*, edited by Russell Burns, Peter Peregrinus Ltd, 1988

Kummritz, Herbert; "On the Development of Radar Technologies in Germany up to 1945," in *Tracking the History of Radar*, ed. by Oskar Blumtritt et al, IEEE-Rutgers, 1994

Muller, G. and H. Bosse; "German primary radar for airborne and ground-based surveillance," in *Radar Development to 1945*, edited by Russell Burns, Peter Peregrinus Ltd, 1988

Muller, Werner; *Ground Radar Systems of the Luftwaffe, 1939-1945*, Schiffer Publishing, 1998

Pendel, Hans S.; Son of Hans N. Pendel, personal communications

"Report On *Flugfunk Forschungsinstitut* Oberpfaffenhofen (F.F.O. Establishments)," Combined Intelligence Objectives Sub-Committee Report Number 156, 1945

Schulze, E.; "German anti-chaff measures," in *Radar Development to 1945*, edited by Russell Burns, Peter Peregrinus Ltd, 1988

Sieche, Erwin F.; "German Naval Radar," 1999; http://www.warships1.com/Weapons/WRGER_01.htp

Suchenwirth, Richard, and Harry R. Fletcher; *The Development of the German Air Force, 1919-1939*, Ayer Publishing, 1970

Swords, Seán S.; *Technical History of the Beginnings of Radar*, Peter Peregrinus Ltd., 1986

Trenkle, Fritz; *Die deutschen Funkmessverfähren bis 1945 (The German radar procedures until 1945)*, in German, Motorbuch Verlag, 1978

Trinkle, Fritz; "History of Radio Flight Navigation Systems." translated by M. Hollmann and Peter Aichner; http://www.radarworld.org/flightnav.pdf

Chapter 6
RADIO-LOCATION DEVELOPMENT IN THE SOVIET UNION

World War II started in the USSR with the June 1941 invasion by Germany. Although the Soviet Union had outstanding scientists and engineers, began research on what would later become radar (*radiolokatsiya*) as soon as anyone else, and made good progress with early magnetron development, they entered the war well behind Great Britain, the United States, and Germany in this capability. To understand this, it is necessary to examine the characteristics of this nation as existing during this period.

HISTORICAL BACKGROUND

Russia basically became a European state under Czar Peter I ("Peter the Great"), who ruled between 1682 and 1725. By his time, Russia was geographically the largest nation in the world, stretching from the Baltic Sea to the Pacific Ocean. It was, however, a backward state, both economically and culturally, as compared with Western Europe. In an effort to improve the science culture, the Russians developed higher-learning institutes, and prominent scientists and other intellectuals were enticed from Europe to staff them. St. Petersburg was then the capital, and the Academic Gymnasium established there set the pattern for future universities throughout Russia.

Emperors following Peter the Great had little of his thirst for developing the backward nation. The universities were outstanding but were primarily for the elite and wealthy, leading to small pockets of outstanding writers, composers, and performing artists, but they had minimal influence on Russia as a whole. The country had little commerce and essentially no industry, and illiterate peasants (actually serfs – a type of slave – until the 1860s) formed the vast majority of the population.

The then-reigning czar was assassinated in 1881, and his successor instigated rigid censorship and police control. The oppression of Jews was particularly severe, with great numbers killed. An intense industrialization effort produced factories in Moscow, St. Petersburg, and Kharkov (in Ukraine). The working and living conditions, however, were miserable, and the socialist philosophy of the German Karl Marx was eagerly accepted by the workers.

The Revolution and the USSR

During the first part of the 20th century, Russia was in continuous strife but remained a constitutional monarchy. The expanding Japanese empire invaded the far-east Port Arthur, resulting in a short-lived war in 1905, and the humiliating defeat of the Russian fleet. Vladimir Illich Ulyanov (later known as Lenin), although in exile much of the time, led the drive for a Socialist government. Strikes and riots in the industrial cities followed.

The czar gave the people a representative assembly – the *Duma*. This was eventually able to enact certain progressive and reform measures, but was all brought to an end in 1914 with the start of World War I. Russia immediately suffered defeats on the Western front that resulted in massive troop desertions; large-scale dissatisfaction with the government spread across the country. Because the name St. Petersburg sounded Germanic, in 1916 the city was renamed Petrograd.

In March 1917, the czar was deposed and a provisional government temporarily took control. The capital was moved from Petrograd to Moscow. This government was overthrown in the October Revolution of 1917, and Lenin was elected the chairman of the Soviet Council of People's Commissars.

After a series of further revolutions, in December 1924, the leaders of the Russian Communist Party established the Union of Soviet Socialist Republics (USSR or Soviet Union). Petrograd (formerly St. Petersburg) was renamed Leningrad. Josif Vissarionovich Dzhugashvili, who took on the name of Stalin during the revolution, became the undisputed master of the Soviet Union by the late 1920s.

The USSR military forces were the Red Army, the Red Fleet, and the Soviet Air Fleet. The *Raboche-Krest'yanskaya Krasnaya Armiya* (RKKA, the Workers' and Peasants' Red Army), first organized early in the revolution, became the official army of the USSR in 1922. The *Raboche-Krest'yansky Krasny Flot* (RKKF, the Workers' and Peasants' Red Fleet) was formed from the Russian Imperial Navy in

Josif Stalin

1918. For the first decade, it existed in a less than service-ready state, but in the 1930s as the industrialization of the Soviet Union proceeded, it was modernized and greatly expanded. Also, the Imperial Russian Air Force was formed into the Workers' and Peasants' Air Fleet. In 1924, this became the *Voenno-Vozdushnye Sily* (VVS, Soviet Air Forces). Like other elements of the Soviet Military, it expanded greatly in the 1930s, and in 1939 was renanamed the VVS of the Red Army (VVS-RKKA).

The years from 1929 to 1939 were a period of massive industrialization and collectivized agriculture, accompanied by internal struggles as Stalin exercised virtually unrestrained power. In a speech, he had said that "we are 50 to 100 years behind the advanced countries and we must make good this distance in 10 years." A program of Five-Year Plans (*Pyatiletka*) was put into place. On the positive side, education and employment opportunities were open to the vast population. Within the first years, electrical power capacity greatly increased, some 1,500 new factories were built, and production of steel and other basic materials quadrupled. Many additional research institutions were formed, all funded by the government and directed to improve the industrial and military status of the USSR.

On the negative side, the most horrible example is found in Ukraine, a region traditionally called "the breadbasket of Europe." Here the movement into collective farming was strongly resisted by the peasants and their production decreased. The Plan, however, called for a production increase, and Stalin ordered that deliveries be made in accord with this plan. Not only were all the products of many communes taken, but also the personal food of the families. As a result, famine swept through Ukraine during 1932-34, and up to eight million people died of starvation.

Between 1934 and 1938, Stalin led the Great Purge; triggered by the assassination of a leading Communist official, millions of people – Communists as well as non-Communist -- were caught up in a nightmare. Characterized by widespread police surveillance and suspicion of being "saboteurs," arrests, torture, slave-labor confinement, and executions followed. The *Narodnyy Komissariat Vnutrennikh Del* (NKVD, People's Commissarit for Internal Affairs) headed by the ruthless Lavrentiy P. Beria, carried this out. Estimates of the number of deaths associated with the Great Purge run to nearly two million.

The Red Army invaded Eastern Poland in September 1939, and then fought a winter war against Finland during 1939-1940. A Nazi-Soviet non-aggression pact was signed in 1939, but on June 22, 1941, Hitler initiated Operation Barbarossa (the invasion of the USSR) with *Blitzkrieg* (lightning war) bombing and border assaults. Within days, three massive, tank-led *Heer* (Army) groups moved in on a 900-mile front with Leningrad, Moscow, and the Ukraine region as objectives. Fast raids by the *Luftwaffe* (Air Force) destroyed 1,200 Soviet aircraft in the first few hours of the war.

There followed what became known to the Soviets as the Great Patriotic War. In August, the German Army reached Ukraine where, at first, some of the people greeted them as liberators. This quickly changed, however, with the arrival of the infamous *Schutzstaffel* (SS, Protective Squadron); the SS rounded up close to 24,000 Jews and executed them all in three days.

Siege of Leningrad

Starting in September, Leningrad was under siege for 900 days. By the middle of October, German troops were within 100 km of Moscow, but Stalin decided not to evacuate; the city neither fell nor was ever under full siege. Beginning in late August 1942, the 90-day battle around Stalingrad was fought; this is often called the bloodiest battle in world history.

Higher Education

When Peter the Great formed the St. Petersburg Academy of Sciences in 1724, the University School was an integral part. In 1747, the School became an independent unit and the first Rector was appointed; this became the St. Petersburg University in 1819. Moscow University was formed in 1755; it is thus debatable as to whether St. Petersburg or Moscow is the oldest Russian "university." In the following years up to the early 1900s, only eight more full universities were started in Russia; one of these was Kharkov University, opened in 1805 in Ukraine. All of the universities had associated research centers.

St. Petersburg University

After completing university undergraduate study, a student might enter a post-graduate period of study and research involving examinations, a dissertation, and scholarly papers, and eventually be awarded a *Kandidat Nauk* (Candidate of Science, C.Sc.) degree (approximately the same as a Ph.D. degree in most other countries). The holder of a C.Sc. might find a position as a lecturer or a docent (equivalent to an assistant professor). Continued research and a second dissertation would be needed to qualify for the higher *Doctor Nauk* (Doctor of Science, D.Sc.) degree. Individuals holding the D.Sc. degree were more often titled "Professor" rather than "Doctor."

Essentially all of the Russian universities had faculties (departments) of physics and mathematics from their start. As the engineering disciplines came into being, this field was not added to the university curricula, but a new class of higher-education institute was formed for this instruction. Initially called vocational schools, then later technical schools, many evolved to have a status near that of a university. The St. Petersburg Electro-technical Institute was founded in 1886 as the engineering college of the Russian Department of Post and Telegraph of Russia. This was the first institute of higher education in Europe for training engineers and researchers in the field of electrical engineering.

Prior to the Revolution, both the Russian Army and Navy had academies providing higher education. These included the Artillery Military Academy in St. Petersburg dating from 1619 and the Russian Naval Academy formed by Peter the Great in 1696. These early academies added engineering as a major course of study. Starting in 1918, the military academies were reorganized as many specialized higher-education institutions for officers who had already completed undergraduate studies.

Emergence of Radio

In Russia, May 7 is called "Radio Day," recognized as the day in 1895 that A. S. Popov invented this technology. Popov was teaching physics at an engineering academy for officers, and on that date gave a presentation at a meeting of the Russian Physical and Chemical Society on a "lightning detector" that he had developed. A journal paper on this was published later in the year.

Aleksandr Stepanovic Popov (1859-1906) was born in a small village in the Ural Mountains. The son of a priest, he first attended a seminary, then studied physics at St. Petersburg University, where he graduated in 1882. A short time later, he took a teaching position with the Torpedo School of the elite Naval Warfare Institute at Kronstadt, which had excellent laboratory facilities and one of the best libraries in Russia. He soon began experiments on electromagnetic radiation, following the work

Aleksandr Popov

of Hertz, and built a coherer detector that improved on the previous designs. The Navy had an interest in predicting storms, and Popov first used this coherer in a receiver for detecting distant lightning strikes, demonstrating it at the school on May 7, 1895.

Continuing with his experiments, Popov developed both a spark-gap transmitter and a further-improved coherer receiver. In March 1996, he gave a demonstration to the Physical and Chemical Society in which the letters spelling out "Heinrich Hertz" were transmitted in Morse code. That year, there were 11 publications on his accomplishments, making the invention accessible to the public. The following year, he demonstrated his apparatus in communicating between two ships in the Baltic Sea; during this, he took note of interference beat caused by the passage of a third vessel.

In Paris, Professor Eugène DuCretet at the College de France heard of Popov's apparatus and in 1899 started a firm building DuCretet-Popov wireless sets. In 1900, Popov's receiver was patented in Russia, England, and France. A 47-km wireless link was set up between islands in the Finnish Gulf and was soon used in saving the lives of many fishermen who were adrift on an ice float. This led to the adoption of wireless communication by the Russian Imperial Navy. In addition to DuCretet-Popov sets from Paris, equipment came from a factory in St. Petersburg opened by the German firms Siemens & Halske under license from Popov. Marconi attempted to establish an operation in Russia, but he was totally unsuccessful.

In 1901, Popov received a professional appointment at the St. Petersburg Electro-technical Institute. There he set up a laboratory for research and also gave lectures in this field. Among other activities, he initiated research in audio modulation. In 1905, he became the first elected Rector of the Institute. Unfortunately, Popov died within a year after taking this position.

Although radio technology had an early start in Russia, its acceptance and adoption under the Tsarist government and poor economic conditions progressed very slowly. Popov himself was a researcher and professor, not an entrepreneur. In fact, there were few true entrepreneurs in the business climate of Russia during the first years of the 20th century. The universities and institutes, however, maintained the level of fundamental radio science and technologies on a par with other European countries.

At the time of the Revolution, the leaders recognized the importance of radio communications. On October 27, 1917, Lenin announced the success through a radio message sent in Morse code from a cruiser anchored at St. Petersburg. In 1920, Lenin sent a letter to M. A. Bonch-

Bruyevich, Professor of Radio Physics at St. Petersburg University, that included the following:

> I take this opportunity of expressing to you my deep gratitude and sympathy for the great work of radio inventions that you are carrying on. The newspaper without paper and "without distances" that you are bringing into being will be a great achievement. I promise to assist you in any and every possible way in this and similar work.

In 1922, as Stalin was rising in power, Lenin sent him a letter concerning using radio as a tool for the government:

> It is technically quite feasible to broadcast human speech over any distance by wireless; furthermore, it is also possible to use many stations that could broadcast speeches, reports and lectures delivered in Moscow to many hundreds of places throughout the Republic.

By 1924, 50 radio stations were spread across the Nation. While there were few homes with receivers, the broadcasts were piped into speakers in apartments, barracks, factories, shops, and street corners.

Research Centers

The *Rossiiskaya Akademiya Nauk* (RAN, Russian Academy of Sciences) was established at St. Petersburg in 1724, under the guidance of the German mathematician Gottfried Leibniz, co-developer (with Isaac Newton) of calculus. Financed by the Imperial State, this outstanding organization remained essentially independent of the State governance throughout the years. It had a small core (usually around 50) of active academicians and a large number of affiliated researchers. While the Academy brought Russian science

Russian Academy of Sciences

into world recognition, it did little to improve the industrial status of the nation.

In early 1918, representatives of the Academy met with Lenin. They agreed that expertise of the Academy would be used to address "questions of State construction," and in return the regime would give the Academy political and financial support. In 1925, the Soviet government recognized the Academy as the "highest all-Union scientific institution" and renamed it the Academy of Sciences of the USSR. In

1934, the Academy headquarters moved from Leningrad to Moscow, together with a number of associated academic institutions.

With the formation of the USSR in 1924, the Party and the government provided means for organizing and operating a number of scientific research institutes. The first of these were the Nizhny-Novgorod Radio Laboratory, founded by M. A. Bonch-Bruyevich and V. M. Leshchinsky, and the Institute of Physics in Moscow, headed by P. P. Lazarev. Then came the Petrograd (later Leningrad) Physical-Technical Institute, led by A. F. Ioffe, and the State Optical Institute, headed by D. S. Rozhdestvensky. Soon afterward, the Central Aerohydrodynamics Institute was set up in Moscow, with N. E. Zhukovsky and S. A. Chaplygin as its leaders. Then came the All-Union Electro-Technical Institute in Moscow, headed by K. A. Krug, and the Ukrainian Physical-Technical Institute in Kharkov, led by I. V. Obreimov.

Abram Ioffe

Abram Fedorovich Ioffe (1880-1960) was generally considered the top Russian physicist of his time. He was educated at the Petersburg Technological Institute, followed by graduate studies at Munich University. He served as a Professor at the St. Petersburg Polytechnic Institute and was elected to the Academy of Sciences in 1918, twice serving as the Vice President. When the Petrograd/Leningrad Physical-Technical Institute (LPTI) was formed, he was selected as the Director.

In the first decade following the Revolution, an estimated 60 research institutes were either formed or nationalized from a few private institutes. All these institutes were rapidly organized and equipped; the Soviet government's appropriations for science were far beyond anything Russia had seen before.

In addition, there were the centers involved with research at universities and those established by the military, particularly the Red Army. The *Glavnoe artilkeriisko upravlenie* (GAU, Main Artillery Administration) was considered the "brains" of the RKKA. It not only had competent engineers and physicists on its central staff but also established a number of scientific research institutes dedicated to special technical areas. These included the *Leningradskii Elektrofizicheskii Institut,* (Leningrad Electro-Physics Institute, LEPI), directed by A. A. Chernyshev. Research in military radio communication was under the *Nauchnoissledovatel skii ispytalel nyi institut suyazi RKKA* (NIIIS-KA, Scientific Research Institute of Signals of the Red Army).

EARLY RADIO-LOCATION DEVELOPMENT

The first known notice of detection by electromagnetic waves was made by radio pioneer Alexander Popov in 1897. In a report concerning communication experiments between two warships on the Baltic Sea, he noted that the passage of another vessel across the path of transmission caused a variation in the received signal. Popov wrote that this phenomenon might be used for detecting objects, but he did nothing more with this observation.

In only a few years after the close of World War I, Germany's *Luftwaffe* had aircraft capable of penetrating deep into Soviet territory. Thus, the detection of aircraft at night or above clouds was of great interest to the *Voiska Protivo-vozdushnoi aborony* (PVO, Air Defense Forces) of the Red Army. As in both Great Britain and America, two methods appeared to have potential: acoustic and optic, both involving devices that would augment human senses.

The detection problem was given to the GAU and assigned to Lt. Gen. M. M. Lobanov. There was a considerable literature base on acoustics, and a sound-locator system was soon developed. Called the *Prozhzvuk* – from *prozhektor* (searchlight) and *zvuk* (sound) – it was made of an assembly of large trumpet-like sound collectors, each feeding a microphone, with the amplified output going to headphones that could be switched between the receptors. The bearing and elevation of the assembly indicated the direction of the sound originator, and these measurements were used to aim a searchlight for illuminating the target.

The first tests of the *Prozhzvuk* were made in early 1932. These showed a success rate of near 50 percent, but under the ideal conditions of still air (no wind) and with only a single target aircraft. Blind operators were brought into the testing under the assumption that their hearing would be more acute, but this failed to improve the detection performance. Nevertheless, systems were built for the PVO.

To pursue the optical solution, Lobanov turned to the All-Union Electro-Technical Institute. Here Professor V. L. Granovski had been investigating the thermal energy radiated by heated objects. The nature of infrared radiation was well known, having been addressed by astronomers and physicists for many years, and the scientific literature included many papers on various types of detectors.

A system was built using a 1.5-meter parabolic reflector (from a searchlight) that focused the received energy onto a detector at its center. The most sensitive infrared detector at that time was the bolometer, and was likely used in the unit. To recognize the presence of received radiation, the input to the detector was interrupted (chopped). The resulting audio signal from the detector was amplified and delivered to headphones used by the operator.

Called the *teplopelengator* (roughly "thermal course and bearing indicator"), the system was first field-tested in 1932. A bomber could be detected at ranges up to 12 km, but only at night (in the absence of radiation from the sun) and when the aircraft was not in or behind clouds. Tests in detecting naval vessels, however, were more successful. Pursuit of the acoustical technique for air defense was essentially abandoned, but was continued for the NKKF (Soviet Navy) in systems for nighttime coastal defense against intruding ships.

The use of optical devices was much more successful for close-in targets. In 1931, a stereoscopic range-finder called *Puazo* for anti-aircraft was introduced, followed by an improved *Puazo-2* in 1934. In use, the operators simply read the range, azimuth, and elevation from mechanical indicators and telephoned this information to the gun battery. In 1939, a new *Puazo-3* provided the information electrically to the fire-control center. Use of *Puazo* was obviously limited to targets in the open and during daylight hours.

Puazo Range Finder

Radio-Location Work in Leningrad

With lack of significant success in acoustical and optical detection, the GAU turned to an unexplored technique: the use of radio. An initial approach for developing such a device was made to the Scientific Research Institute of Signals of the Red Army (NIIIS-KA). There the GAU was told that <u>*radiolokatory*</u> (radio-location) techniques were not feasible. Without making any investigations, they said that radio signals from engine ignition sparks could never be received on the ground, and detection by receiving reflections from an aircraft illuminated by radio beams would require an unachievable level of transmitter power.

The Nizhny-Novgorod Radio Laboratory had been moved to Leningrad in 1929 and enlarged to become the *Tsentral'naya radiolaboratoriya* (TsRL, Central Radio Laboratory), with D. N. Rumyantsev as Director and M. A. Bonch-Bruyevich as Technical Director. After being rebuffed by the NIIIS-KA, the GAU approached the TsRL in August 1933. Here Rumyantsev agreed to start a research program on radio-location and assigned Yu. K. Korovin to make an initial investigation.

Korovin, who earlier studied radio physics under D. A. Rozhansky at Kharkov University, was conducting research on VHF communications, and had built a 50-cm (600-MHz), 0.2-W transmitter using a Barkhausen-Kurz tube. The transmitting and receiving antennas were backed by two-meter parabolic reflectors. The antennas were arranged to aim along the flight path of an aircraft, and on January 3, 1934, a Doppler signal was received by reflections from the aircraft at some 600-m range and 100-150-m altitude. This experiment showed that radio reflections from objects could be detected, refuting the position of the NIIIS-KA and indicating that radio-location was feasible. An agreement was soon made by the GAU under which the TsRL would develop an experimental radio-location system for aiming searchlights in supporting anti-aircraft guns.

In the economic and political climate of the USSR at that time, competition was seldom allowed, much less instigated. The potential value of radio-location to the military, however, led the GAU to enter a separate agreement with the Leningrad Electro-Physics Institute (LEPI), for a similar radio-location system. This technical effort was led by B. K. Shembel.

While the GAU had earlier been pursuing the acoustic and optical approaches, engineers at the PVO also came up with the concept of radio detection for extending the reconnaissance range. P. K. Oshchepkov was assigned to prepare a paper for the Defense Commissar, asking that a special research unit be set up for this technology. The proposal was accepted in June 1933, and a program with high priority was funded for developing a *razvedyvlatl'naya elektromagnitnaya stantsiya* (reconnaissance electromagnetic station).

Upon receiving the go-ahead from the Defense Commissariat, Oshchepkov was sent to Leningrad to discuss the program with leaders in the Academy of Sciences. The concepts were warmly received, and Oshchepkov was transferred to Leningrad to be in charge of a Special Construction Bureau (SCB), as well as being responsible for the related *experino-tekknicheskii sektor* (technical expertise component) of the PVO.

On January 16, 1934, a major conference was held in Leningrad on detection methods for air-defense. Arranged by the Academy of Sciences for the PVO and chaired by A. F. Ioffe, it was attended by top radio specialists, acoustical and optical experts, academicians from related fields, and officials from the PVO command. It is interesting, however, that the participants did not include representatives from Kharkov University or the Ukrainian Institute of Physics and Technology (UIPT), both, as described later, already with research activities in magnetrons and their applications in radio detection.

The conference opened with Oshchepkov reviewing the state-of-the-art in various air-defense detection technologies. Proponents for sound and infrared detection urged that future research center on these areas, but the general consensus was that radio detection was feasible and must be emphasized. A case was made by Oshchepkov for pressing research into decimeter and centimeter wavelengths to facilitate forming better beams. Ioffe strongly argued that longer wavelengths should be used, mainly because higher power from existing transmitters would far overshadow any potential advantages of yet-to-be-developed shorter-wavelength components.

Overall, the conference did little in establishing the direction for future research. To give the findings to a wide audience, the proceedings were published in the journal *Sbornik PVO* in February 1934. This included the then-existing information on radio-location in the USSR, available (in Russian language) to researchers in this field throughout the world.

Although the conference did not produce concise guidance for continued development, Oshchepkov was given permission by the PVO to place various types of radio-location work in Ioffe's Leningrad Physics and Technology Institute (LPTI) and the Ukrainian Institute of Physics and Technology (UIPT) in Kharkov (for magnetrons), his own SCB, and the Svetlana factory in Leningrad (a radio manufacturer and the largest supplier of vacuum tubes in the USSR). Since the GAU-funded work at the LEFI and the TsRL was for the PVO, Oshchepkov had some monitoring responsibilities in this also. As a consequence, like Watson Watt in Great Britain, Oshchepkov is often credited with initiating radar development in the Soviet Union

Pavel Oshchepkov

Pavel (sometime listed by nickname Piotr) Kondratyevich Oshchepkov (1908-1992) began life as one of the many street waifs generated by the war and strife in Russia at that time. He was 12 before receiving his first education at a Shalashinsk Commune school, but in 1928 he was able to enter the Plekhanov Institute in Moscow to study the economics of electrical engineering. His early performance at this Institute was excellent, and he was allowed to transfer to Moscow University, where in 1931 he completed his undergraduate education in the physics curriculum.

Oshchepkov was initially employed as an electrical engineer in a power station, but before the end of 1932, was on the Moscow engineering staff of the PVO, working on improved optical instruments for aircraft detection. Here his technical and leadership abilities were quickly recognized

Fundamentally, two types of radio-location equipment were needed. The GAU wanted equipment for support of searchlights for anti-aircraft guns as well the coordination of defensive fighter aircraft. This required the determination of three coordinates: azimuth, elevation, and range, particularly for nighttime and cloudy conditions when the existing *Puazo* range-finder was ineffective. On the other hand, the PVO was more interested in reconnaissance that required only azimuth and range; this was sometimes called *radioobnaruzehenie* (radio-detection).

At this stage, little theoretical analysis of requirements for radio equipment was needed in the different applications. Despite Ioffe's reservations, most of the researchers believed that any ultimate success would involve operating at high frequencies. Little support was available from other research institutes and factories; essentially all prior studies and equipment was for continuous wave (CW) communications equipment operating at long wavelengths.

Initial Radio-Location Systems

Earlier at the LEPI under GAU sponsorship, Shembel's team had built a transmitter and receiver to study the radio-reflection characteristics of various materials and targets. This equipment was readily made into an experimental bi-static (transmitter and receiver stations separated by a significant distance) radio-location system called *Bistro* (Rapid).

The *Bistro* transmitter, operating at 4.7 m (64 MHz), produced near 200 W and was frequency-modulated by a 1-kHz tone. A fixed transmitting antenna produced a broad coverage of what was called a radio screen. A regenerative receiver, located some distance from the transmitter, had a dipole antenna mounted on a hand-driven reciprocating mechanism, sweeping the cone of reception over the screened zone.

An aircraft passing into the screened zone would reflect the radiation, and the receiver antenna would collect this reflected signal when it swept the zone. The detection was a result of the interference beat between a portion of the transmitted signal and the reflected signal. The aiming direction of the receiver antenna when the signal was detected would roughly show the direction (azimuth) of the invading aircraft.

Bistro was first tested during the summer of 1934. With the receiver up to 11 km away from the transmitter, the set could only detect an aircraft entering a screen at about 3-km range and a height up to 1,000 m. With minor technical and operational improvements, 75 km was determined to be the maximum range.

There was no means for *Bistro* to indicate the target range; only the presence and a coarsely indicated direction of the target could be obtained. While the set was completely unsuitable for anti-aircraft applications, the PVO, including Oshehepkov, believed that it might have some value as a reconnaissance set. In October 1934, the PVO placed an order with the Svetlana Factory for five additional *Bistro* sets for experimental purposes.

In 1935, both LEPI and TsRL were made a part of *Nauchno-issledovatel institut-9* (NII-9, Scientific Research Institute #9), a new GAU organization opened in Leningrad. Professor M. A. Bonch-Bruyevich, the renowned radio physicist previously with TsRL, was named the NII-9 Scientific Director. In addition to radio-location, NII-9 was engaged in projects involving a wide range of disciplines, including military television.

M. A. Bonch-Bruyevich

Mikhail Alexandrovich Bonch-Bruyevich (1888-1940) was possibly the leading authority on radio physics at that time in the USSR. He received his undergraduate education at St. Petersburg University, graduating in 1909, then continuing with advanced studies, research, and teaching. Lenin had personally sought him out for advice on radio and its uses, and he been elevated to Professor. He was one of the founders of Nizhny-Novgorod Radio Laboratory, established in 1922 as the first State research institute.

As described in the next section, the UIPT in Kharkov had developed a series of magnetrons. In 1936, one of these producing about 7 W at 18 cm (1.7 GHz) was used by Shembel at the NII-9 as a transmitter in a *radioiskatel* (radio-seeker) called *Burya* (Storm). The type of receiver was not described. In a bi-static configuration, the set used dipole antennas with parabolic reflectors – likely producing a relatively broad beam for transmission and a narrower beam for receiving – and operated similarly to *Bistro*. The range of detection was about 10 km, and provided azimuth and elevation coordinates estimated to within 4 degrees. This performance, however, was again insufficient for directing AA guns.

Plans for improving *Burya* included a more sensitive receiver and, as proposed by Bonch-Bruyevich, the division of the transmitter output into several antennas to produce a flattened radiation pattern. To improve the angular resolution, Shembel proposed three receiving antennas arranged for horizontal and vertical pairing and using lobe-switching technique. Sets using the different antennas were designated *Burya-2* and *Burya-3*, sometimes simply *B-2* and *B-3*.

None of these sets went further than the experimental laboratory. No attempts were made to make *Burya* a pulsed system; Bonch-Bruyevich strongly believed in the merits of CW radio-location, although he had earlier used a pulsed transmitter in research on the ionosphere.

While work by Shembel and Bonch-Bruyevich on CW systems was taking place at NII-9, Oshehepkov at the SCB and V. V. Tsimbalin of Ioffe's LPTI were pursuing a pulsed system. Pulsing the transmitter had two great advantages. First, the output would be much higher in pulsed power (the tube "rested" between pulses, then concentrated the power into the pulse). Second, the receiver would possibly be open for reception of the returned signal during the off-time of the pulse. (The powerful transmitted signal entering a close-by receiver would saturate the circuit and obscure reception of the weak reflected signal.)

The PVO's vacuum-tube laboratory had developed a transmitter tube producing up to 50 watts continuous power. In 1936, Oshchepkov and Tsimbalin used this tube in an experimental pulsed radio-location system operating at 4 m (75 MHz.). It had a peak power of about 500 W and a 10-µs pulse duration. The overall configuration of this experimental set was similar to that of the CW *Bistro* and *Burya*. Initial tests using separated transmitting and receiving sites resulted in an aircraft being detected at 7 km. In April 1937, with the peak-pulse power increased to 1 kW and the antenna separation also increased, test showed a detection range of near 17 km at a height of 1.5 km.

Although this was a good beginning for pulsed radio-location, the system was still bi-static and incapable of directly measuring range (the technique of using the pulses for determining range had not yet been developed). It was the intent of Oshchepkov and the LPTO to obtain approval for transforming this into a radio-location systems with all of the desired characteristics, but these plans and the project soon went away.

The PVO conducted the first large-scale exercise of Soviet air defense in early 1937. This clearly showed that sound and infrared detectors would be of essentially no benefit in extending the range of aircraft detection. The preliminary accomplishments in radio-location systems were brought to the attention of A. I. Sedyakin, newly named as the PVO chief. The Red Army Signals Command had believed that radio-location was not possible, and tried to influence Sedyakin against supporting the effort. However, Sedyakin asked the NII-9 to prepare a proposal for an improved radio-location system meeting all requirements that might be submitted to the "highest level" (presumably Stalin).

From the start of the radio-location development, the NIIIS-KA (Red Army Signals Research) had been strongly opposed to this effort and repeatedly attempted to have it terminated. With the prospect of a larger effort at NII-9, the NIIIS-KA was able to have established a commission to investigate the organization and its activities. The resulting findings – very biased by senior officers from the Signals Command being participants – highly criticized the NII-9 operations.

In June 1937, all of the work in Leningrad on radio-location suddenly stopped. The infamous Stalin Purges swept over the military high commands and the supporting scientific community. The PVO chief was executed, as were many other military officials at his level. The report from the investigation commission led the NKVD to arrest the director of NII-9 and to the dismissal of Shenbel. Through the influence of Bonch-Bruyevich, who had been a favorite of Lenin in the prior decade, NII-9 as an organization was saved, and Bonch-Bruyevich was named the new director.

The other radio-location activities in Leningrad were even more severely affected; the SCB was closed and most of the senior personnel arrested. Oshchepkov was charged with "high crimes" and sentenced to 10 years at a *Gulag* penal labor camp. Although the two had often been at technical odds, Ioffe plead without success for Oshchepkov to be restored. Over the years of Oshchepkov's internment, Ioffe assisted in his survival by providing food packages and letters of encouragement.

Relative to the purge and subsequent effects, General Lobanov later made the following comments:

> The baseless repressions – the purges – eliminated valued scientific and engineering cadres and produced a climate in which fear of repression for failure and for taking scientific-technical risk led to indecisiveness. . . . Yet the intrusion [of the NIIIS-KA] was far from a regressive development.

This latter was in reference to the various radio-location activities being placed under a single organization, and the rapid reorganization of the work to accomplish needed results.

Radio-Location Development in the Soviet Union

Having accomplished the decimation of prior radio-location organizations and activities in Leningrad, the NIIIS-KA then did an about face, pressing hard for speedy development of such systems. They took over Oshchepkov's laboratory and were made responsible for all existing and future agreements for research and factory production.

Scientists at NII-9 under Bonch-Bruyevich developed two types of very advanced microwave generators. In 1938, a linear-beam, velocity-modulated vacuum tube (a klystron) was developed by N. D. Devyatkov, based on theory given earlier by D. A. Rozhansky at Kharkov University. This device produced about 25 W at 15-18 cm (2.0-1.7 GHz) and was later used in experimental systems. The next year, he followed this with a simpler, single-resonator device (a reflex klystron).

At this same time, D. E. Malyarov and N. F. Alekseyev were building a series of magnetrons, also based on designs from Kharkov; in 1938, the best of these produced 300 W at 9 cm (3 GHz). As will be noted later, all of the Soviet-built magnetrons suffered from frequency instability and were never fully satisfactory in their radio-location applications.

Also at NII-9, D. S. Stogov was placed in charge of the improvements to the *Bistro* system. Redesignated as *Reven* (Rhubarb), it was tested in August 1938, but was only marginally better than the predecessor. With additional minor operational improvements, it was made into a mobile system called *Radio Ulavlivatel Samoletov* (*RUS*, Radio Catcher of Aircraft), soon designated as *RUS-1*. This CW, bi-static system had a truck-mounted transmitter operating at 4.7 m (64 MHz) and two truck-mounted receivers.

[A note should be made concerning designations for the Soviet systems. The names of systems under development frequently changed even when only minor improvements were made. On the other hand, once they had been "accepted" by a factory and placed into production, the same designation was often retained even with significant changes.]

Although the *RUS-1* transmitter was in a cabin on the rear of a truck, the antenna had to be strung between external poles anchored to the ground. A second truck carrying the electrical generator and other equipment was backed against the transmitter truck. Two receivers were used, each in a truck-mounted cabin

RUS-1

with a dipole antenna on a rotatable pole extended overhead. In use, the receiver trucks were placed about 40 km apart; thus, with two positions, it would have been possible to make a rough estimate of the range by triangulation on a map (the crossing point of the directions).

The *RUS-1* system was tested and put into production in 1939, then entered service in 1940, becoming the first deployed radio-location system in the Red Army. A total of about 45 *RUS-1* systems were built at the Svetlana Factory before the end of 1941, and deployed along the western USSR borders and in the Far East. Without ranging capability, however, the military found the *RUS-1* to be of little value.

At the NIIIS-KA, A. I. Shestako had continued work with the 4-m (75-MHz) pulsed-transmission system started by Oshchepkov. Through pulsing, the 40-W transmitter tube produced a peak power of 1 kW, the highest level thus far generated. In July 1938, a fixed-position, bi-static experimental system detected an aircraft at 30-km range at heights of 500 m, and at 95-km range for high-flying targets at 7.5 km altitude.

The project was then taken up by Ioffe's LPTI, resulting in the development of a mobile system designated *Redut* (Redoubt). An arrangement of new transmitter tubes was used, giving near 50-kW peak-power with a 10-µs pulse-duration. Yagi antennas, separated by a few hundred meters, were adopted for both transmitting and receiving.

The *Redut* was first field tested in October 1939, at a site near Sevastopol, a port in Ukraine on the coast of the Black Sea. Sevastopol was a highly strategic port, and this testing was in part to show the NKKF (Soviet Navy) the value of early-warning radio-location for protecting this port. With the equipment on a cliff about 160 meters above sea level, a flying boat was detected at ranges up to 150 km. The Yagi antennas were spaced about 1,000 meters; thus, close coordination was required to aim them in synchronization.

In addition to having greater power and thus a longer range, *Redut* also gave better direction-angle measurements than the *RUS-1*. In late 1939, 10 sets were ordered built in a crash effort, just in time for the Russo-Finish War. Both *RUS-1* and *Redut* sets were used in this conflict, but they were considered to be of no value to the military; to a large measure, this was because of the separation of the transmitting and receiving stations. Nevertheless, 13 additional *RUS-1* sets were later ordered.

During 1940, experiments were made at the LPTI on different pulse rates and durations; from this, the very important capability of range measurement resulted. A cathode-ray display, made from an oscilloscope, was used to show the range information. The R&D Institute of Radio Industry No. 20 (NII-20), located in Moscow, was brought in for the final design. In July 1940, the system was designated *RUS-2*.

Radio-Location Development in the Soviet Union

Engineers at NIII-KA, LPTI, and NII-20 coordinated in a major effort on a transmit-receive device (a duplexer) to allow operating with a common antenna; this was finally accomplished in February 1941. All of this testing was done at an experimental station at Toksovo (near Leningrad). The range-finding capability was immediately added as a modification to the *RUS-2* (the name did not change), and an order was placed with Svetlana for 15 systems (this was later changed to the Aviapribor plant in Moscow). In April, one of these was adopted for shipboard conditions and tested on the light cruiser *Molotov*.

The final *RUS-2* had pulse-power of near 40 kW at 4 m (75 MHz). The equipment was in a cabin on a motor-driven platform, and with a seven-element Yagi antenna mounted about five meters above the roof. The cabin, and thus the antenna, could be rotated over a large sector to aim the transmit-receive pattern. The detection range was 10 to 30 km for targets as low as 500 m and 25 to 100 km for higher-altitude targets. The variance was about 1.5 km for range and 7 degrees for azimuth.

RUS-2

Other activities at the NII-9 in this period included the development under A. E. Suzant of an experimental *radiodal'nomer* (radio range-finder) for anti-aircraft applications. This pulsed system operated at 80 cm (375 MHz) with a power of 16 kW. Called *Strelets* (Shooter), it was capable of detecting an aircraft at 20 km and accuracies of 160-m range and 3-degree angle of elevation. Other experimental sets for adding azimuth measurements were built, including the previously noted *Burya-2* and *Burya-3*, the latter with an accuracy of 1 degree.

During 1940, an experimental gun-laying system designated *Luna* (Moon) was designed by NII-9 incorporating the combined features and accuracies of *Burya-3* (a CW set for azimuth) and *Strelets* (a pulsed set for elevation and range). This would also have likely included the developments from LPTI on pulse-based range measurement and the transmit-receive device allowing use of a common antenna.

Bonch-Bruyevich, the Director of NII-9, died in March and there was no strong leader to push the *Luna* project. Production was originally scheduled to start at Svetlana in August 1941, but the design would not have been ready. The swift German invasion and the evacuation of industries sealed the demise and the project was dropped.

Hitler began the invasion of the Soviet Union on June 22, 1941. There was no warning, but the people of Leningrad had realized that it would come soon. The evacuation of vital Leningrad operations began immediately, with the LPTI and NII-9 packing their laboratories and moving eastward in July. The Svetlana factory was also evacuated.

Radio-Location Work in Kharkov

Historically, Ukraine and its people were not held in high regard by most of the rest of Eastern Russia. Except for the cities of Kharkov, Kiev, and Sevastopol, the region was originally agricultural, mainly occupied by serfs and later by peasant farmers. Under the Soviets, the land was organized into huge collective, but unproductive, farms. A major exception was the area around Kharkov; beginning in the late 1800s, this became an industrial center. At the start of the Revolution, the area had some 5,000 individually owned firms. These were then all nationalized and placed under a relatively small number of State operations.

Kharkov University

Opened by imperial decree in 1805, Kharkov University (KU) was *the* intellectual center of Ukraine. From the start, KU included a Faculty (Department) of Physical and Mathematical Sciences and was soon recognized as one of the most prestigious schools of higher education in Russia.

Dmitry A. Rozhansky (1882-1936) graduated in physics from St. Petersburg University in 1904, then worked at the St. Petersburg Electro-Technical Institute under A. S. Popov. After graduate studies in Germany, he joined the KU Physics Faculty in 1911, where he was soon named Professor and Department Head. Rozhansky initiated work in high-frequency generation, especially developing theories for velocity-modulation devices. Two of his students were A. A. Slutskin and Yu. B. Korovin, who would separately make major contributions in future radio-location developments. (Korovin's demonstration showing the feasibility of radio-detection was described in the previous section.) In 1921, an independent Physics Research Department was established under Rozhansky's guidance at KU, giving birth to what would later be called the Kharkov radiophysics community.

Dmitry Rozhansky

Abram A. Slutskin (1881-1950) graduated from KU in 1916, and remained there doing research. In this, he and D. S. Shteinberg worked with Rozhansky in developing magnetron-type oscillators that generated signals in the 40- to 10-cm (750-MHz to 3.0-GHz) range. For two years he was at Barkhausen's laboratory in Germany, then returned to KU where, in 1930, he was awarded the D.Sc. degree. Slutskin received the title of Professor and was named head of the Physics Research Department.

Abram Slutskin

Since the beginning of the USSR, the government had centralized its advanced research at the Leningrad Physical-Technical Institute (LPTI). A. A. Ioffe, then Director of the LPTI, persuaded the government that some decentralization of certain activities was needed and suggested that a new institute be established at Kharkov. The Ukrainian Institute of Physics and Technology (UIPT) was opened in 1929; this operated in close cooperation with KU and shared many senior personnel. Although supported by, and reporting through, government bureaus, both LPTI and UIPT were affiliates of the Academy of Sciences.

A basic staff of scientists, including the first director Professor Ivan V. Obreimov, were transferred from the LIPT. Both Slutskin and Shteinberg were on the initial staff, but also kept their KU positions. Rozhansky left to take Obreimov's place at the LIPT, but remained an advisor to the UIPT, as well as mentoring students at KU. Nine departments were initially started at the UIPT. The department-level Laboratory of Electromagnetic Oscillations (LEMO) was headed by Professor Slutskin; all of the other departments were led by scientists transferred from Leningrad.

The UIPT received substantial funds for purchasing equipment from abroad, and offered higher salaries to recruit exceptional researchers. The Institute very soon came to be known outside the USSR, and attracted visits from world-recognized physicists such as Niels Bohr and Paul Dirac. The UIPT adopted a highly unusual practice: it published its research papers in German to give wider readership to the world scientists.

The future Nobel Laureate Lev Davidovich Landau joined the UIPT in 1932 and headed the Theoretical Department for the next five years. He also served on the Physics Faculty at KU, and was a major influence in Kharkov becoming regarded as the center of theoretical physics in the USSR.

With a staff of about ten scientists and engineers, the LEMO was devoted to research in electrical engineering and electromagnetic waves. On a personal basis, Slutskin continued with his research in magnetrons and developed the theory of a magnetron oscillator operating in the dynamic mode. By 1933, the development of new magnetrons was well established. Led by Aleksandr Y. Usikov, a team designed a number of segmented-anode magnetrons covering 80 to 20 cm (0.37 to 1.5 GHz), with output power between 30 and 100 W. They, however, had a short lifetime, usually less than 50 hours. These were put into small-scale production at the LEMO for use elsewhere; this was considered an insulting imposition by the scientists. In 1934, the contributors published journal papers on their achievements.

Aleksandr Usikov

Pavel P. Lelyakov was the leading expert on magnetrons at the LEMO. He invented a device with a hollow anode that was water-cooled from the inside. After a series of improvements by Semion Y. Braude, this became a glass-cased magnetron producing up to 17 kW with 55 percent efficiency at 80 cm (370 kHz); the primary advantage was that it was tunable over a wavelength change of 30 percent. He also investigated the pulsing of magnetrons, and developed a high-power, pulsed device. Both Usikov and Braude were awarded the C.Sc. (equivalent of Ph.D.) from KU in 1937.

Semion Braude

In March 1937, the NIIIS-KA contracted with the LEMO for developing a pulsed radio-location system for anti-aircraft applications. The project was code-named *Zenit* (a popular football team at the time) and was under the guidance of Slutskin. The transmitter development was led by Usikov. The unit used a 60-cm (500-MHz) magnetron pulsed at 7-10-μs duration and providing 3-kW pulsed power, later increased to near 10 kW.

Braude led the development of the receiver. This was a super-heterodyne unit initially using a tunable magnetron as the local oscillator. The magnetron, however, lacked stability and was replaced with a circuit using an RCA type 955 Acorn triode. The returned pulses were displayed on a cathode-ray oscilloscope. The *Zenit* had separate transmitting and receiving antennas, built with dipoles backed by 3-meter parabolic reflectors; these formed a beam with a width of about 16 degrees. The full transmitter and receiver units were attached on the backs of the respective reflectors in hermetically sealed cases. They were set about 65 meters apart, with selsyns used to synchronize their movements.

Zenit was first tested in October 1938. In this, a medium-sized bomber was detected at a range of 3 km, and needs for improvements were determined. A new agreement was made in May 1939, specifying the required performance and calling for the system to be ready for production by 1941. After the changes had been made, a demonstration was given to all of the interested customers in September 1940. It was shown that the three coordinates (range, altitude, and azimuth) of an aircraft flying at heights between 4,000 and 7,000 meters could be determined at up to 25-km distance. As compared with barometric measurements, the height error averaged about 5 percent. With the antennas aimed at a low angle, there was a dead zone of some distance; this was caused by interference from ground-level reflections.

Zenit Antenna

While this performance was not satisfactory for immediate gun-laying applications, it did show the way for future systems. Another characteristic, however, rendered Zenit completely unsuitable for use with fire-controllers on anti-aircraft guns. A null-reading method was used for analyzing the signals; this made it necessary to determine azimuth and elevation coordinates independently, requiring a sequence of antenna movements that took 38 seconds for the three coordinates.

As the Zenit project was underway at the LEMO, there was considerable turmoil throughout the parent UIPT. Many scientists there passionately believed that the Institute should not engage in military projects; this led to an unsuccessful movement by some of the scientists, including Lev Landau, to have the LEMO become a totally separate organization.

The radio-location work was also of great secrecy. In general, the other departments and their directors did not even know the nature of the work, and this was in contradiction to "freedom of scientific research." In addition, it was revealed that the personnel of LEMO were paid substantially more than the average in UIPT. All of this created open hostility throughout the organization.

As Stalin's Great Purges reached Kharkov, the UIPT was an immediate target of the NKVD. There was no shortage of informers;

many persons were arrested under charges of sabotage and two were executed by a firing squad. Landau was accused of "trying to spoil defense work in UIPT," and only his international reputation saved him. He transferred to another institute in Moscow, but he was then arrested in 1938 and spent a year in prison.

The situation within UIPT improved in 1939, and work at the LEMO continued on *Zenit*, particularly in converting it into a single-antenna system designated *Rubin*. This effort, however, was disrupted by the invasion of the USSR by Germany in June 1941. In a short while, all of the critical industries and other operations in Kharkov were ordered to be evacuated far into the East.

WARTIME RADAR IN THE USSR

As the German tank-led armies swept into the USSR and bombing of cities began, in July 1941 the *Komitet Oborony* (Defense Committee – the small group of leaders surrounding Stalin) made a decision to concentrate the defense at Moscow but evacuate to the East the critical operations in other large cities in the West. Initially, some of the movements were to Moscow, but within a few months many operations in this city were also relocated.

Between July and November, over 1,500 factories and institutes were moved eastward. Gorky (Nizhny Novgorod before 1932) was the major city closest to Moscow to which evacuations were initially made. About 400 km southeast of Moscow, it was located at the strategic confluence of the Volga and Oka Rivers. Factory No. 183 relocated from Kharkov to Gorky to become Factories No. 92 and 112, producing the highly effective T-34 medium tanks.

Kazan was the second closest city to Moscow to receive initial evacuations. Located about 800 km southeast of Moscow on the Volga River in the Tarter Republic, it had a population of about 400,000. Many factories from Leningrad, including Svetlana, were moved there.

Kubyshev, in the Urals some 850 km southeast of Moscow, had two extremely important wartime roles. First, the defense contingency plan called for the evacuation of the Government to this city in the event Moscow was lost. Second, Factory No. 115 was evacuated from Moscow to Kubyshev, where thousands of Yak fighter aircraft designed by A. S. Yakovlev were built. The importance of this city was such that it was closed to foreigners during the war and for some time thereafter.

The furthest city receiving major evacuees was Alma-Ata, roughly 3,000 km from Moscow. This was the capital of the Kazakh Republic on the southwest corner of the USSR and had a population of about 250,000.

Although a great distance away, industries and institutes involving over 25,000 persons were evacuated to this city.

Omsk was another large city in the East that received many evacuated industries. Located in southwest Siberia some 2,200 km from Moscow and with a population of about 500,000, Omak became home to so many industries producing heavy military equipment that, like Kubyshev, the city was closed to foreigners for many years.

The Uzbek Republic, in the south-central part of the USSR, had two cities involved in the evacuations: Tashkent, the capital, 2,800 km from Moscow at the far northeast corner of Uzbek; and Bukhara, some 425 km to the southwest of Tashkent. Although Tashkent was already industrialized, many additional operations relocated there, increasing the population to over a million. In contrast, Bukhara was much smaller than Tashkent and had almost no industry. Dating from the 1st century AD and on the ancient "Silk Road, " Bukhara was a provincial city with a strong Islamic heritage.

In a very short time, Moscow became the best-defended city in the world. It had three lines of defense, with a huge number of soldiers and massive emplacements of anti-aircraft guns. Nevertheless, near the end of September an overwhelming number of German troops pressed on the city from the north and south, and on October 12, Stalin ordered the bulk of the Government offices and remaining critical factories to move. About half of the population was either relocated or fled the city. By the end of November, the Germans were at the last line of defense, only 22 km from Moscow.

Then the tide of battle turned; large numbers of Soviet troops from Siberia and the Far East were rushed to defend Moscow. These troops had been held in reserve under the assumption that Japan would join Germany and also attack the USSR, but in late November a Soviet spy in Tokyo learned of an impending attack on the United States. Believing then that Japan would not attempt a second-front invasion, Stalin immediately ordered the reserve troops brought to Moscow, and the city was saved.

A winter 1941-42 counterattack drove the Germans emplacements from close positions around the city. It would be early 1943, however, before the front was moving westward, and October of that year before Moscow was again fully secured.

Ground-Based Radar Systems

The anti-aircraft batteries around Moscow performed well; the *Luftwaffe* was never able to make major strikes against the city. Dating from the early 1800s when Napoleon's invasion was turned, the Russians had depended on massive fire-power as opposed to precision aiming of large weapons. Around Moscow there was a huge number of anti-aircraft guns in batteries of up to 100 weapons each. By concentrating their fire with fused shells into even an even larger volume, a lethality of almost 100 percent could be achieved. Factories in the area operated continuously in producing the needed ammunition.

Moscow Munitions Factory

Before the war, the military had not yet accepted radio-location as having any real value. Human spotters gave warnings and the optical *Puazo* range-finder assisted in aiming (laying) guns on close-in targets. The rudimentary radio-location sets, however, were used beneficially as the German air strikes started.

The early *RUS-2* test unit that had been set up at the Toksovo experimental site near Leningrad had the two antennas mounted on separate 20-meter towers. When air attacks on Leningrad began, this system was pressed into tactical operation and provided early-warning of *Luftwaffe* raids, often in formations of up to 100 bombers. It was initially operated by the staff of LPTI, then turned over to the military as the research personnel evacuated the city.

After two months of intensive fighting involving nearly two million combatants, the Germans got the upper hand in the region around Leningrad and the siege on the city began in early September. Within the first year of the siege, many thousands of aircraft were detected by the *RUS-2* at Toksovo. With a range up to 100 km, this gave a good warning to the city, as well as to notifying the defending fighter aircraft. This gained the attention of military leaders who previously had shown little interest in radio-location equipment.

In mid-July, the LPTI and the NII-9 were evacuated from Leningrad to Moscow. A *RUS-2* system was set up near Moscow and manned by recently moved LPTI personnel; it was first used on July 22, when it detected at night an incoming flight of about 200 German bombers while they were 100 km away. This was the first air attack on Moscow, and it immediately led to three rings of anti-aircraft batteries being built around the city, all connected to a central command post.

Several transmitters and receivers built for *RUS-2* systems were quickly adapted by the NIII-KA for fixed radio-location stations around Moscow. Designated as *RUS-2S* and also *P2 Pegmatit*, these had their

Yagi antenna mounted on 20-meter steel towers and could scan a sector of 270 degrees. For building additional equipment, in January 1942, Factory 339 in Moscow became the first manufacturing facility in the Soviet Union devoted to radio-location sets (soon officially called radar). During 1942, this facility built and installed 53 *RUS-2S* sets around Moscow and other critical locations in the USSR.

Factory 339 was the part remaining in Moscow of the Aviapribo Plant, the other portions having been evacuated to Kazan. Aviapribor had been originally founded in 1917 by the noted scientist Alexander Fridman (a crater on the Moon was given his name) and was the largest supplier of aircraft instrumentation in the Soviet Union.

The Aviapribor Plant had an outstanding research and engineering staff; this had earlier been administratively separated and designated as the Scientific Institute of Radio Industry No. 20 (NII-20). Victor V. Tikhomirov, a pioneer in domestic aircraft radio engineering, was the Technical Director. Factory 339 and the associated NII-20 dominated radar equipment development and fabrication in the USSR throughout the war.

Victor Tikhomirov

About 700 of a number of different versions of the *RUS-2* were built during the war. While providing early warning, these sets suffered from the deficiency of not providing target height (elevation angle). Thus, they were mainly used in conjunction with visual-observation posts, with humans using optical devices for estimating altitude and identifying the type of aircraft.

From the time of the first efforts in radio-location, the question had been raised as to how the aircraft identification could be made – was it friendly or an enemy? With the introduction of *RUS-2*, this problem required an immediate solution. At NII-20, E. N. Genishta led in developing a unit to be carried on an aircraft that would automatically respond as "friendly" to a radio illumination from a Soviet radar. The greatest problem was in determining the optimum antenna position on different types of aircraft

A transponder, designated as *SCH-3* and later called an Identification Friend or Foe (IFF) unit, was placed into production at Factory 339 in 1942. This unit initially responded only to the 4-m (75-MHz) signal of *RUS-2*. While thousands of IFF sets were built for aircraft and ships in America and Great Britain, only a relatively small number of *SCH-3* and successor units were built in the USSR.

The *RUS-2* was sponsored by the PVO and intended for early warning. The GAU still wanted a gun-laying system capable of supporting the anti-aircraft batteries. Upon arriving in Moscow, the radio-location group of the NII-9 continued working for the PVO on this problem, returning to the magnetrons built earlier by Malyarov and Alekseyev and the *Burya* experimental microwave sets. Within a few weeks, a team led by Mikhail L. Sliozberg and with the cooperation of NII-20, developed a bi-static CW set designated *Son* (Sleep) using a 15-cm (2.0-GHz) magnetron.

In early October, the experimental *Son* set was tested in combat by an anti-aircraft battalion near Moscow. The performance of the radio-based *Son* was poor as compared with that of the existing optics-based *Puazo*. The project was discontinued, and no further attempts were made to use magnetrons in radio-location sets. The available magnetrons were simply too unstable in frequency for either the transmitter or local oscillator in the super-heterodyne receiver.

After this failure, NII-9 was no longer involved in radio-location activities. Later references to NII-9 concerned work at other locations on missile control systems and equipment for the atomic program; thus they were apparently again evacuated. A portion of the radio-location group, including Sliozberg, remained in Moscow working for NII-20.

In July 1940, the Government had created a special commission on uranium separation; that was the first official act in the Soviet Union's Atomic Bomb program. Abram Ioffe was Deputy Chairman of the commission and from this point onward edged the LPTI toward activities that supported this effort. (Ioffe later turned down heading the overall bomb development.) After the war started, the LPTI first left Leningrad for Moscow, but soon evacuated again and did not continue with radio-location work; as with NII-9, however, some of the team joined NII-20 in Moscow.

Shortly after Germany invaded the USSR, a delegation of Soviet military officers visited Great Britain seeking assistance in defense hardware. From their intelligence sources, the Soviets were aware of Britain's gun-laying RDF (radar) system, the GL Mk II, and asked for this equipment to be tested in the defense of Moscow. In early January 1942, Churchill agreed to send one of these systems to Russia, but with the provision that it would be totally secured under British officers and operated by British technicians.

When the ship carrying the equipment arrived at Murmansk, a seaport off the Bering Sea above the Arctic Circle, there was a winter storm and unloading had to wait overnight. The next morning, it was found that the entire GL Mk II system – mounted on three trucks – had disappeared. The British Embassy made an immediate protest, and after several days the officers were informed that the equipment had been taken to Moscow for security.

It indeed had gone to Moscow – directly to NII-20 and Factory 339, where intelligence experts gave it a total examination and Mikhail Sliozberg led a team in quickly reverse-engineering the hardware. In mid-February, the NII-20 announced that it had developed a new radio-location system designated *Son-2a*. It was essentially a direct copy of the GL Mk II.

Operating at 5 m (60 MHz), *Son-2a* used separate trucks for the transmitting and receiving equipment, and a third truck carried a power generator. In use, a dipole-array transmitting antenna giving a broad pattern was fixed in position atop a grounded pole. Separated from the transmitter by about 100 meters, the receiving station was on a rotatable cabin with wing-like antennas mounted on each side. A mast above the cabin held a pair of antennas that were used with a goniometer for height-finding.

Like its "parent" in Great Britain, the *Son-2a* was not of great assistance in directing searchlights and anti-aircraft guns. Nevertheless, it was put into production and released to the Red Army in December 1942. Over the next three years, about 125 of these sets were built. Under the Lend-Lease Program, over 200 GL Mk II systems from Great Britain were also provided, making the combination the most-used radar equipment in the Soviet Union during the war.

Ukraine had been the third objective of the invading German Army. By late July 1941, the mechanized forces were approaching this region, and, following orders from the Defense Committee, the UIPT in Kharkov made evacuation preparations. For this, the LEMO was split from the UIPT, and the two organizations would be sent to different cities: Alma-Ata for the main operation and, separated by 1,500 km, Bukhara for the LEMO. The war had accomplished what the scientists had earlier failed to do: fully dividing the Institute's activities.

While the preparations for moving were going on, the LEMO was directed to bring the experimental *Zenit* equipment to Moscow for testing by the NIIIS-KA. In the middle of August, A. Y. Usikov, S. Y. Braude, and several other LEMO staff members went to Moscow, where they were attached to the NIIIS-KA, and the *Zenit* system was installed in the Moscow outskirts.

The NIIIS-KA offered the Moscow Defense Command the opportunity to test the *Zenit* system in combat. It was noted that, while the accuracy of the system was not sufficient for precise aiming, it was satisfactory for barrage firing. It could also be used as a supplement to the *RUS-2* surveillance system in guiding fighter aircraft.

The offer was accepted and connections were made directly with the command post. The defense officials were initially skeptical as to the value of this system; that changed, however, when a bomber that had left the main formation was detected, reported to the command post, and shot down by anti-aircraft fire, all in few minutes.

In September, the team made field modifications to the *Zenit* and more tests were run. It was found that the detection range had been doubled, but the dead zone increased by a like amount. The NIIIS-KA believed that the prospects were good for this to be developed into a suitable system, but laboratory conditions were necessary for making the improvements. Thus, the *Zenit* and all of the NIIIS-KA staff were sent 3,200 km away to Bukhara, joining the remainder of the LEMO as it also moved.

It is noted that when the German army approached Kharkov and the LEMO was evacuated, Pavel Lelyakov, their leading expert on magnetrons, stayed behind and then left with the Germans when they retreated in 1942. It is also noteworthy that the UIPT in Alma-Ata was an early and important participant in the Soviet atomic program.

The relocation of the LEMO to Bukhara possibly reflected the attitude of the Government toward radio-location at that time. While very safe from the German threat, the provincial city offered no scientific or technology infrastructure, and the closest equipment fabrication facilities were at Trchkent, 425 km to the northwest across deserts and mountains. No real expectation could have been held that the scientists and engineers in this environment would make any significant contributions.

In late October 1941, trains brought the laboratories and personnel from Kharkov and the *Zenit* equipment and other personnel from Moscow, but it was into 1942 before adequate facilities were completed in Bukhara and the LEMO and NIIIS-KA staffs could began meaningful work. Still under funding from the GAU, their objective was to develop a radio-location system with sufficient range, azimuth, and elevation accuracies to support anti-aircraft batteries.

Because of the null-reading method of analyzing the signals, the *Zenit* system suffered from slowness in measurements (38 seconds for determining the three coordinates) as well as accuracy. It also had a large dead zone caused by ground returns. While still at Kharkov, work had started on *Rubin*, a system intended to correct *Zenit* deficiencies. With

Abram Slutskin as LEMO Director, this project continued at Bukhara under Aleksandr Usikov's leadership.

Major contributions in the development of *Rubin* were made by Ivan D. Truten. He had been with the LPTI since 1932, when he was an undergraduate student. As a graduate student at Kharkov University, he studied under Professor Slutskin and completed his dissertation while at Bukhara. He later made significant advancements in magnetrons and was recognized by being named a Lenin Prize recipient.

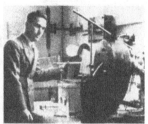
Ivan Truten

The basic transmitter and receiver of the *Rubin* system started with those of the *Zenit*. A somewhat improved magnetron operated at 54 cm (470 MHz) with a pulse-power increased to 15 kW. The superheterodyne receiver was improved, with two intermediate-frequency transforms and high-frequency blocking. Truten developed a gas-discharge T-R device that isolated the receiver from the direct transmitter pulse, thus allowing the use of a common transmitting-receiving structure. (A similar development had been made for the *RUS-2* common antenna, but this would not have been suitable for the microwave *Rubin*.)

Several techniques for replacing the null-reading methods were considered, with the final selection making use of a fixture to provide a stationary dipole against which the directional position of the antenna could be continuously determined. Range, azimuth, and elevation were shown on a CRT display. There was no provision, however, for feeding this information into an automatic unit for aiming searchlights and guns.

Rubin Antenna

Separate transmitting and receiving dipoles were at the focus of a 3-meter paraboloid reflector. A reflector from the *Zenit* was initially used, but the wind load was too great. The team built a segmented reflector made of a wire net stretched on a frame, carefully soldering the wires spaced in 20-mm squares. The antenna assembly, with remote controls, could rotate 0-90 degrees vertically and 0-400 degrees horizontally. The width of the main beam was 16 degrees equatorial and 24 degrees meridian.

Rubin Deployed

The system was carried on two trucks, the electronics and control console in one and the power generator in the other. Both the transmitter magnetron and front-end portions of the receiver were in sealed containers attached to the rear of the reflector. The antenna assembly was on rails and could be rolled out to near the truck.

Testing of units was done in the desert outside Bukhara. Here an unusual phenomenon was observed: occasional reflections from very distant objects so strong that they obscured nearby targets. Usikov concluded that it was the result of extremely low electromagnetic attenuation along an atmospheric channel over the desert. This was later called the "surface-ducting effect," through which greatly increased ranges were observed.

By August 1943, the prototype *Rubin* system was completed. (It is again noted that all of the work had to be performed by the small LEMO and NIIIS-KA staffs.) The system was transported to Moscow where Usikov, Truten, and others conducted further tests and gave non-combat demonstrations. By this time, the British GL Mk II and its Soviet replication, *Sun-2a*, were also available and were possibly used in direct comparison with the *Rubin*; if so, the *Rubin* would not have faired well.

Rather than releasing the prototype for production, the Army made arrangements for the *Rubin* to be tried by the Red Fleet Command. At the beginning of 1944, the system was transported to Murmansk, the only non-freezing port in the Soviet Arctic. Here, despite the cold, Usikov continued with tests and demonstrations under better conditions than in the still chaotic Moscow.

Test aboard a ship showed aircraft detection at 60 km and reliable measurement starting at 40 km. The mean errors were no more than 120-m in range and 0.8-degree in azimuth and elevation angles. The time for determining the angular coordinates never exceeded 7 seconds, and the dead zone was down to 500 m. Similar accuracies were found for detecting all types of surface vessels, but with the *Rubin* antenna at deck level, the detection range was understandably much less than that for aircraft.

During the last year of the war, *Rubin* was used by the NKKF (Navy) for air and surface surveillance in the polar sector. If the GL Mk II and its clone, *SUN-2a*, had not become available, the *Rubin* would likely have been completed much earlier and gone into production. Although never

Radio-Location Development in the Soviet Union

put into regular service, this system provided a good foundation for future magnetron-based radars in the Soviet Union.

Airborne Radar Systems

A number of new fighter and bomber aircraft were being designed in the years before the war. Vladimr Petlyakov led a Soviet Air Forces (VVS) design bureau, responsible for developing a twin-engine attack-dive bomber that was eventually designated *Pe-2*. Having gotten behind in the schedule, Petlyakov was charged with sabotage and thrown into a technical *Gulag* for a year, actually doing a large part of his design there.

In late 1940, the VVS developed the requirement for an on-board enemy aircraft detection system. The radio-location group at NII-9 in Leningrad was directed to design such a set for the *Pe-2*. Most of radio-location equipment at that time was large and heavy, and for this aircraft, a small, lightweight set was needed. Also, limitations on antenna size drove the design to frequencies as high as possible. The reflex klystron (as it was later called) had just been developed by N. D. Devyatkov. Using this, the design was started on a set designated *Gneis* (Origin) and operating at 16 cm (1.8 GHz).

When the NII-9 was evacuated to Moscow in July 1941, this greatly affected the schedule. Also, the reflex klystron had not been put into production and its availability in the future was doubtful; therefore, the project was terminated. The need, however, for an airborne radio-location set was now even more important; the *Pe-3*, a heavy fighter aircraft and successor to the *Pe-2*, was in production. Some of these aircraft were being configured as night-fighters, and the radar (as it was now called) was urgently needed. The NII-20 and Factory 339 took up the design, led by the Technical Director, Victor V. Tikhomirov.

The new set, designated *Gneis-2*, operated at 1.5 m (200 MHz). The *Pe-3* fighter was a two-place aircraft, with the pilot and the rear gunner/radio operator seated back to back. The radar was designed as another piece of equipment for the radio operator.

The antennas were mounted above the top surface of the wings, a broad-pattern transmitting array on one wing and two Yagi receiving antennas on the other. One Yagi was directed forward and the other, a few feet away, aimed outward 45 degrees. The fuselage of the aircraft provided a shield between the transmitting and

Pe-3 with Gneis-2

receiving antennas. The system had a range of about 4 km and could give the target's azimuth relative to the fighter's flight path.

The *Gneis-2*, the first aircraft radar in the Soviet Union, was proven in combat at Stalingrad during December 1942. About 230 of these sets were built during the war. A few were installed on Yak-9 and (out of number sequence) Yak-3 aircraft, the advanced fighters that eventually gave the VVS parity with the *Luftwaffe*. Other sets with *Gneis* designations were developed at Plant 339 for experimental purposes, particularly with Lavochkin La-5 fighters and Ilyushin Il-2 ground-assault aircraft, but none of these sets were placed into production.

Naval Radar Systems

During the 1930s, the RKKF (Red Navy) had major programs in developing radio communications. Starting in 1932, this activity was headed by Aksel Ivanovich Berg (1893-1979, Director of the NIIIS-KF, Red Fleet Signals Research) and later given the rank of Engineer-Admiral. He was an electrical engineering graduate from the Naval Institute, and completed graduate studies with the Electrotechnical Faculty. Berg was also a Professor at Leningrad's universities and closely followed the early radio-location progress at the LPTI and NII-9. He started a research program in this technology at the NIIIS-KF, but was interrupted by being arrested in 1937 during the Great Purge and spent three years in prison.

Aksel Berg

Berg was released in early 1940 and reinstated in his positions. After reviewing the tests of *Redut* conducted at Sevastopol, he obtained an *RUS-2* cabin and had it adapted for shipboard testing. Designated *Redut-K*, it was placed on the light cruiser *Molotov* in April 1941, making this the first warship in the RKKF with a radio-location capability. After the start of the war, only a few of these sets were built.

In mid-1943, radar (*radiolokatsiya*) was finally recognized as a vital Soviet activity. A Council for Radar, attached to the State Defense Committee, was established; Berg was made Deputy Minister, responsible for all radar in the USSR. While involved with all future developments in this activity, he took special interest in Navy systems. It should also be noted that Berg was later mainly responsible for introducing cybernetics in the Soviet Union.

Other indigenous Soviet Navy radars developed (but not put into production) during the war included *Gyuis-1,* operating at 1.4 m with 80-kW pulse power. This was a successor to *Redut-K* for early warning; the prototype was installed on the destroyer *Gromkii* in 1944. Two fire-control radars were simultaneously developed: *Mars-1* for cruisers and *Mars-2* for destroyers. Both were tested just at the close of the war, and later placed into production as *Redan-1* and *Redan-2,* respectively.

CLOSURE

Until the surrender of Germany on May 7, 1945, the Soviet Union lost about 27 million people, 14 percent of its population.

Although essentially all of the USSR's industry was devoted to the war, much materiel was received from the U.S., Great Britain, and Canada under the Lend-Lease Program. In this, the materiel was "loaned" with the provision of post-war payment. Among other items, the Soviet Union received about 19,000 aircraft, 12,000 tanks, 442,000 vehicles, 600 naval vessels, 43,000 radios, and 2,600 radars.

Most of the radars used by the Soviet Navy during the war were British and American units, provided under the Lend-Lease Program. The shipments began in late 1942, and by the end of the war, essentially all large vessels were fitted with one or more of these systems.

Radars from Great Britain included the following Types: 271, 281, 282, 284, 285, 286, and 291. The British did not sequentially number their systems, so the numerical order does not necessarily relate to when the sets were supplied. The Type 271, a 10-cm (3.0-GHz) surface-search system, was the first microwave radar placed into full service in the USSR.

American-built naval systems provided included the SG, SF, SL, and SO, all 10-cm surface-search sets; the SK, a 1.5-m (200-MHz) air-search set; and the FC (Mark 3), a 40-cm (750-MHz) fire-control set. Most of these radars were included aboard fully equipped vessels. A relatively small number of air-defense systems were provided, including several of the high-performing SCR-586, less the M-9 Predictor-Corrector Unit.

Most of the scientists and engineers involved in developing radar continued in important roles after the war. Pavel Oshchepkov was released from the *Gulag* camp in 1946. Returning to academic studies, he eventually earned doctoral degrees and was given the title Professor. He never returned to radar research, but found entirely new avenues for his creativity: thermal physics and material science.

The Leningrad Physical-Technical Institute (LPTI) reopened and Abram Ioffe continued as Director. During a campaign in 1950 by Stalin against Jews, Ioffe was removed from this position but remained there as a scientist, primarily working in semiconductor development. A large crater on the far side of the Moon was named in Ioffe's honor.

The Ukranian Institute of Physics and Technology (UIPT) reopened in Kharkov (now often spelled Kharkiv), but the Laboratory of Electromagnetic Oscillations (LEMO) remained a separate department under Abram Slutskin. Continued research in magnetrons was conducted by Ivan Truten, resulting in a series of mm-wavelength-band magnetrons. A closely related operation, the Radio Wave Propagation, was formed in 1950 under Aleksandr Usikov. In 1955, these departments were combined to form the Institute of Radio-Physics and Electronics (IRE), directed by Usikov. Later, this Institute was named for Usikov.

Semion Braude remained with the LEMO and then the IRE. He was a USSR State Prize (1952) and Ukrainian National Prize winner (1997), and was awarded a title of Honored Scientist of Ukraine and the A. S. Popov Gold Medal of the USSR Academy of Sciences. His accomplishments included the design of the world's largest decameter-band radio telescope, the UTR-2.

Although the USSR had exhausted its resources during the war, Stalin immediately set this nation on a path that attempted to dominate in military power. Included in this effort was the reverse-engineering of many of the British and American radars for production in the USSR. The major efforts, however, were in missiles and nuclear bombs. As earlier noted, many of the research institutes were converted from radar and electronics to these new activities.

Russia had a long heritage in rocket research, actually pre-dating both Germany and the U.S. During the war, many types of short-range unguided rockets were produced, and a number of the research institutes became engaged in guidance technologies; however, no operational guided rockets were used by the USSR during the war.

Immediately with the fall of Germany, captured guided rockets were sent to Russia, along with many scientists and engineers who had developed them. This included a number fully operational V-2 rocket systems. In April 1949, the decision was made to produce the R-1, basically a copy of the V-2. The first R-1 systems were placed into service in November 1950. Sergei Pavlovich Korolyov was the leader of the Soviet missile program.

The Soviet nuclear weapons program actually began in 1943, initiated by reports about the rapidly growing Manhattan Project in the U.S. During the last two years of the war, it was primarily an intelligence operation, assessing information provided by spies at "highly secret" American facilities.

Radio-Location Development in the Soviet Union

When the two atomic bombs were dropped on Japan and ending the war in August 1945, Klaus Fuchs, a senior physicist at the Manhattan Project's scientific laboratory in Los Alamos, New Mexico, had already given the USSR the full design of one of these bombs. Under the leadership of physicist Igor Vasilievich Kurchatov, the Soviet Union's first atomic bomb was completed and detonated in August 1949. This was followed by the thermonuclear bomb in August 1953, just 10 months after the detonation of a similar bomb by America.

Bibliography and References for Chapter 6

Brown, Lewis; *A Radar History of World War II – Technical and Military Imperatives*, Inst. of Physics Pub., 1999

Chernyak, V. S., I. Ya. Immoreev, and B. M. Vovshin; "Radar in the Soviet Union and Russia: A Brief Historical Outline," *IEEE AES Magazine*, vol. 19, December, p. 8, 2003

Erickson, J.; "The air defense problem and the Soviet radar programme 1934/35-1945," in *Radar Development to 1945*, ed. by Russell Burns, Peter Peregrinus Ltd., 1988

Erickson, John; "Radio-location and the air defense problem: The design and development of Soviet Radar 1934-40," *Social Studies of Science*, vol. 2, p. 241, 1972

Gallagher, Matthew; *The Soviet History of World War II: Myths, Memories, and Realities*, Greenwood Press, 1976

Gregorovich, Andrew; "World War II in Ukraine," *Forum: A Ukrainian Review*, No. 12, Spring, 1995

Ioffe, A. F.; "Contemporary problems of the development of the technology of air defense," *Sbornik PVO*, February 1934 (in Russian)

Kobzarev, Y. B.; "The First Soviet Pulse Radar," *Radiotekhnikn*, vol. 29, No. 5, p. 2, 1974 (in Russian)

Kostenko, Alexei A., Alexander I. Nosich, and Irina A. Tishchenko; "Development of the First Soviet Three-Coordinate L-Band Pulsed Radar in Kharkov Before WWII," *IEEE AP Magazine*, vol. 43, June, p. 31, 2001

Kostenko, A. A., A. I. Nosich., and I. A. Tishchenko; "Radar Prehistory, Soviet Side," *Proceedings of IEEE APS International Symposium 2001*, vol. 4, p. 44, 2002

Lobanov, M. M.; *The Beginning of Soviet Radar*, Sovetskoye Radio, 1975 (in Russian)

Lobanov, M. M.; *Development of Soviet Radar Technologies*, Voyenizdat, 1982 (in Russian)

Oshchepkov, P. K.; *Life and Dreams: Notes of Engineer, Inventor, Designer and Scientist*, Mosltovsky Rabochy, 1967 (in Russian)

"Russian Radar Equipment in World War II," *Taifun Magazine*, Feb. 2002; http://www.navweaps.com/Weapons/WNRussian_Radar_WWII.htm

Shenibel, B. K.; *At the Origin of Radar in USSR*, Sovetskoye Radio, 1977 (in Russian)

Siddiqi, Asif A.; "Rockets Red Glare: Technology, Conflict, and Terror in the Soviet Union; *Technology & Culture*, vol. 44, p. 470, 2003

Tvrnov, O. F. and B. G. Yemets; "Fifty years of Kharkov University's Department of Radio Physics," *Proc. of the IEEE International Crimean Conference*, p. 824, Sept. 2003

Chapter 7
RRF DEVELOPMENT IN JAPAN

At the start of World War II, the Japanese Imperial Navy, and especially its Air Division, was on a par with that in any other country. The Imperial Army was also highly competent, having shown its might in the defeat of China several years earlier. From December 7, 1941, until June 6, 1942, Japan appeared to be invincible – scoring victory after victory.

Then came the Battle of Midway. In that turning point, a larger Japanese naval force was devastated by the United States, with the Japanese losing four carriers and 300 planes to American losses of a single carrier and 100 planes. While many factors contributed to this difference, radar was certainly one; the U.S. warships had radar but the Japanese aircraft carriers had none.

This deficiency was by no means caused by any lack of scientific and engineering capability; the Japanese warships and aircraft clearly showed high levels of technical competency. They were ahead of Great Britain in the development of magnetrons, and their Yagi antenna was the world standard for VHF systems. It was simply that the top military leaders failed to recognize how the application of radio in detection and ranging – what was often called the Radio Range Finder (RRF) – could be of value, particularly in any offensive role; offense not defense, totally dominated their thinking.

HISTORICAL BACKGROUND

For 700 years, the island nation of Nippon (Japan) was ruled by a *shogun* (military dictator) and essentially closed to the western world. In 1853, Commodore Matthew Perry sailed an American fleet into Tokyo Bay and forced Japan to open up for trade; thus began Japan's transition from a medieval to a modern power. In a few years, the feudal world of the shoguns was abolished and an emperor returned to the throne. In 1889, a parliamentary government with a Prime Minister was formed.

After establishing a *Teikoku Rikugun* (Imperial Army) and *Teikoku Kaigun* (Imperial Navy), Japan initiated an aggressive expansion with the objective of controlling the Far East. A brief war with China (1894-95) gained them Formosa (Taiwan) and southern Manchuria. They defeated the Russian Navy (1904-05) and took control of Port Arthur and other eastern ports on China's Liaodong Peninsula, as well as certain rail-rights, all of which were previously leased by Russia. Chosen (Korea) was annexed in 1910. Japan declared war on Germany in 1914 and seized Germany's Pacific islands and certain leased areas in China. At the 1922 Peace Conference, however, Japan agreed to respect Chinese national integrity.

The growth of modern Japan, especially on the main island of Honshu, was thwarted by an 8.4-magnitude earthquake and accompanying tsunnami on September 1, 1923. Tokyo, Yokohama, and all the large cities were devastated, mainly through firestorms that swept across the predominantly wooden-constructed houses and buildings, driven by winds from an off-shore typhoon. Upwards of 38,000 people who packed into the huge Army Parade Ground in downtown Tokyo were incinerated. Rioting occurred throughout the nation and the civil government was slow to react, helping forge the path to eventual military dominance. It was years before the physical and economic recovery was complete.

Japan had party governments in the 1920s, but, as the Great Depression came, there was major political unrest and the military gained power. In 1931, without the knowledge of the civilian government, the military invaded Manchuria and set up this area as a puppet state. In 1932, party cabinets were abolished and the Army was increasingly in control. The Prime Minister was selected by the Army and, with their backing, exercised essentially total control of national affairs. There was still an emperor, but his role was symbolic. A major effort was placed on preparing for a war of aggression. On November 25, 1936, Japan joined Germany and Italy in a Tripartite Pact. The invasion of China came the next year, followed by the attack on the U.S. and southeastern Pacific areas beginning December 7, 1941.

MAGNETRONS AND OTHER EARLY DEVELOPMENTS

As a sea-faring nation, Japan had an early interest in wireless (radio) communications. Message exchanges with ships outside visual and audio contact quickly came to rely on this new technology. The Japanese alphabet (*katakana* or *kana*) has 48 characters, and with the earlier advent of the wired telegraph, the International Morse Code with 26 Roman characters had been extended to the *Kana*-based Code to accommodate Japanese communications. The first known use of wireless telegraphy in warfare at sea was by the Japanese Imperial Navy fighting the Russian Imperial Fleet in 1904; this used a set developed by Shunkiti Kimura at the Tokyo Imperial University.

In addition to equipment for exchanging messages, there was an early interest in equipment for radio direction-finding, for assistance in both navigation and military surveillance. The Imperial Navy developed the Type 10 Direction Finder in 1921, and soon most of the Japanese warships had this system. Knowing that similar equipment was available to other navies, the Japanese often maintained "operational silence" for security. This led to the reluctance of both the Army and Navy to consider radio for active surveillance (radar), fearing that the radiated signals would disclose their location.

Radio engineering was strong in the institutions of higher education, especially the Imperial (government-financed) universities. This included undergraduate and graduate study, as well as academic research in this field. Special relationships were established with foreign universities and institutes, particularly in Germany, with Japanese teachers and researchers often going overseas for advanced study.

The academic research tended toward the improvement of basic technologies, rather than their specific applications. For example, there was considerable research in high-frequency and high-power oscillators, such as the magnetron, but the application of these devices was generally left to industrial and military researchers. One of the best-known academic researchers from the 1920s–30s era was Professor Hidetsugu Yagi.

Hidetsugu Yagi (1886-1976) received his degree in electrical engineering from Tokyo Imperial University in 1909, then studied and performed research in Germany under Heinrich G. Barkhausen at the Dresden Technische Hochschule. As World War I loomed, Yagi left Germany and went to Great Britain, where he studied electrical engineering with John A. Fleming at University College, London; then to America and further study under George W. Pierce at Harvard. Returning to Japan, he began his academic career at Tohoku Imperial University and was awarded the engineering doctorate there in 1919. Two of his first doctoral students, Shintaro Uda and Kinjiro Okabe, would remain at Tohoku after graduation and participate with Yagi in extremely important projects.

Hidetsugu Yagi with Antenna

Yagi's research at Tohoku was supported by a private foundation and centered on antennas and oscillators for high-frequency communications. With assistance from Uda, a radically new antenna emerged. First called a wave projector, it had a number of parasitic elements (directors and reflectors) and would come to be known as the Yagi or Yagi-Uda antenna. They published the first paper on this antenna in the February 1926 issue of a Japanese journal, and applied for Japanese and American patents. The U.S. patent, issued in May 1932, was assigned to RCA. To this day, this is the most widely used directional antenna worldwide.

Shintaro Uda

The magnetron was another of Yagi's interest at Tohoku University. This high-frequency (~10-MHz) device had been invented in 1921 by Albert W. Hull at General Electric, but Yagi saw it as a potential VHF and even UHF generator. In 1927, under Yagi's guidance, Kinjiro Okabe developed a split-anode device that ultimately generated oscillations at wavelengths down to about 12 cm (2.5 GHz). Okabe was awarded the D.Eng. degree in 1928, and a paper by him on the subject was published in the April 1929 issue of the *Proc. of the IRE*.

A summary of the radio research work at Tohoku University was contained in a seminal paper by Yagi in the June 1928 issue of the *Proc. of the IRE*. Information on the wave projector was given, as well as a description of the system using a 40-cm magnetron with a wave-projector antenna that he, Uda, and Okabe had tested to achieve a transmission distance of about 1 km.

In the early 1930s, Yagi moved to the Osaka Imperial University, where he was appointed Director of the Radio Research Laboratory. Okabe accompanied Yagi to Osaka and continued his research on split-anode magnetrons, but, as later noted, in 1936 he changed his research to VHF equipment and applications.

Researchers at other Japanese universities and institutions also started projects in magnetron development, leading to improvements in the split-anode device. These included Dr. Kiyoshi Morita at the Tokyo Institute of Technology, Dr. Yoji Ito at the Naval Technical Research Institute (NTRI), and Dr. Tsuneo Ito at Tohoku Imperial University. An industrial effort was also in process under Dr. Shigeru Nakajima at Japan Radio Company (JRC).

For use in his doctoral research, Morita had designed an improved split-anode magnetron. He provided detailed drawings to the JRC and asked them to build the device; this was done at JRC's Tube Department in 1933. After graduating, Morita remained at Tokyo Institute of

Technology and was a major participant in the future development of magnetrons and especially in their laboratory applications.

The JRC Tube Department was headed by Shigeru Nakajima. Many General Electric patents for tubes were licensed by the competing firm, Tokyo Shibaura Denki (trade named Toshiba), and JRC was attempting to find devices that would circumvent these patents. Nakajima had earlier examined without success the Barkhausen-Kurz tube, but, after being introduced to Morita's device, he saw the commercial potential of magnetrons. With the medical diathermy market as an objective, Nakajima began the further development and subsequent very profitable production of magnetrons.

Yoji Ito (1901-1950), an older brother of Shigeru Nakajima, was of central importance in the development of Japanese radar. Their father was a schoolmaster in a small fishing village and encouraged them to achieve science and mathematics excellence. After graduating in electrical engineering from the Tokyo Imperial University, Ito was commissioned in the Imperial Navy. After several years in assignments at sea, he was sent to Germany for graduate study where, like Yagi, he was a student of Heinrich Barkhausen at the Dresden Technische Hochschule. Upon completing his engineering doctorate there in 1929, he was promoted to the rank of Commander and assigned as a researcher at the recently opened Naval Technology Research Institute (NTRI) at Neguro in Tokyo.

Yoji Ito

At NTRI, Ito was involved with analyzing long-distance radio communications, and wanted to investigate the interaction of microwaves with the Kenelly-Heaviside Layer (the ionosphere). He started a project using a Barkhausen-Kurz tube, then tried a split-anode magnetron developed by Okabe but the frequency was too unstable. In late 1932, believing that the magnetron would eventually become the primary source for microwave power, he started his own research in this technology, calling it the magnetic electric tube. Since both JRC and NTRI were then engaged in magnetron research, a cooperative effort ensued and continued for many years.

Looking ahead at Ito's contributions, he remained a career naval officer, was named a Department Director at the NTRI, and promoted to the rank of Captain. His technical accomplishment gained him respect in the scientific community, and his leadership skills soon established him as the primary technical liaison between the Navy, industries, and university research institutes in radar development. As later described, in 1940, he was a key member of a delegation to Germany for a technology exchange.

As a doctoral research project at Tohoku Imperial University, in 1936, Tsuneo Ito (no relationship to Yoji Ito) developed an 8-split-anode magnetron that produced about 10 W at 10 cm (3 GHz). Based on its appearance, it was named *Tachibana* (or Mandarin, an orange citrus fruit). After completing his doctorate, Tsuneo Ito also joined the NTRI and continued his research on magnetrons in association with Yoji Ito. In 1937, they developed the technique of coupling adjacent segments (called push-pull), resulting in frequency stability, an extremely important magnetron breakthrough.

By early 1939, NTRI/JRC had a 10-cm (3-GHz), stable-frequency Mandarin-type magnetron (No. M3) that, with water cooling, could produce 500-W power. In the same time period, magnetrons were built with 10 and 12 cavities in what was called the Cosmos or Rising Sun configuration (again based on appearance) and operating at wavelengths as low as 0.7 cm (40 GHz). Subsequent developments provided progressively higher powers.

Magnetron M3 - 1939

The configuration of the NTRI Mandarin-type magnetron was essentially the same as that used in the frequency-stabilized, 8-segment, Boots and Randall magnetron developed at Birmingham University in early 1940; however, unlike the high-power magnetron in Great Britain, the device from the NTRI generated only a few hundred watts.

THE MILITARY AND RADAR EVOLUTION

Until shortly before the start of World War II, there were almost no applications of radio for other than communications in either the Imperial Navy or the Imperial Army. The need for augmenting human senses for detecting the enemy was certainly recognized, but apparently no thought was given to using radio for this. Sound-detection devices for relatively long-range warning of approaching aircraft appeared to satisfy their needs.

Listening Systems - 1930

In both of the Army and Navy, there were a few visionaries in radio technology, and a few activities – such as the previously mentioned radio-propagation studies in the NTRI – that should have led to significant accomplishments. There were, however, two major hindrances: organizational ego and enmity.

For a long time, the Japanese had believed they had the best fighting ability of any military force in the world. This belief was further engendered by their victories over the giants Russia and China. The military leaders, who were then also in control of the government, sincerely felt that the weapons, aircraft, and ships that they had built were fully sufficient and, with these as they were, the Japanese Army and Navy were invincible.

Certainly in the pre-war years, and to a large measure extending throughout the war, there was essentially no cooperation between the two military organizations concerning technology advancements. Further, there was a gap between the military and non-military personnel. In a document commonly called the "Compton Report," prepared immediately after the war by the United States Scientific Intelligence Commission, Hidetsugu Yagi is quoted as follows:

> The Army and Navy certainly never co-operated with each other. Each acted as if it would be preferable to lose the war rather than co-operate. . . . They treated [university] scientists exactly as if they were 'foreigners.' . . . It seems that they attempted to keep information relating even to the enemy's weapons secret from scientists.

To back up Yagi's accusation, it is noted that the Yagi-Uda antenna was used throughout the rest of the world on many early radars, but not in Japan. It was only after capturing American and British equipment using these antennas that the Japanese used them extensively on their own systems. Further, there is no indication of university scientists being intimately involved in any of the radar developments.

The 1921 Conference settling World War I limited the number of naval vessels of Japan. In their plan to become a world power, Japan gave emphasis to quality rather than quantity. In the implementation of this plan, the Naval Technical Research Institute (NTRI) was formed in 1922 and became fully operational in 1930. Here first-rate scientists, engineers, and technicians were engaged in activities ranging from designing giant submarines to building new radio tubes. Included were all of the precursors of radar, but this did not mean that the heads of the Imperial Navy accepted these accomplishments.

The Imperial Army also had significant research and development, but mainly contracted these activities to a small number of highly select industries. It was not until 1942 that a central operation, the Tama Technological Research Institute (TTRI), was formed to promote Radio Range Finder (RRF) development.

In late 1940, Japan arranged for two missions to visit Germany and exchange information about their developments in military technology. In December, a 20-person group representing the Imperial Army traveled by the Trans-Siberian railway through then-neutral Russia. In the following January, 22 persons representing the Imperial Navy went by ship through the Panama Canal to Portugal, then by train to Berlin. Lieutenant Colonel Kinji Satake represented the Army's interest in radar, and Commander Yoji Ito did the same for the Navy.

For the next several months, the Japanese were afforded a wide range in Germany, including military installations and factories. They exchanged significant general information, as well as limited secret materials in some technologies, but very little directly concerning radar. Neither side even mentioned magnetrons; it is possible that both sides, at least initially, believed that the other had no knowledge of this technology.

In March 1941, the British Royal Navy essentially destroyed a large part of the Italian Navy during a nighttime engagement at Cape Matapan off the coast of Greece. German intelligence immediately attributed this to new radars carried by the British warships, allowing blind firing without searchlights; this information was passed on to the visiting Japanese.

The Germans eventually showed the Japanese a wrecked British GL Mk 1 radar and an MRU (the earliest searchlight-control radar), both having been left behind during the Dunkirk evacuation, and told them what they knew about the British use of pulsed systems. Ito, who was fluent in German, talked his way into seeing a *Würzburg* radar – called the most secret of Germany's weapons – and was even given a brief description of the system. The delegations returned home in August, convinced that Japan was far behind in radar development.

Satake was allowed to stay in Germany and gathered much useful information. When Japan began the war with the United States and Great Britain, Germany decided to share its radar secrets. Satake was given drawings and other documentation on the *Würzburg*, but Japan needed an actual system to expedite building their own version. A *Würzburg* was provided, but shipping it by a surface vessel was considered too risky. Therefore, a huge submarine, the *I-30* of the B-Class, was sent to transport to Japan the full system and documentation.

Japan had by far the most diverse submarine fleet ever assembled. This ranged from one-man suicide vessels to giant I-Class underwater aircraft carriers that could transport up to four floatplane bombers. Their B-Class submarines were 108-meters (356-feet) long, over 9-meters (30-feet) high, and had a forward hanger for carrying a seaplane. They could make untended voyages of up to 14,000 nautical miles.

The *I-30* reached the German U-boat base at Lorient, France, in early August, 1942. A few days later, it departed with the *Würzburg* system and related documentation in its water-tight hanger. After a six-week trip, in mid-October it reached the captured naval base at Penang on the Malay Peninsula. Following a brief refueling stop, it headed for Japan. A few days later, it struck a British-planted mine offshore from Singapore. Although it sank in relatively shallow water, the radar and documentation were not recoverable. Some items, however, were salvaged; these included a *Metox* warning receiver.

Newspaper Picture of I-30 at Lorient

Satake, who had not left Germany with the *I-30*, assembled another set of documentation and obtained another *Würzburg*. Arrangements were made with Italy to provide the transportation, again by submarine. The Italian submarines were of ordinary size, so the radar had to be disassembled and divided between two vessels. In the summer of 1943, they departed for Japan, this time with Satake and a German radar engineer from GEMA (the firm that built the *Würzburg*).

Again there was misfortune – only the submarine carrying Satake and the GEMA engineer made it. However, with the portions of the radar that they carried, they were eventually able to re-engineer the system. Two prototypes, designated *Tachi*-24 by the Army and Mark 2 Model 3 (Type 23) by the Navy, were completed in March 1944, but neither was ever put into serial production.

Developments by the Imperial Army

In 1938, engineers from the Research Office of Nippon Electric Company (NEC) were making coverage tests on high-frequency transmitters using a portable field-strength meter when rapid fading of the signal was observed. This was noted whenever an aircraft passed over the line between the transmitter and receiving meter. On an earlier trip to London, Masatsugu Kobayashi, the Manager of NEC's Tube Department, had noted a similar fading on a television receiver, and now recognized that this was due to the beat-frequency interference of the direct signal and the Doppler-shifted signal reflected from the aircraft.

Kobayashi had relationships with the Army Science Research Institute and suggested to them that this phenomenon might be used as an aircraft warning method.

Although the Army had rejected all earlier proposals for using RRF techniques, this one had appeal because it was based on an easily understandable method and would require little developmental cost and risk to prove its military value. Also, Kinji Satake of the Research Institute's staff had observed a similar phenomenon in 1937. NEC proceeded to develop a system called the Bi-static Doppler Interference Detector (BDID). The major drawback was that the technology could not provide range or travel direction of the intruding aircraft.

Japan had recently occupied parts of China, and there was little air traffic there; thus, for testing the prototype system, it was set up along the China coast. It operated between 4.0-7.5 MHz and involved a number of widely spaced stations; this formed a radio screen that could detect the presence (but nothing more) of an aircraft at distances up to 500 km. It was the Imperial Army's first deployed RRF system, placed into operation in early 1941. Since it predated the TTRI, it did not have a *Ta* designation but was called *Ko*. Later, these and similar CW detection systems were classified as Type A.

While this was being established, a similar system was developed by Satake for the Japanese homeland. This was implemented in three sectors: Tokyo, Osaka, and Fukuoka. An information center in each segment received oral warnings from both human observers and the operators at radio detector stations. The spacing of stations was usually between 65 and 240 km, with the longest leg being about 650 km, outside the mainland between Formosa to Shanghai. The transmitters operated with only a few watts power; the low power was to reduce homing vulnerability, a great fear of the military. Although originally intended to be temporary until better systems were available, they remained in operation throughout the war.

In the weeks following the December 7, 1941, air attack on Pearl Harbor, the Japanese military swept through Southeast Asia. Singapore fell in February 1942, and Corregidor the following May. The remains of what turned out to be a British GL Mk-2 radar and a Searchlight Control (SLC) radar were found at Singapore. At Corregidor, the captors found two U.S. Army radars, an SCR-268 in operating condition and a heavily damaged SCR-270.

Emplaced SCR-268

RRF Development in Japan

The Imperial Army sent Shigenori Hamada of Toshiba and Masatsugu Kobayashi of NEC to examine these sets and ship them back to their laboratories for analysis and reverse engineering. Along with the hardware, a set of hand-written notes was found, giving details of the theory and operation of the SLC. All four of the captured radars were pulsed systems, a technology that had not yet been introduced in Japanese sets. With these radars and the notes (22 pages when typed in English), the Imperial Army had full information to develop similar pulsed systems.

On April 18, 1942, four months after the start of the war, the aircraft carrier USS *Hornet* brought 16 B-25 Mitchell bombers to within 1,050 km (650 miles) of Japan. Commanded by LTC James H. Doolittle, the planes took off from the carrier and flew directly to the mainland, where they bombed Tokyo and three other cities with minimal opposition, then safely escaped; 15 going to coastal China and the remaining one (short of fuel) to Russia. While over Japan, they encountered only a few fighter aircraft and minimal anti-aircraft fire.

The relative ease with which this attack was accomplished led Japanese officials to realize the vulnerability of their islands. They had believed that American planes could not reach them from available airfields, and apparently never thought of the use of aircraft carriers for bombers. The bi-static CW defense network was inadequate. In response, the Imperial Army reassigned many aircraft to home fields and initiated a major effort to develop pulse-modulated radars to augment the existing BDID equipment in the defense system.

The Tama Technology Research Institute (TTRI or Tama), located in a highly industrialized section of Tokyo, was formed to lead the Army's RRF developments. TTRI was staffed with competent personnel, but most of their developmental work was done by contractors at the Toshiba Research Institute and the NEC research laboratory of Sunitomo Communications.

The TTRI established a system for designating the Army radar equipment, based on its use. The prefixes were Ta-Chi (written herein as Tachi) for land-based systems, Ta-Se for shipborne systems, and Ta-Ki for airborne systems. The "Ta" denoted Tama, the "Chi" was from tsuchi (earth), the "Se" means mizu (water) rapids, and "Ki" was from kuki (air).

In a rare cooperative Army and Navy effort, the radars captured at Singapore and Corregidor were jointly reverse-engineered under TTRI and NTRI direction. The Navy development related to this effort is described in the next section.

NEC had established a large research laboratory at its factory in Tamagana, just west of Tokyo. NEC was originally formed as a joint venture between Sumitomo – one of the largest firms in Japan – and the U.S. giant Western Electric. Because of this American connection, at the start of the war the Tamagana factory was taken over by the Imperial Army and, under government management, was devoted to radio research and production. The wartime products were shown as built by Sumitomo Communications; herein, however, this will continue to be called NEC.

In June 1942, both NEC and Toshiba started projects based on the captured SCR-268. The American system operated at 1.5 m (200 MHz) with a peak power of 75 kW from a ring oscillator. It had a very complex set of three antennas on a horizontal, rotatable boom and used lobe-switching.

A project led by Masatsugu Kobayashi at NEC was for a target-tracking system designated Tachi-1, essentially a copy of the SCR-268. The duplication of this system was found to be too difficult, and Tachi-1 was soon abandoned. At Toshiba, Shigenori Hamada led a project that was also for a target-tracking system. Designated Tachi-2, this was to incorporate many simplifications to the SCR-268, particularly in the antenna system. This was also abandoned before reaching serial production, primarily because tests showed that it would be too fragile for field operation.

Much was learned in these two initial projects, and this was carried into more ambitious activities. The British GL Mk 2 was much less complicated than the SCR-268 and was easily reverse engineered; in addition, the notes on the MRU, the predecessor of the GL radar series, were available. At NEC, Masanori Morita led the development of Tachi-3, a ground-based tracking radar. Although based on the GL, this included many significant changes to the original British system. Foremost were a change to a fixed-location configuration and a totally different antenna system.

Tachi-3 Transmitter Antenna

The Tachi-3 transmitter operated at 3.75 m (80 MHz), produced about 50-kW peak power, with 1- to 2-µs pulse width and 1- or 2-kHz PRF. The transmitter was designed for enclosure in an underground shelter. It used a Yagi antenna that was rigidly mounted above the shelter and the entire unit could be rotated in azimuth. By phasing the antenna elements, some elevation change could be attained.

The receiver for Tachi-3 was located in another underground shelter about 30 m distance from the transmitter. Four dipole antennas were mounted on orthogonal arms, and the shelter and antennas rotated to scan in azimuth. The receiving antenna provided a conical scan, and lobe-switching was included. The maximum range was about 40 km. NEC (Sumitomo Communications) built about 150 of these sets, and they finally entered service in early 1944.

Tachi-3 Receiver Antennas

The follow-on project at Toshiba was designated Tachi-4 and again led by Shigenori Hamada. This was for a ground-based tracking radar, again using the SCR-268 as a pattern. Still with the original 1.5 m (200 MHz) operation, this set performed reasonable well and about 70 sets were produced. These began service in mid-1944; however, by then the Tachi-3 was available and was superior in performance.

Engineers at Toshiba had already begun work on a pulse-modulated system. With the arrival of the highly damaged SCR-270, certain portions were incorporated into this ongoing development. This project went very well; a prototype of a fixed, early-warning system designated Tachi-6 was completed near the end of 1942. The transmitter operated in the 3- to 4-m (100- to 75-MHz) band and had a peak power up to 50 kW. It used a dipole-array antenna fixed atop a tall pole; this array could radiate either 360 degrees or in a selectable 90-degree segment.

Multiple receiver stations were spaced around the transmitter at a distance of about 100 m. Each of these had a hand-rotated pole with Yagi antennas at two levels, allowing azimuth and elevation measurements to about 5 degrees accuracy. One receiver station could be used to track an aircraft while the others were searching. Ranges up to 300 km were attained, shown on a CRT display ("A" Scope) with an accuracy of about 7 km. This went into service in early 1943, before the other lower-numbered Tachi sets; some 350 Tachi-6 systems were eventually built.

A transportable version of this early-warning system was added. Designated Tachi-7, the primary difference was that the transmitter with a folding antenna was on a pallet. About 60 of these were built. This was followed in 1944 with the Tachi-18, a much lighter, further simplified version that could be carried with troops. Several hundred of these "portable" sets were built, and a number were found as the Japanese vacated distant occupied territory. All of these continued to operate in the 3- to 4-m band.

Other land-based radars developed by the Imperial Army included two height-finder sets, Tachi-20 and Tachi-35, but they were too late to be put into service. There was also Tachi-28, a radar-based aircraft guidance set, but there is no information on this system. As previously noted, the TTRI developed the Tachi-24, their slightly modified version of the German *Würzburg* radar, but this was never put into production.

It was early recognized that identification of Japanese aircraft (am IFF system) was needed. Although JRC had worked on a set for several years, it was well into the war with the mounting threat of homeland invasion before a system was even tested. The Taki-17 ground set had a 184-MHz, 10-kW transmitter for interrogating, and the companion Taki-15 aircraft transponder used a 175-MHz, 100-W transmitter for replying; a provision for communicating using Kana-based code was included. The system was not mutually operational with the Navy's M-13 set that was designated for the same purpose. Only 50 sets were built, and information concerning implementation (if ever) is not available

The Imperial Army had its own ships; these included attack motorboats, large landing-craft carriers, many sizes and types of landing craft, troop and supply transport ships, and even transport submarines. For these many vessels, they developed only two types of shipboard systems: Tase-1 and Tase-2, both anti-surface radars. The Imperial Army also had its own Air Divisions with fighters, bombers, transports, and reconnaissance aircraft. Only two systems were developed for these aircraft: Taki-1, an airborne surveillance radar in three models, and Taki-11, an airborne ECM set. Information on these shipboard and airborne developments is not available.

Developments by the Imperial Navy

In the mid-1930s, some of the technical specialists in the Imperial Navy became interested in the possibility of using radio to detect aircraft, and turned to Professor Yagi of Osaka Imperial University for consultation. Yagi suggested that this might be done by examining the Doppler frequency-shift in a reflected signal; this originated from his earlier experiments in Germany on supersonic signaling.

Funding was provided to the Radio Research Laboratory at Osaka University for experimental investigation of this technique. Dr. Kinjiro Okabe (1896-1984, the inventor of the split-anode magnetron) led the effort. Theoretical analyses indicated that the reflections would be greater if the wavelength was approximately the same as the size of aircraft structures. Thus, a VHF transmitter and receiver with Yagi antennas separated some distance were used for the experiment.

In 1936, Okabe successfully detected a passing aircraft by the Doppler method; this was the first recorded demonstration in Japan of aircraft detection by radio. With this success, Okabe's research interest switched from magnetrons to VHF equipment for target detection. This, however, did not lead to any significant funding. The top levels of the Imperial Navy believed that any advantage of using radio for this purpose were greatly outweighed by enemy intercept and disclosure of the sender's presence. Time would see this position greatly change; the 1944 Order of Culture award for significant advancements in science and technology was presented to Okabe by the Emperor of Japan in recognition of these early efforts.

Kinjiro Okabe

Historically, warships in formation used lights and horns to avoid collision at night or when in fog. Newer techniques of VHF radio communications and direction-finding might also be used, but all of these methods were highly vulnerable to enemy interception. At the NTRI, Dr. Yoji Ito proposed that the UHF signal from a magnetron might be used to generate a very narrow beam that would have a greatly reduced chance of enemy detection.

Development of microwave radar for collision avoidance started in 1939, when funding was provided by the Imperial Navy to JRC for preliminary experiments. In a cooperative effort with Yoji Ito of the NTRI and Shigeru Nakajima of JRC, an apparatus using a 3-cm (10-GHz) magnetron with frequency modulation was designed and built. The extremely short wavelength was selected to make the beam as narrow as possible. Details were not given, but it is likely that dipole antennas with parabolic reflectors were used.

The equipment was used in an attempt to detect reflections from tall structures a few kilometers away. This experiment gave poor results, attributed to the very low power from the magnetron. The initial magnetron was replaced by one operating at 16 cm (1.9 GHz) and with considerably higher power. The results were then much better, and in October 1940, the equipment obtained clear echoes from a ship in Tokyo Bay at a distance of about 10 km. Nothing more was done at this time, however; there was still no commitment by top Japanese naval officials for using radar aboard warships.

After receiving the reports from the Technical Exchange Commission in Germany, as well as intelligence reports concerning the success of Great Britain with blind-firing using radar, the Naval General Staff reversed itself and accepted this technology. On August 2, 1941, even before Yoji Ito returned to Japan, funds were allocated for the initial development of pulse-modulated radars. Commander Chuji Hashimoto of the NTRI was responsible for initiating this activity.

A prototype set operating at 4.2 m (71 MHz) and producing about 5 kW was completed on a crash basis. With the NTRI in the lead, the firm NEC and the Research Laboratory of Japan Broadcasting Corporation (NHK) made major contributions to the effort. Kenjiro Takayanagi (1899-1990), Chief Engineer of NHK's experimental television station and called "the father of Japanese television," was especially helpful in rapidly developing the pulse-forming and timing circuits, as well as the receiver display. In early September 1941, the prototype set was first tested; it detected a single bomber at 97 km and a flight of aircraft at 145 km,

Kenjiro Takayanagi

The system, Japan's first full radar, was designated Mark 1 Model 1. (This type of designation is shortened herein to the numbers only, e.g. 1,1 or more clearly Type 11.) Contracts were given to three firms for serial production; NEC built the transmitters and pulse modulators, Japan Victor the receivers and associated displays, and Fuji Electrical the antennas and their servo drives. The system operated at 3.0 m (100 MHz) with a peak-power of 40 kW. Dipole arrays with mat-type reflectors were used in separate antennas for transmitting and receiving.

In November 1941, the first manufactured Type 11 was placed into service as a land-based early-warning radar at Katsu-ura, a small town on the Pacific coast about 100 km from Tokyo. A large system, it weighed close to 8,700 kg. Some 30 sets were built and used throughout the war. The detection range was about 130 km for single aircraft and 250 km for groups.

Type 12, another land-based early-warning system, followed during 1942. It was similar to its predecessor but lighter in weight (about 6,000 kg) and on a movable platform. Three versions were made; they operated at either 2.0 m (150 MHz) or 1.5 m (200 MHz), each with a peak-power of only 5 kW. The lower power limited the range to 50 km single and 100 km for groups. About 50 sets of all versions of these systems were built; these were primarily used at captured locations and to fill gaps in the Type 11 coverage.

Another similar system was the Type 21. Fundamentally, it was the 200-MHz version of the Type 12 redesigned for shipboard use and weighing only about 840 kg. The first sets were installed on the battleships *Ise* and *Hyuga* in April 1942. About 40 sets were eventually built.

Type 12 Installation

Type 21 Antenna

In this same time period, the more use-flexible Type 13 was also being designed. Operating at 2.0 m (150 MHz) and with a peak power of 10 kW, this set included a major advancement. A unit (diplexer) had been developed to allow the use of a common antenna. With a weight of near 1,000 kg (a small fraction of that of the Type 11), this system could be readily used on shipboard as well as at land stations. Its detection range was about the same as the Type 12. It was placed into service in late 1942, and by 1944 it had also been adapted for use on surfaced submarines. With some 1,000 sets eventually being built, the Type 13 was by far the most used air- and surface-search radar of the Imperial Navy.

Land Version Type 13

The Type 14 was a shipboard system designed for long-range, air-search applications. With a peak power of 100 kW and operating at 6 m (50 MHz), this could detect single aircraft at 250 km and large flights at 360 km. Weighing a huge 30,000 kg and with large Yagi antennas, only two of these systems were built. They were placed in service in May 1945, just before the end of the war.

The Imperial Navy built two radars based on the captured SCR-268. The Type 41 was electronically like the original, but with two large dipole array antennas with mat reflectors and configured for shipboard, fire-control applications. About 50 of these were built, and it went into service in August 1943. The Type 42 had more revisions, including a change to using four Yagi antennas. Some 60 were built and put into service in October 1944. Both systems had a range of about 40 km.

The *Würzburg*, Germany's prized land-based radar, operated in the 60-cm (500-MHz) band and used a 3-m (10-ft) parabolic reflector. After finally receiving one of these sets from the Germans, the Imperial Navy made minimal changes, mainly converting to a magnetron oscillator from the special vacuum tube that had been designed for this system. Resulting was the Type 23 anti-ship, fire-control radar intended for cruisers and larger ships. With the change to a magnetron, the output was approximately halved to a peak-power of about 5 kW; this gave a disappointing range of only 13 km for most surface ships. Although the prototype was completed in March 1944, only a few sets were built and it was never put into serial production.

In parallel with the development of VHF systems, the JRC was given a contract to work with the NTRI in designing and building a microwave surface-detection system for warships. Designated Type 22, this used a pulse-modulated, 10-cm (3.0-GHz) magnetron with water-cooling and producing a peak-power of 2 kW. Since it was the first full set using a magnetron, Yoji Ito was made responsible and gave it special attention.

Type 22 Antennas

The receiver was a superheterodyne type with a low-power magnetron serving as the local oscillator. Separate horn antennas were used for transmitting and receiving. These were mounted on a common platform that could be rotated in the horizontal plane. Since the antennas could not be rotated vertically, the set could not track close-in aircraft.

The prototype for the Type 22 was completed in October 1941; tests showed that it detected single aircraft at 17 km, groups of aircraft at 35 km, and surface ships at over 30 km (depending on the height of the antenna above the sea). Different versions for surface ships and submarines were put into serial production. The submarine version had the horn antennas mounted on a pressure-proof unit containing the transmitter and receiver; it thus weighed considerably more than the surface-ship version.

The first factory-produced Type 22 sets were put into service on destroyers *Kazegumo* and *Makigumo* in March 1942, making them the first

Japanese warships equipped with radar. The next month, these were also installed on the battleships *Ise* and *Hyuga*, giving these vessels two types of radar. The Type 22 radar was in wide use on surface vessels and submarines by late 1944, with about 300 sets being built during the war.

Type 13 and 22 Antennas on Submarine

Although not designed for this purpose, the Type 22 radar could provide relatively accurate range and azimuth coordinates for gunnery control. For larger guns, the Japanese had the finest optical fire-control system afloat, including infrared systems for night-fighting capabilities; the Type 22 was used to augment, not replace, these systems.

With the disappointment of the Type 23 (the *Würzburg* copy), development was started on three microwave systems for fire-control applications. The Type 31 operated at 10 cm (3 GHz) and used a somewhat smaller parabolic reflector. While the prototype could detect larger ships at up to 35 km, it was not completed until March 1945 and was never placed into production.

The Type 32 was another 10-cm system, this one having separate square-horn antennas. Detection range for large ships was about 30 km. It became operational in September 1944, and some 60 sets were produced. Type 33 was still another 10-cm set; this one used separate round-horn antennas. The prototype was completed in August 1944. Like the Type 23, detection range was only 13 km; thus, it was not put into production.

Type 32 Installation

The Imperial Navy had a large number of aircraft. At various times during the war, they had 26 aircraft carriers, serving as floating bases for up to 90 combat planes. In addition, they had land bases both at their homeland and on occupied territory. As previously noted, they even had a number of giant submarines capable of carrying several *Sei ran* floatplane bombers to very near their targets (the name literally means "storm out of a clear sky"). It was a year after the start of the war, however, before any Imperial Navy planes were equipped with radar.

At the same time that the NTRI was designing the Types 13 and 21, the first airborne radar was being developed at the Oppama Naval Air Technical Depot (ONATD) located at Yokosuka. Initially designated Type H-6 with a number of experimental sets built, this was eventually produced as the Type 64 and began service in August 1942. The greatest developmental problem was in bringing the weight down to that allowable for an aircraft; 110 kg was eventually achieved.

Intended for both air- and surface-search, the Type 64 operated at 2 m (150 MHz) with a peak power of 3 to 5 kW and a pulse width of 10 μs. It used a single Yagi antenna in the nose of the aircraft and dipoles on each side of the fuselage, and could detect large surface vessels or flights of planes at up to 100 km. This set was initially used on Kawanishi H8K-class 4-engine flying boats, then later on a variety of mid-sized attack planes and torpedo bombers. It was by far the most used airborne radar, with about 2,000 sets being produced.

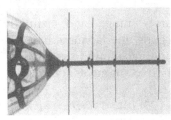
Type 64 Antenna

Development continued on lighter-weight systems at the ONATD. The Type N-6 weighing 60 kg was available in October 1944, but only 20 sets were built. This was a 1.2-m (250-MHz), 2-kW experimental set intended for a single-engine, 3-place (pilot, gunner, and radar operator) fighter aircraft.

Another was the Type FM-3; operating at 2 m (150 MHz) with 2-kW peak-power, this weighed 60 kg and had a detection range up to 70 km. It was specifically designed for the Kyuysu Q1W1 *Tokai*, a new 2-engine 3-place anti-submarine aircraft. For this purpose, the radar used two Yagi antennas on each side of the fuselage. Although 100 sets were built, they did not go into service until just days before the end of the war.

Kyuysu Q1W1 Tokai

With assistance from the NTRI and Yoji Ito, the ONATD also developed Japan's first airborne microwave radar. Designated FD-3, this was a magnetron-based, 25-cm (1.2-GHz), 2-kW set weighing about 70 kg. Tests showed that it could detect aircraft at 3 km and ships at some 10 km, sufficient for close-range night-fighter aircraft such as the

RRF Development in Japan

Nakajima J1N1-S Gekko with FD-3

Nakajima J1N1-S *Gekko*. It used four Yagi antennas mounted in the nose area; separate elements for transmit and receive were skewed for searching. Unlike in the air warfare in Europe, there were few night-fighter aircraft used by Japan; consequently, it was mid-1944 before the Type FD-3 was put into use. Some 100 sets were manufactured.

In mid-1942, after the Type 21 system had been placed in operation on several types of Japanese vessels, the operators sometimes found that the receiver was picking up signals from some other radar transmitter. Unknown to them at the time, this would have been an American SA, SC, or SK radar, all operating on 1.5-m (200 MHz, the same as the Type 21, but with a different PRF).

When the Type 21 transmitter was aimed in the direction of the unknown signal, there was usually no received reflection. This was because the travel path of the reflected signal was twice that of the direct signal, and the strength of an electromagnetic wave is inversely proportional to the square of the distance traveled. Also, only a small portion of a wave striking the target is reflected. This led to the concept of a warning receiver that could indicate the presence of a radar at a far greater distance than the radar's detection range.

As the first Japanese radar countermeasure, the NTRI developed the E-27 warning indicator. The receiver could be tuned between 0.75 and 4.0 m (400 and 75 MHz), the overall band of known and potential VHF radars. When used with an omni-directional antenna, the E-27 could warn that an enemy radar was somewhere in the region; the change in received signal strength could indicate an approaching or receding emitter.

A directional antenna for the E-27 would provide additional information, and two or more separated receivers could be used in triangulation to approximate the position of the radar. Limited by the curvature of the Earth, detection ranges in the order of 300 km were possible, twice or more that of the emitting radar because the signal was direct, not reflected.

The E-27 warning receivers were beginning to be used by the end of 1942, and eventually more than 2,000 sets were on ships and submarines. Somewhat earlier, the German Navy had a warning receiver called R600A or *Metox* (the name of the manufacturer in France). As noted earlier, one of the *Metox* sets had been salvaged in October 1942 from the cargo of a sunken submarine off Singapore, and was possibly the model for the E-27.

The FT-B and FT-C were slightly different warning receivers for aircraft. These were put into service on bombers and large reconnaissance planes in early 1944. They would automatically sweep the 0.45- to 2.7-m (670- to 110-MHz) band, and give an audio signal into the aircraft intercommunications headsets. The sets weighed 20 kg and could detect radar signals at up to 300-km range. About 400 total sets were built.

As more of the Allied radars operated in the microwave region, a warning receiver (type-number unknown) was developed to cover the 3- to 75-cm (10- to 0.4-GHz) range. The *Naxos* was a German warning receiver that used an untuned crystal diode receiver to detect all microwave signals. Although the *Naxos* never worked reliably in Germany for detecting radars, the Japanese might have used it in developing an improved version. About 200 microwave warning receivers were built and placed on ships and ground stations near the end of the war; there is no indication of their success.

Other radar countermeasures attempted by Japan included chaff and jammers. Although chaff (strips of aluminum foil) was occasionally dropped from aircraft to confuse the radar returns, aluminum was in very short supply and little was allocated for this purpose. Experiments were made with jamming transmitters, possibly some using designs obtained from Germany, but none were adopted for combat.

It has been previously noted that the Japanese Army and Navy had independent developments of systems for allowing aircraft to electronically identify themselves (IFF systems). The M-13 transponder was used on Navy aircraft, primarily for responding to Type 13 radars operating at around 2 m (150 MHz). The M-13 receiver would sweep the 145- to 155-MHz band to find the radar interrogation signal, and then give one of five coded responses on the same frequency. This would show as a coded increase in intensity on the CRT display used by the radar operator. Only about 100 sets were built and installed beginning in early 1945.

All of the radars deployed in Japan used the single-trace ("A" scope) CRT display. The contributions of Kenjiro Takayanagi, Chief Engineer of NHK's experimental television station, were noted in the development of Type 11, Japan's first full radar system. Takayanagi continued to assist the Navy, and in early 1945, he and Commander Shoji Baba of the NTRI developed a PPI for Japanese radar. Although too late for use in the war, ten experimental radar sets with PPI CRTs were completed in July 1945.

RRF Development in Japan

Even before radar was developed for detection, a similar technology was used for altitude determination. Several types of radio altimeters were worked on in Japan, but only one type was put into service. The FH-1, operating at 88 cm (340 MHz) with very-low power was developed for use in large flying boats on sea-skimming flights. Although not specified, it is likely that frequency-modulation, rather than pulsing, was used. With this unit, the planes could safely fly as low as 15 meters above the surface while searching for submarine periscopes. About 100 sets were built, and the system became operational in early 1945.

CLOSURE

Following the conclusion of the war in Europe, the Allied Forces (not including the USSR) moved relentlessly toward the Japanese homeland. The Ryukyus Islands, that included Okinawa, were the last obstacle to the invasion of the mainland islands. Between March 26 and July 2, 1945, the campaign to capture the Ryukyus, particularly Okinawa, was the most fiercely fought of the Pacific war. About 12,500 Americans were killed or missing, but this was a small fraction of the 110,000 Japanese military men who died in battle and the 150,000 Okinawans – one-third of the native population – who perished.

There followed a brief pause in the intensive land and sea fighting, during which time U.S. President Harry S. Truman demanded that Japan agree to an unconditional surrender. This was rejected by Japan, and Truman then had two options: exercise Operation Downfall, with the massive invasion of the mainland islands, first Kyushu and followed by Honshu, or use the just developed atomic bombs on selected mainland targets to show the consequence of continued resistance.

The first atomic bomb used in warfare was dropped on the city of Hiroshima at 8:15 a.m. on August 6, 1946. The demand to surrender was again issued, and again rejected. Three days later, another atomic bomb fell on the city of Nagasaki. Japan's leaders finally accepted the unconditional terms of surrender on August 15, and the signing on September 2 brought an official close to World War II. The USSR finally declared war on Japan on August 8.

The Japanese death toll from the two bombings was approximately 214,000, but the intent was not to take lives but to save them. The number killed in the two bombings was significantly less than the Japanese military and civilian losses in fighting on the Ryukyu Islands, and the total for the mainland invasion was projected to be in the millions.

Contrary to the widespread belief that the United States initiated atomic weapons, a number of other countries had projects pursuing this technology before and during WWII. In Japan, the Imperial Navy strongly felt that the Japanese would need some revolutionary and powerful weapon to be successful in the war. Toward this, research was conducted in atomic energy, intercontinental bombers, beamed-energy devices, and several other areas directed toward advanced weapons with great destructive potential.

These highly secret activities were under what has sometimes been called Project Z, but it is likely that the actual, seldom-used designation was the vowel "*E*" in the Japanese *katakana* alphabet, which looks similar to the Roman character Z. For this work, a large research facility had been built in Shimada City, at the foot of Mount Fuji about 200 km southwest of Tokyo.

By 1942, Japan was aware of the Manhattan Project (development of the atomic bomb) in America. Future Nobel Laureate Hideki Yukawa and other physicists convinced the Navy that Japan could never be the first with such a weapon. This work in Japan was then officially discontinued, although in late 1944 they were still collecting a stock of U-238 and were building centrifuges capable of isotope separation.

The intercontinental bomber, called *Fugaku*, was specified to have a range of 19,400 km and carry fuel and payload weight of 80,000 kg. With six 5,000-hp engines, it was intended to take off from Japan, bomb the continental United States, and fly on to land in Germany. When development of the atomic bomb was cancelled, the bomber program was slowed but continued until the end of the war; at that time all plans and documentation were destroyed.

The *Ku-go* (Death Ray) was another matter. An article on a death-ray device invented by Nikola Tesla had been in *The New York Times* (July 11, 1934), and was picked up by the Japanese press. In this, Tesla was quoted as saying that his beam would "drop an army in its tracks and bring down squadrons of airplanes 250 miles away." Magnetron research at the NTRI and JRC appeared to indicate that a beam device with tremendous output power might be possible.

In 1943, work began at the Shimada City research facility on developing a high-power magnetron that, if not as capable as Tesla had boasted, could at least incapacitate an aircraft. A number of Japan's leading physicists were involved in this activity, including Sin-Itiro Tomonaga, another future Nobel laureate. By the end of the war, their effort had produced a 20-cm magnetron with a continuous output of 100 kW, far short of the desired 500 kW, which itself would likely have been

RRF Development in Japan

insufficient for the mission. Like other Project Z efforts, documentation was totally destroyed before Japan surrendered.

With the close of the war, all organizations, facilities, and projects related to the military in Japan were disbanded. This included the NTRI and ATRI, as well as all parts of private firms engaged in weapons research and manufacturing. About 7,250 radar sets of 30 different types had been developed for the Army and Navy; except for a small number converted by universities for use in radio astronomy, these were all destroyed.

Scientists and engineers, as well as military technical officers, engaged in communications and radar formed the base for Japan's future electronics industry. Captain (Dr.) Yoji Ito was one of these. With the hope of making a peaceful contribution of technologies cultivated in his naval days, he founded the Koden Electronic Company. Among his successful products was a fish-finder that revolutionized the Japanese commercial fishing industry. All such firms and their products had to be approved by the General Headquarters, Supreme Commander of Allied Powers.

Bibliography and References for Chapter 7

Boyd, Carl, and Akihiko Yoshida; *The Japanese Submarine Force and World War II*, Naval Institute Press, 2002

Brittain, James E.; "Scanning the Past: Hidetsugu Yagi," Proc. IEEE, vol. 81, p. 934, 1993; http://ieee.cincinnati.fuse.net/reiman/05_2004.htm

Brown, Lewis; *A Radar History of World War II – Technical and Military Imperatives*, Inst. of Physics Pub., 1999

Coffey, Thomas M.; *Imperial Tragedy: Japan in World War II, the First Days and the Last*, World Publishing Co., 1970

Compton, K. T.; "Mission to Tokyo," *The Technology Review*, vol. 48, no. 2, p. 45, 1945

Goebel, Gregory V.; "Japanese Radar Technology at War," in "The Wizard War: WW2 & The Origins of Radar," a book-length document; http://www.vectorsite.net/ttwiz.html

Nakagawa, Yasudo; *Japanese Radar and Related Weapons of World War II*, translated and edited by Louis Brown, John Bryant, and Naohiko Koizumi, Aegean Park Press, 1997

Nakajima, S.; "Japanese radar development prior to 1945," *IEEE Antennas and Propagation Magazine*, vol. 34, Dec., p. 18, 1992

Nakajima, S.; "The history of Japanese radar development to 1945," in *Radar Development to 1945*, ed. by Russell Burns, Peter Peregrinus Ltd., 1988

Nakajima, Shigeru; Oral history #212 conducted by William Aspray for the IEEE History Center, Rutgers University, 1994; http://www.ieee.org/web/aboutus/history_center/oral_history/oh_japan_menu.html

Okabe, Kinjiro; "On the Applications of Various Electronic Phenomena and the Thermionic Tubes of New Types," *Japanese IEEE*, vol. 473 (Suppl. Issue), p. 13, 1927 (in Japanese)

Sato, Genzo; "The Secret Story of the Yagi Antenna in World War II," *The Radioscientist*, vol. 2, no. 4, p. 71, 1991

Swords, S. S.; *Technical History of the Beginnings of Radar*, Peter Peregrinus Ltd., 1986

"The Pacific War," *A Brief History of U.S. Army in World War II*, U.S. Army Center of Military History; http://www.history.army.mil/books/wwii/11-9/wpac.htm

Wilkinson, Roger I.; "Short survey of Japanese radar – Part I," *Trans. AIEE*, vol. 65, p. 370, 1946

Wilkinson, Roger I.; "Short survey of Japanese radar – Part II," *Trans. AIEE*, vol. 65, p. 455, 1946

Yagi, H.; "Beam Transmission of Ultra-Short Waves", *Proc. of the IRE*, vol. 16, p. 715, 1928

Chapter 8

EARLY RADAR DEVELOPMENT IN OTHER COUNTRIES

In addition to the five countries discussed in the preceding chapters, there was early development of radio-based detection and ranging systems in three other nations: The Netherlands (Holland), France, and Italy. These efforts, however, were cut short by World War II, with little else done until later years.

In early 1939, several of the British Commonwealth Nations were invited to send representatives to England to receive information about a 'new secret technique of a most important character concerning defence.' This was, of course, Great Britain's highly secret Range and Direction Finding (RDF) technology. Based on this, Australia, Canada, New Zealand, and South Africa started their own development of RDF systems.

Activities in Hungary must also be noted. At the start of WWII, this nation attempted to be both in the Axis Powers and still maintain relationships with the rest of the world. As a consequence, they were on their own in defense developments, and in 1942 started a completely independent activity in radio-based detection.

This chapter provides a brief summary of the early radar developments in all eight of these countries. As in the previous treatments, the emphasis is on the people, organizations, and activities.

THE NETHERLANDS

After Christian Hülsmeyer of Germany announced his *Telemobilskop* device in 1904, he was invited by the Holland American Line to give a demonstration on board the ship-tender *Columbus*, near Rotterdam. While the demonstration was reasonably successful and the need for such an apparatus was acknowledged, Hülsmeyer's equipment was found to be cumbersome to operate and could not determine distance to a target. In addition, the ship lines were making large investments in wireless communications and gave this a higher priority. Thus, no action on Hülsmeyer's invention was taken.

When vacuum tubes came into being, the Philips Company in Eindhoven soon became the largest tube manufacturer in Europe. In 1914, they established the *Natuurkundig Laboratorium* (Natlab) for fundamental research related to their products. In the early 1930s, researcher Klass Posthumus, who had just developed negative feedback to greatly broaden the bandwidth of amplifiers, was investigating tubes for very-high frequency applications. Following the trend to divide the anode, he developed a magnetron with the anode split into four elements. This functioned up to 1 GHz (30 cm) and generated about 10 watts. By 1934, the tube was put on the market by Philips as a 50-cm (600-MHz), 50-watt magnetron. The detailed analysis of this type of magnetron was published by Posthumus in *Wireless Engineer* in 1935.

In a project led by C. H. J. A. Staal at Natlab, the new magnetron was being tried in a voice communication system. There were difficulties in audio-modulating the magnetron, so rapidly switching the power on and off (pulse modulation) was used. In 1935, to simplify the field testing, Staal had the parabolic transmitting and receiving antennas set side-by-side and tried using a large plate set at some distance along the beam to reflect the signal back to the receiver. It was found that a strong reflected signal was received.

Parabolic Reflector

Recognizing the potential importance of this as a detection device, Natlab arranged a demonstration for the *Koninklijke Marine* (Royal Dutch Navy). This was conducted in 1937 across the entrance to the main naval port at Marsdiep. Reflections from sea waves obscured the return from the target ship, but the Navy was sufficiently impressed to initiate sponsorship of the research.

The equipment was improved, making use of pulse modulation to omit reflections except those at a selected distance. In 1939, at a new facility in Wijk aan Zee, the minesweeper *Jan van Brakel* was detected at a range of 3.2 km. Based on this, a prototype system was built by Philips, and plans were started by the firm Nederlandse Seintoestellen Fabfrek for building a chain of warning stations to protect the primary ports. Some field testing of the prototype was conducted, but this was discontinued in May 1940 with the invasion by Germany. Within the Natlab, however, the project was continued in great secrecy until 1942.

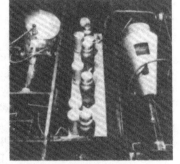

1-GHz Electronics

Early Radar Development in Other Countries

During the late 1920s and early 1930s, there were widespread rumors of a "death ray" being developed. The July 11, 1934, issue of the *New York Times* specifically attributed such a ray to Nikola Tesla. The Dutch Parliament set up a Committee for the Applications of Physics in Weaponry under Dr. G. J. Elias to examine this potential, but the Committee quickly discounted death rays. The Committee did, however, established the *Laboratorium voor Fysieke Ontwikkeling* (LFO, Laboratory for Physical Development), dedicated to supporting the Netherlands Armed Forces.

In order not to arouse suspicions, the LFO was called the *Meetgebouw* (Measurements Building). The facility was located on the Plain of Waalsdorp and was shared with the Dutch Military Weather Service. J. L. van Soest led the initial research efforts; this was primarily on sound devices for detecting aircraft, a radio-meteography system, and an infrared detection apparatus.

Measurements Building

J. L. W. C. von Weiler (1902-1988) joined the LFO in 1934 and, with S. G. Gratama, began research on a 1.25-m (240-MHz) communication system to be used in artillery spotting. In 1937, while tests were being conducted on this system, a passing flock of birds disturbed the signal. Realizing that this might be a potential method for detecting aircraft, the Minister of War ordered continuation of the experiments. Weiler, assisted by Gratama, set about developing a system for searchlight directing and anti-aircraft gun aiming.

The experimental "electrical listening device" operated at 70 cm (430 MHz) and used pulsed transmission at 10 kHz. A transmit-receive blocking circuit was developed to allow a common antenna; the blocking resulted in a minimum range of 0.45 km. The received signal was displayed on a CR tube with a circular time base (a PPI).

Model of Dutch 430-MHz System

This set was demonstrated in April 1938 to an inspector-general of the Army and detected an aircraft at a range of 18 km. The set was rejected, however, because it could not withstand the sand and water environment of Army combat conditions.

The Navy was more receptive. Funding was provided for final development, and Max Staal was added to the team. To maintain secrecy, they divided the development into parts. The transmitter was built at the University of Delft and the receiver at the University of Leiden. Ten sets would be assembled under the personal supervision of J.J.A. Schagen van Leeuwen, head of the firm Hazemeijer Fabriek van Signaalapparaten. Only 37 persons were initially given clearance to perform the work. This would be followed by an order for 50 sets, with firms NEDALO and NEDISCO to be added for production.

The prototype had a peak-power of 1 kW, provided by four RS-297 tubes from Telefunken. The pulse length was 2 to 3 μs, with a 10- to 20-kHz PRF. The pulses were modulated by a 1-kHz tone to allow audio detection of the signal. The receiver was a super-heterodyne type using Acorn tubes and a 6-MHz IF stage. The antenna consisted of 4 rows of 16 half-wave dipoles backed by a 3- by 3-meter mesh screen. The operator used a bicycle-type drive to rotate the antenna, and the elevation could be changed using a hand crank.

430 MHz Electronics

Several sets were completed, and one was put into operation on the Malievelt in The Hague just before the Netherlands fell to Germany in May 1940. The set worked well, spotting enemy aircraft at up to 120-km range during the first days of fighting, but there were no associated anti-aircraft guns so the operation was frustrating. As the country fell, essentially all assemblies and plans for the radar were destroyed.

CRT Display

Von Weiler and Max Staal fled to England on the HMS *Wessex*, one of the last ships able to leave, carrying two completed sets with them. Later, Gratama and van Leeuwen also escaped to England. All eventually worked on equipment for the Dutch Navy through the Research Department of HM Signal School in Portsmouth. One of their important contributions was the integration of their system for range measurements on the Bofors 40-mm anti-aircraft gun, a project that they had started at the LFO in Holland. Vickers built 200 of these systems, and it was also adopted by the U.S. where thousands were produced.

After the war, von Weiler returned to the LFO as Testing Department Head and was also appointed Physics Professor at the Delft University of Technology. Max Staal became the Technical Director of Signall. The pictures below are from the post-war era.

J. L. W. C. von Weiler   *Max Staal*

ITALY

The "father of radio," Guglielmo Marconi, initiated the research that led to radar in his native Italy. He presented a paper to the Institute of Radio Engineers in 1922 that included reference to using reflected radio waves for navigation purposes. In 1933, while participating with his Italian firm in experiments with a 600-MHz communications link across Rome, Marconi noted transmission disturbances caused by moving objects adjacent to its path. This led to the development at his laboratory at Cornegliano of a 330-MHz (0.91-m) CW Doppler detection system that he called *radioecometro*. Barkhausen-Kurz tubes were used in both the transmitter and receiver.

In May 1935, Marconi demonstrated his system to the Fascist dictator Benito Mussolini and members of the military General Staff. (In 1923, Marconi had joined the Italian Fascist party, and in 1930, Mussolini appointed him President of the *Accademia d'Italia*, which made Marconi a member of the Fascist Grand Council.)

The output power of the *radioecometro* was insufficient for military use, but Marconi's demonstration raised considerable interest. The press promoted the fantasy of a "death ray." Little more was done by Marconi on this apparatus, and he died two years later in Rome.

Mussolini directed that radio-based detection technology be further developed, and it was assigned to the *Regio Instituto Electrotecnico e delle Comunicazioni* (RIEC, Royal Institute for Electro-technics and Communications, or "Marinelettro"). The RIEC had been established in 1916 by the Royal Navy in Livorno, a major port some 350 km above Rome on the western side of the peninsula; the Institute was physically located within the campus of the Italian Naval Academy. Under the technical leadership of Professor/Admiral Giancarlo Vallauri, the RIEC was the center of communications research and development in Italy. Vallaur immediately asked that Ugo Tiberio, a Lieutenant in the Naval Reserves, be called to active duty at the RIEC to head the project.

Ugo Tiberio

Ugo Tiberio (1904-1980) graduated in civil engineering from the University of Naples. In 1932, he completed electrical engineering graduate studies at the School of Engineering in Rome. At this same time, he was fulfilling his required military service by teaching and doing research at the *Istituto Militare Superiore Trasmissioni* (Institute of Higher Military Transmissions). After transferring to Livorno in 1936, he officially served as a physics and radio-technology instructor at the Academy and conducted the RDT research at RIEC on a part-time basis.

For several years, Tiberio had been examining the application of radio for detection applications. Fully familiar with the work of Marconi and his *radioecometro*, Tiberio prepared a report outlining further research in both CW Doppler and pulsed technologies to be pursued, following the lines used in Great Britain and America in Kennelly-Heaviside Layer research. The report, submitted in mid-1936, included what was later called the radar range equation. The system was called *telemetro radiofonico del rivelatore* (RDT, Radio-Detector Telemetry).

Nello Carrara, a physics instructor at the Academy, started collaborating with Tiberio. Cararra, who had been doing research at the RIEC in microwaves for 10 years, was mainly responsible

RDT Research Laboratory at the RIEC

for the development of the RDT transmitter. Several petty officers and technicians were also assigned to the project.

Before the end of 1936, Tiberio's team had demonstrated the EC-1, the first Italian RDT system. This had an FM transmitter operating at 200 MHz (1.5 m) with a single parabolic cylinder antenna. It detected by mixing the transmitted and the Doppler-shifted reflected signals, resulting in an audible tone.

The EC-1 did not provide a range measurement; to add this capability, the development of a pulsed system was initiated. In 1937. Captain Alfeo Brandimarte joined the group and primarily designed the first pulsed system, the EC-2. This operated at 175 MHz (1.7 m) and used a single antenna made with a number of equi-phased dipoles. The detected signal was intended to be displayed on an oscilloscope. There were many problems, primarily concerning the receiving techniques, and the system never reached the testing stage. Brandimarte left the RIEC and returned to regular duty.

Alfeo Brandimarte and Nallo Carrara (right)

Work then turned to developing higher power and operational frequencies. Carrara had followed the literature on development of magnetrons, and turned to this device for improving the RDT transmitter. In cooperation with the firm Fabbrica Italiana Valvove Radio Elettriche (FIVRE), a magnetron-like device was developed; this was composed of a pair of triodes connected to a resonate cavity and producing 10 kW at 425 MHz (70 cm). It was used in designing two versions of the EC-3, one for shipboard and the other for coastal defense.

Italy entered WWII in September 1939, without an operational RDT. A prototype of the EC-3 was built and tested from atop the laboratory building, but most RDT work was stopped as direct support of the war took priority. In June 1940, three naval officers, including Captain Brandimarte, were sent to Germany in an attempt to obtain information on their radio-based detection systems. They were shown some of the German equipment, including a *Freya* system, but were given very little detailed information.

On the night of March 28, 1941, the Regia Marina Italiana (Royal Italian Navy) suffered greatly in an engagement off Cape Matapan with radar-equipped British warships. The importance of RDT systems was immediately reevaluated, and the Navy again turned to the RIEC and

Radar Origins Worldwide

Tiberio's team. In April, the EC-3 prototype was installed on the vessel *Carini* to demonstrate to senior military officials the benefits of RDT equipment. The demonstration was very successful, with the detection at ranges up to 12 km for ships and 80 km for aircraft.

Even before the demonstrations were completed, the Ministry of Defense directed that the EC-3 be placed into production. With the assistance of FIVRE and Radio Marelli (another Italian firm), prototypes of two versions of the EC-3 were made. The *Folaga* (Coot) was for coastal defense and the *Gufo* (Owl) for the Italian Navy shipboard applications. Both transmitters used push-pull type 1628 triodes with a resonant cavity and a pulse-width of 4 μs. The receivers were regenerative designs using RCA 955 Acorn tubes.

Folaga RDT System

The *Folaga* operated at around 200 MHz (1.5 m) and used a pair of parabolic-reflector antennas. Targets were shown on a CRT with a circular-range display.

The *Gufo* operated at 425 MHz (70 cm) and used two horn antennas, with the transmitter and receiver mounted directly behind the horns. The horns could be rotated 90 degrees; these were located at different levels above the bridge but synchronized electrically. Detection depended on receipt of a 600 Hz audio signal. In late 1941, the *Gufo* prototype was made permanent and placed aboard the battleship *Littorio*, making this the first vessel in the Italian Navy with radar.

The firm Marelli received an order from the Navy for 150 *Folaga* RDTs, while SAFAR was to build 50 of the *Gufo* sets. However, due to the lack of skilled production personnel and damages to the facilities from Allied bombing, only 12 sets were actually delivered before the war ended.

Gufo Control Unit

Gufo Horn Antenna

The Italian Army and Air Force had also ordered RDTs to be designed and built for land and airborne applications. Participating firms included Allocchio Bacchini, SAFAR, Philips Italiana, IMCA Radio, Officine Marconi, and Telefunken Italiana. However, except for a small number of the *Lince* (Lynx), a ground-based air-search system developed at SAFAR for the Air Force by Arturo Castellani and Francesco Vecchiacchi, no other systems went beyond prototypes.

Lince Antenna

A mention should be made of the development of intercept receivers and jammers. This task was assigned to the Regio Istituto Radio-tecnico (Royal Radio-technical Institute) of the Air Force, located at Guidonia. The development group was supervised by Professor Gaetano Latmiral from the University of Naples. The only product known to have been fielded was a 200-MHz jammer used in Sicily against the British radar station in Malta.

The Allies invaded Southern Italy on September 3, 1943, and the Italian government surrendered on September 8. The German military remained in the country and fighting continued. Rome was finally liberated in June 1944.

All activities in Italy relating to radar development ceased when the Nation surrendered. For protection during the continued fighting by the Germans, the Naval Academy moved temporarily to Brindisi; here Tiebrio was promoted to Lieutenant Colonel of Naval Weapons. Alfeo Brandimarte had been killed in naval action in 1942.

Ugo Tiberio and Nallo Carrara remained as professors and researchers at the Academy for many years. Commonly called the "father of Italian radar," Tiebro was eventually named to head the Faculty of Radio at Pisa University.

FRANCE

In the late 1920s, Pierre David at the *Laboratori National de Radioelectricite* (LNR, National Laboratory of Radioelectricity) in Bagneux experimented with reflected radio signals at about a meter wavelength. During the same time, Professor Camille Gutton of the *Faculté de la Science à l'université de Nancy* (Faculty of Science at Nancy University) studied reflections at 16 cm. Both activities gained the interests of Maurice Ponte, Research Director at the firm *Société Française Radioélectrique* (SFR, French Radioelectric Company) in Paris. The SFR was headed by its founder, Emil Girardeau. These were the individuals and organizations at the foundation of radar in France.

France was a nation with a rich heritage in science and technology; thus, there had certainly been earlier interests in using electromagnetic waves for detection purposes. The very elementary work by Christian Hülsmeyer and his *Telemobilskop* in Germany was well known by Girardeau and others who were interested in detection means for vessel safety. The ever-present threat of invasion by the resurgent Nazi military led the LNR and similar governmental research operations to give significant attention to means for giving early warning of approaching aircraft.

Professor René Mesny at the University of Paris had led many students in conducting research on electromagnetic waves. In 1925, one of his students, Pierre David, began an investigation at the LNR on radio emissions from sparks in aircraft engines. After considerable study, David concluded that this means would not be practical for long-distance detection.

Pierre David

After consultation with Mesny, David suggested that radio beams impinging on an aircraft might be reflected to allow detection. His analysis of the reflections of HF radio signals then became the center of his thesis research. He was awarded the *Docteur de la Faculté des Sciences de Paris* in 1928); however, he did not pursue this experimentally for several years.

In 1933, researchers at the Bell Telephone Laboratories in America published a paper in the *Proceedings of the IRE* on their studies in radio-wave reflections. (See Precursors of Radar in Chapter 1.) Upon reading this paper, David returned to his earlier proposal and initiated a letter to the Minister of War recommending that the LNR conduct an experimental study on radio detection of aircraft.

David's proposal was approved, and in mid-1934, experiments were conducted at Le Bourget Airport in Paris. A 50-W CW transmitter operating at 75 MHz (4 m) and a super-regenerative receiver were set up about 5 km apart The experiments were immediately successful; aircraft crossing the path at altitudes up to 5,000 m were readily detected by the Doppler interference.

This detection technique was called *barrage électromagnétique* (electromagnetic curtain). While this could indicate the general location of penetration, precise determination of direction and speed was not possible. David did further studies and devised a method to give close estimates of these parameters. He called this the *maille en Z* (Z-network); it used multiple beam paths in a Z-configuration.

While the work by David at the LNR was taking place, a totally different approach to radio-based detection was underway. Professor Camille Gutton at the University of Nancy developed a retarded-field triode (similar to the Barkhausen-Kurz tube) to generate 16-cm (1.9-GHz) signals, but with a power of only about 0.2 W. His assistant, Emile Pierret, built an apparatus using a parabolic reflector to form a beam. In 1927, reflected signals were observed as the beam was moved across nearby metal objects.

Emile Girardeau

Emile Girardeau (1882-1970), a graduate of *École Polytechnique*, then considered the oldest and most prestigious engineering school in the world, had founded SFR in 1910 to compete with the Marconi Cmpany and Telefunken. Within a few years, they supplied the equipment for a French chain of communication stations that stretched around the world, and he formed the *Compagnie Générale de Télégraphie Sans Fil* (CSF) in 1918 as the parent organization. The SFR was the research center, investigating new concepts and developing advanced commercial products.

Maurice Ponte

Maurtius (Maurice) Ponte (1902-1983) earned undergraduate degrees in mathematics and in natural science from the *Ecole Normale Superieure* (National University, then was Rockefeller Foundation graduate student in physics at London University. He joined the SFR in 1926 while working on his thesis, and received the *Docteur de la Faculté des Sciences* in 1930. At the SFR, he quickly gained the confidence of Emil Girardeau and was made Director of the Research Center.

While both Girardeau and Ponte had outstanding capabilities in the radio field and made personal contributions to the research projects, they also brought in highly capable personnel to head most of the activities. One such person was Dr. Henri Gutton (1905-1984), who led the SFR tube development. The son of Professor Camille Gutton at Nancy, Henri Gutton recognized the potential advantages of microwaves for detection and turned to Dr. M. Robert Warneck in his department seeking a means to increase the generator power at these wavelengths. They examined the paper by Japanese scientist Dr. Kinjiro Okabe on a split-anode in the April 1929 issue of the *Proc. of the IRE*. From this, in 1932, Warneck developed magnetrons producing several watts at around 80 cm (470 MHz). At the same time, Henri Gutton improved on his father's 16-cm device by adding a quarter-wave antenna within the glass envelope; this was designated as type UC16. When this was placed at the focus of a parabolic reflector, a narrow beam was produced.

Warneck's Magnetron

Ponte was aware of the early work on the electromagnetic curtain by David at the LNR. With Henri Gutton's assistance, Ponte set up a similar experiment using both the 16-cm UC16 and an 80-cm magnetron. The objective was to develop a system for maritime safety, and satisfactory tests were made in the second half of 1934. These were still CW systems and depended on Doppler interference for detection, but the antennas were co-located. A patent in Ponte's name was filed by SFR in July on a *Système de réperages d'obstacles* (System of locations of obstacles); obstacles cited in the patent were ships, icebergs, and aircraft.

During tests at the SFR, a Doppler signal was observed from a passing aircraft. Based on this, Ponte prepared a proposal in early 1935 for "the location of mobile objects by microwaves and its immediate applications to national defense." The Army and Navy showed some interest, but doubted that the needed power of several kilowatts at the proposed 4- to 6-cm wavelength could be achieved.

In March 1935, the Navy conducted comparative tests using David's 4-m apparatus and Henri Gutton's UC16 device from SFR. Power from the latter was insufficient for the tests, but the 4-m system detected aircraft over a 21-km path. SFR then abandoned its microwave approach to aircraft detection and proposed a project using lower frequencies. Government funds were allocated in April for testing three transmitters and six receivers in a 75-MHz (4-m) Z-network.

Still believing that microwaves would be best in detection systems for naval safety applications, Giardeau and Ponte approached the owners of the recently launched SS *Normandie*, the largest liner of that day, to outfit the ship with a SFR obstacles-locating system. The proposal was accepted, and preliminary tests were conducted with both 16-cm and 80-cm sets on the cargo ship *Oregon* as it steamed along the English Channel. The tests were inconclusive, and it was decided that Henri Gutton and Maurice Bridge would develop a set for the *Normandie*.

The resulting set operated at 16-cm (1.9 GHz) and used two 0.75-meter parabolic reflectors with UC16 tubes attached at the backs. The reflectors gave beam-widths of some 12 degrees, and were mechanically synchronized to scan 40 degrees on each side of the bow. A super-regenerative receiver was used, with another UC16 as the local oscillator. The transmitter was modulated with a 7.5-kHz audio signal, and the send and received signals compared on an oscillograph. The maximum range for detection was earlier tested to be between 3 and 5 km for ships and up to 10 km for the coastline. The equipment was ready in time for the *Normandie's* second trans-Atlantic voyage in July 1935.

Obstacle Locater Antennas on the Normandie

Equipment on the Normandie, Henri Gutton (standing)

On the initial part of the voyage, it was found that side signals from the transmitter made it difficult to identify reflected signals. Also, few other large vessels passed within range. Thus, performance of the equipment could not be determined. Then, the weather turned very bad, with waves reaching to the reflectors. When the ship arrived in New York, the reflectors were found to have been damaged beyond repair. Upon return of the *Normandie* to Le Havre, the equipment was removed from the ship.

On the military side, the Défense Aérienne de Territoire (DAT, Defence of Air Territory) set up David's simple electromagnetic curtain near Gien. In tests during the summer maneuvers of 1936, the line detected many entering aircraft, but some were missed.

At SFR, the tests of the 75-MHz (4-m) screen had been promising, but it was decided to use a lower frequency to obtain greater power. They proposed to develop 30-MHz (10-m) equipment and run tests in several screen configurations. Funds were received for building 12 stations; these were set up in the Reimers-Argonne region for maneuvers in June 1937. Different configurations were tested, with the Z-network giving the best results. Using David's methods, estimates were made of aircraft speed to within 10 percent, flight direction to between 10 and 20 degrees, and altitude to within 1,000 meters.

Undaunted by unsatisfactory initial trials on the *Normandie*, SFR continued the development of microwave equipment. Henri Gutton and Robert Warneck improved their magnetron by using multi-segmented anodes. By 1937, an 8-segment device designated M16 was generating a few watts at 16 cm (1.9 GHz). The oxide-coated cathode gave an efficiency of about 15 percent. This magnetron was made public in a 1938 bulletin from SFR. The write-up makes no reference to the problem of frequency stability, something that plagued other magnetron developers.

To increase the peak power, Maurice Elie at SFR developed a means of pulse-modulating transmitter tubes. In late 1937, a new 16-cm system was developed; this used pulsed transmission with a pulse-width of 6-µs, giving a peak-power of near 10 W. There was also audio modulation at 800 Hz. The transmitting and receiving antennas were backed by parabolic reflectors one meter in diameter. A super-heterodyne receiver was used with a type UC16 as the local oscillator; the pulses were shown on a CRT indicator.

The most significant advancement in the system was the use of returned pulses to determine range, thus making this the first full system developed in France for radio detection and ranging (radar). The CRT display provided a range-determination precision of some 200-300 meters.

Antennas at La Havre

The new apparatus was set up at near the docks at La Havre, allowing testing and demonstrations to the Navy without going to sea. By early 1939, tests showed detection of cargo vessels at up to 10 km and smaller boats at 3 km. It could not, however, detect aircraft at a usable range. Likely because of this deficiency, the Navy did not adopt the system. In 1939, a paper on the 16-cm system was published in a French journal, and French and American patent applications were filed in December in the names of Elie, Gutton, *et al* for a *Système pour la détection d'objet et la mesure de distance* (System for object detection and distance measurement). The system was planned to be sea-tested aboard the *Normandie*, but this was cancelled at the outbreak of war.

Development of David's curtain system continued at the LNR. Several mobile, 62-MHz (4.8 m) versions were ordered for the military's 1938 maneuvers. They operated well, leading to 30 type E62 emitters (transmitters) and 60 type R62 receivers being purchased for Z-network systems being installed in 1939 for the protection of several ports in France and one in Tunisia. They were also set up along the main air approaches from Germany to Paris. The transmitters were built by the firm SADIR and receivers by Le Matériel Téléphonique (LMT).

As war in Europe drew closer, France and Great Britain held a technical exchange for defense information. Under this exchange, Robert Watson Watt and Arnold Wilkins, two of the early developers of the British RDF, visited France in April 1939. Almost nothing is known about this technical exchange – Watson Watt does not even mention it in his extensive autobiography. It is evident, however, that the advantages of pulsed-transmission were discussed.

David already recognized these advantages; in October 1938, he had sent a note to the Comité d'Expériences Physiquers outlining the need for monostatic, pulsed systems for aircraft detection. After the visit by the British experts, David had SADIR develop a 50-MHz, pulse-modulated system with a peak-pulse power of 12 kW.

France declared war on Germany on September 1, 1939. The Maginot Line (France's "impenetrable" defenses along the Eastern border) deterred the invasion of land forces, but the threat of bombing by Germany, and perhaps Italy, was great. Thus, all efforts were concentrated on early-warning detection systems. The new set from SADIR was taken to near Toulon, where it could detect aircraft at a range up to 55 km.

SFR brought its pulsed system with ranging into the defense activity. They installed a set in Paris operating at 100 MHz (3 m) and producing a peak-power of 25 kW. This could detect, and measure the range of, aircraft as far away as 130 km. Before the end of 1939, SFR moved their 15-cm pulsed set to near Brest; as previously at La Havre, it gave good detection of ships, but was of little value in warning of approaching aircraft.

The pulse-transmission technology improved rapidly. In the first few months of 1940, the firm LMT developed a system operating at around 50 MHz (6 m) that generated near 350-kW peak-power. Pulse frequency could be selected at several values between 50 and 1,250 Hz with a pulse width of either 10 or 25 µs. The CRT display had a scale up to 130 km, but targets were often detected beyond this value.

In April 1940, Henri Gutton and Warneck at SFR developed a 16-cm magnetron producing about 500-W power. Time did not permit this being incorporated in the existing microwave system (now at Brest), but on May 9, just days before the Germans arrived, Maurice Ponte left for Great Britain with several of the new magnetrons. These were taken to Dr. Eric Megaw of the GEC Wemberly Research Laboratory; there their oxide-coated cathodes were quickly duplicated to improve the 10-cm Boots and Randall magnetron. The result was the well-known E1189, a high-power device that became the heart of British radars.

Major bombing of France took place in May and June 1940; the detection systems functioned well, but the French Air Force was no match for the German *Luftwaffe*. The German Army then circumvented the Maginot Line by invading from the north through Belgium, and the Italians came in from the south. As the Axis troops arrived, the French destroyed all of their detection systems.

An Armistice was signed with Germany on June 22, 1940. France was primarily divided into a German-occupied zone in the north and west, and an unoccupied or "free" zone in the south. The free zone became known as Vichy France and had a collaborationist puppet government. The only significant open activity in radar until after France was liberated was at SADIR and LMT; these firms supported the Vichy government by continuing to develop defense electronic equipment.

In the occupied zone, Girardeau created a factory to build communication equipment for underground Free French forces. After the war, he was responsible for rebuilding radio communications in France. Ponti returned to France and eventually became the head of the SFR, then head of the parent company CFS. Henri Gutton remained a leading scientist at SFR until 1957.

AUSTRALIA

In February 1939, the British Government requested that representatives from the more technically advanced Commonwealth nations come to England to be extensively briefed on the highly secret development of RDF (radar). Dr. David Martyn of the University of Sidney represented Australia.

David F. Martyn (1906-1970) was a native of Scotland and earned a doctorate in physics from the University of London. In 1929, when the Australian Radio Research Board was formed under the Australian Council for Scientific and Industrial Research (CSIR), Martyn moved to Australia to become a Board member. Three years later, he joined the Electrical Engineering Department at the University of Sidney as a Research Professor. His work there centered on experimental and theoretical studies of the ionosphere. On a visit to England in 1936, he became generally acquainted with the work on RDF being done by Robert Watson Watt and his team; he then urged that a similar project be started in Australia. Thus, Martyn was a natural selection to be sent to England for the 1939 briefings.

When Martyn returned to Australia in August, he convinced Dr. John P. V. Madsen, chairman of the Radio Research Board, to establish the Radiophysics Laboratory (RPL) as a unit of the CSIR, devoted to developing RDF systems for Australia. On September 3, 1939, Australia declared war on Germany, then on September 8 with Japan.

David Martyn

When the RPL was formed, Martyn was named director and immediately set up highly secure facilities on the campus of the University of Sidney. Dr. John H. (Jack) Piddington (1910-1997) was made responsible for system development. A native of Australia and a Sidney University Fellow, Piddington had earned his doctorate at the University of Cambridge under Professor E. V. Appleton.

Dr. Joseph L. Pawsey, an Australian who had previously been with EMI's Television Group in England, was hired to lead technical research. The first efforts at the RPL were in assessing the British RDF for application in Australia. It was concluded that the large coast of Australia needed a different type of system.

John Piddngton

In 1938, Piddington had built a pulse-transmission system intended for ionoshperic studies, but his appointment to the RPL changed the direction of his work. His first RPL project was developing a 200-MHz (1.5-m) shore-defense RDF system for the Australian Army. Designated ShD, the system was tested at Dover Heights in Sidney in September 1941. Sets were eventually installed at 17 ports around Australia.

ShD Radar at Dover Heights

The Air Council of Canada agreed to open a school for training RDF personnel from throughout the Commonwealth. This school (RAF #31) opened in Clinton, Ontario, in May 1941, with a number of officers and enlisted personnel from Australia in attendance. Realizing the need for indigenous training, Australia established its own programs in September 1941. These included six-month courses for signal officers at Sidney University and for maintenance mechanics at Melbourne Technical College.

With the Japanese attack on Pearl Harbor, the need for an air-warning system became urgent. For the Royal Australian Air Force (RAAF), Piddington's team designed and built a 200-MHz (1.5-m) system in five days. Designated AW Mark I, the set was briefly tested using the existing antenna at Dover Heights, then sent to Darwin for installation. Unfortunately, it was not yet operational when the Japanese devastated that city in their initial attack on Australia on February 19, 1942. Before the end of March, the set was in operation and detected a large incoming raid. While still 20 miles offshore, the raiders were successfully intercepted by U.S. fighter aircraft. Eventually, the RAAF established an integrated network of AW sets and successfully repelled the Japanese attacks.

After a few AW sets were made by the RPL, production was turned over to HMV Gramophone (an EMI company). The New South Wales (NSW) Railways Engineering Group developed a lightweight antenna, making the set air transportable. Called simply the LW/AW Mark II, 56 of these sets were used by the Australian forces, 60 by the U.S. Army

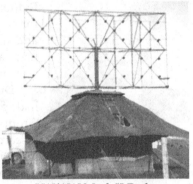

LW/AW Mark II Radar

in the early island landings in the South Pacific, and 12 by the British in Burma. The antenna was hand-cranked, and the operators sat in a tent under the structure.

In early 1942, U.S. Army troops (including the author's brother-in-law) started arriving in Australia. Several SCR-268s brought by these troops were turned over to the Australians, who rebuilt them to become Modified Air

LW/AW Mark II Uncovered

Warning Devices (MAWDs). These 200-MHz systems were deployed at 60 sites around Australia. In time, many other radar systems were supplied from Great Britain and America. The Royal Australian Navy had essentially all types of British-built radars on their vessels.

In June 1942, Martyn became the head of a special operational research group examining radar problems; Dr. Frederick W. G. White, who had recently moved from New Zealand, then became Director of the RPL. During 1943-44, the RPL involved a staff of 300 persons working on 48 radar projects, many associated with improvements on the LW/AW. Height-finding was added (LW/AWH), and complex displays converted it into a ground-control intercept system (LW/GCI). There was also a unit for low-flying aircraft (LW/LFC). Near the end of the war in 1945, the RPL was working on a microwave height-finding system (LW/AWH Mark II).

Dr. Edward G. Bowen, one of the three original leaders of RDF development in Great Britain, came to Australia in January 1944 and joined the RPL as chief of the Radiophysics Countermeasures. Projects to detect, locate, and jam Japanese radars ensued.

Following the war, Dr. Bowen led a major effort in establishing a comprehensive research effort in radio astronomy in Australia. Dr. Martyn became the Director of the RRP's Atmospheric Research Station in New South Wales. There he made fundamental improvements in knowledge of the ionosphere. Like Bowen, Dr. Piddington turned to work in radio astronomy, first developing microwave interferometers, then later in theoretical analysis. Dr. Pawsey also became a renouned radio astronomer and was President of the International Radio Astronomy Commission. Dr. White eventually became Chairman of the CSIR and was knighted in 1962.

NEW ZEALAND

Ernest Marsden, Secretary of the Department of Scientific and Industrial Research (DSIR), represented New Zealand at the 1939 briefings in England.

Marsden (1889-1970) was the son of a cotton weaver in Lancashire, England. While attending Victoria University, he came under the wing of the New Zealand experimental physicist Ernest Rutherford, a recently appointed professor of physics, and assisted him in experiments that ultimately led to a Nobel Prize for his mentor. In 1915, on Rutherford's recommendation, Marsden came to New Zealand to be a professor of physics at Victoria University College in Wellington, and was soon the most recognized physicist in the nation. In 1926, he led in establishing

Ernest Marsden

the DSIR and was named the first Secretary, a position that he held for over 20 years.

Following the RDF briefings, Marsden remained in Great Britain for several months, gaining knowledge of the equipment and operations, and collecting a large shipment items needed for future projects in New Zealand; these included a variety of radio components, the electronic sections of two television sets, and portions of an Air-to-Surface Vessel (ASV) Mk 1 airborne RDF set.

Shortly after returning to New Zealand in October 1939, Marsden established two developmental facilities – one at Canterbury University College in Christchurch, and the other at the Radio Section of the Central NZ Post Office in Wellington. Like in Great Britain, the Post Office had general responsibilities for radio activities in New Zealand, and their Radio Section was the best in the Civil Service organizations. Both facilities were set up with the highest possible security.

Dr. Frederick White, a professor of physics at Cantrbury was asked to lead the Christchurch activities. He was also asked to establish year-long courses in radio physics at Auckland and Canterbury Universities for preparing future researchers and maintenance engineers.

Frederick W. G. White (1905-1994) was a native of Wellington, New Zealand. He graduated in physics from Victoria University College, then studied in England at Cambridge under Ernest Rutherford. A research position with Professor E. V. Appleton followed, and he was awarded his doctorate in 1934. After teaching at Kings College in London for two years, he returned to New Zealand in 1936 as Professor of Physics at Canterbury University College and was also a senior scientist performing ionospheric experimental studies at the DSIR.

Frederick White

Marsden gave different objectives to the two groups. At Christchurch, they were to develop a set for shipboard detection of aircraft and other vessels, and a companion set for directing naval gunfire. For this effort, White was initially assigned one engineer and two technicians from the DSIR. The engineer, Congreve J. Banwell, was highly talented and invaluable for the initial work; however, he found the RDF development so interesting and challenging that he left in less than a year to join the much larger TRE effort in Great Britain; some of his accomplishments there are noted in Chapter 2.

The objective of the Wellington group was to develop land-based and airborne RDF sets for detecting incoming vessels and a set to assist in gun-directing at coastal batteries. As a small country with a large coastline in a huge ocean, New Zealand's primary threat would be from enemy warships, submarines, and raiders; consequently, the work at Wellington was given first priority. Charles Watson-Munro from the DSIR was assigned by Marsden to lead the activity; in addition, there were several engineers and technicians assigned from the Post Office.

Charles Norman Watson-Munro (1915-1991) studied physics as an undergraduate, followed by a master's degree in geophysics, both at Victoria University College. While completing graduate studies, he joined the DSIR in 1936 and worked directly under Marsden, primarily in geophysics research. After leading the RDF project at Wellington, he was elevated in his position with the DSIR and was sent to the United States to learn microwave technology; for this, he worked at the Radiation Laboratory at MIT.

Charles Watson-Munro

Before the end of 1939, the Wellington group had an experimental set that was tested from atop the Post Office Research Facility. An available 180-MHz (1.6-m) transmitter designed for Morse code was modified to produce 2-μs pulses with a PRF at 2 kHz; using two RCA Type 834 tubes, it produced about 1-kW peak power. The receiver was built around a 45-MHz IF unit from a Pye television set brought from England by Marsden. Separate dipole antennas with reflectors were used for transmitting and receiving. Large vessels were detected at ranges up to 30 km.

A number of additional personnel were added to the group, and by mid-1940 the experimental set had been developed into a system suitable for deployment. Improvements included VT-90 Micropup tubes obtained from the Admiralty, giving a much increased output power. A 6-inch CRT was used to display the pulses and indicate the range. The antenna was a broadside array, containing separate transmitting and receiving portions. This system was designated a type CW (Coast Watching), and the first unit was placed at a high elevation near the Auckland Naval Base. With this location and higher power, the range increased to over 45 km.

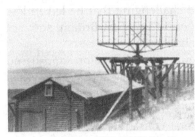

Typical CW or CD RDF Station

An almost identical system was developed at the same time for fire control. Designated CD (Coast Defense), the only significant differences were in the use of lobe-switching on the two sections of the receiving antenna and incorporation of a larger 12-inch CRT display. Horizontal lobe-switching was used to obtain better azimuth resolution, and the elevation angle was determined by vertical lobe-switching. Two 5-inch CRTs were needed for these functions. An experimental set was tested at Motutapu Battery near Auckland using Yagi antennas. These were replaced by the broadside arrays when the first CD set was deployed at Palmer Head Battery at Wellington in late 1940. This was near the Post Office facility; thus, the system remained in regular use while it was improved over the next year. Eventually, resolutions as small as a quarter-degree were possible in azimuth readings.

Type CD RDF

A partially assembled ASV Mk-1 set was included in the materials sent from England by Marsden. This set, operating at 200 MHz (1.5 m), was adapted by the Wellington group for use on aircraft of the Royal New Zealand Air Force (RNZAF). A central dipole antenna was used for transmitting, and receiving dipoles were on each side of the aircraft; switching between the skewed receiving dipoles was used to determine target azimuth.

In early 1940, New Zealand's first airborne radar was flown on an American-made Waco aircraft; the set had a range of about 20 km. Following this, about 20 sets were built and put into service on Vincent and Oxford aircraft; however, they were replaced when the much better ASV Mk-2 became available in mid-1941.

Here it might be noted that the RNZAF was not a great supporter of the development efforts of the DSIR. In fact, they had ordered equipment for an overseas version of the Chain Home system and a Ground Control

Intercept (GCI) system from Great Britain. The DSIR was only involved in the installation efforts.

At Christchurch, White had started the hardware development as well as the radio-physics instructional activity. For the ship-based set, another available Morse-code transmitter was modified, this one operating at the higher frequency of 450 MHz. With a smaller staff, the work went slower than at Wellington, but the initial staff was greatly augmented by the assignment of Lieutenant S. D. Harper of the Royal Navy. Harper introduced a number of innovations, including a shift from racks to cabinets with roll-out drawers for the equipment, greatly facilitating maintenance. Earlier, Harper had been a research engineer in England with the British Post Office.

By July 1940, the Christchurch group had developed an experimental fire-control set that for trials was installed on the Armed Merchant Cruiser *Monowai*. This was designated SS-1 and operated at 66 cm (450 MHz) with a pulse-power of 5 kW at a PRF of 2 kHz; type VT-90 tubes were used. The receiver was built around a Pye television set: the display used a 4-inch CRT with 1,000-yard range-markers. Two dipole antennas were used, backed by a single cylindrical parabolic reflector.

A series of modifications were made, including using HF110 tubes and changing to 430 MHz (70 cm), the highest frequency attainable with available components. The PRF could be switched between 800 and 1,200 Hz, and the pulse-width was 2 μs. The antenna was improved by using 14-element Yagis. The range was up to 7 km with an accuracy of 50 m, and the azimuth accuracy was about one degree.

Two of these systems, designated SWG (Ship Warning, Gunnery) were built at Christchurch and in August 1941 were installed for regular service on the *Archilles* and *Leander*, Cruisers transferred to the newly formed Royal New Zealand Navy (RNZN). Eight SWG systems were eventually built for the RNZN.

The same basic equipment was used by the Christchurch group in developing a ship-based air- and surface-warning system. The primary difference was that the SW antennas could be directed in elevation for aircraft detection. Designated SW (Ship Warning), it was usually installed together with the SWG.

A factory was set up by a commercial firm in Auckland to manufacture the SW and SWG sets, eight of each type were eventually accepted by the RNZN. A number of SWGs were also built for the British fleet stationed in Singapore; some of these with their manuals were captured by the Japanese in early 1942.

A Radio Physics Board (RPB), later known as the Radio Development Board, had been established in April 1941 to formulate policy regarding all radar questions, including priority in development and production. The Chiefs of Staff of the three military services were included. Lieutenant Harper was appointed as RDF officer responsible for the coordination of naval radar.

At the end of 1941, the RPB consolidated the administration of the program, establishing a Radio Development Laboratory (RDL) as a unit of the DSIR with Watson-Munro as Director. Disappointed that he had not been named to direct the new RDL, in early 1942 Frederick White accepted an invitation to give assistance in Australia – initially for a year, but then it became permanent. As an exchange, Dr. Oliver Owen Pulley came from Australia to lead the Christchurch group.

In late 1941, Watson-Munro went to the United States as a Scientific Liaison Officer to learn microwave technology. After spending several months at the MIT Radiation Laboratory, he returned in May 1942 with a shipment of magnetrons, klystrons, and other critical components; these were obtained from the U.S. and Canada under a "Lend-Lease" arrangement.

A microwave development activity was then started at the RDL laboratory at the Wellington Post Office. It was decided to develop mobile 10-cm (actually 9.2-cm) systems for coast-watching and surface-fire-control that might be used throughout the Pacific. With a great demand for such systems, an experimental unit was developed and tested before the end of 1942.

A major portion of the follow-on development involved designing equipment that could withstand the mechanical vibration and tropical environment in a mobile system for the South Pacific islands. The system was mounted in the cabin of a 10-wheel truck and a second truck carried the power generator and workshop. A single parabolic antenna was on the roof of the equipment cabin, capable of laying flat when in transport. A plan-position indicator CRT was used, the first such in New Zealand.

Designated ME, these truck-mounted microwave systems were built by the RDL, mainly in Wellington but with some elements in Christchurch. The first of these saw service in early 1943 in support of a U.S. torpedo-boat base in the Solomon Islands; the RNZN and the RDL provided the maintenance personnel.

Starting in 1943, some of the microwave radars were used to replace 200-MHz CW sets, and several systems were built for

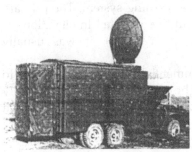

ME Truck (Camouflaged)

operation on RNZN minesweepers in the Pacific. The last six of these systems were constructed for tracking radiosonde balloons of the Meteorological Service.

As the Allies progressed upward in the Pacific, a need arose for a long-range warning set that could be quickly set up following an invasion. The RDL took this as a project in late 1942, and in early 1943 six Long-Range Air Warning (LWAW) systems were available for service. These operated at 100 MHz (3 m) and, like the microwave sets, were mounted in trucks.

Interior of ME Truck

A single Yagi antenna was used, demountable for transport but ready to operate in a half-hour. There was also a broadside array that could be used when a more permanent operation was established. The range using the Yagi was near 150 km; this increased to over 200 km with the broadside.

From the start in late 1939, 117 radar sets of all types were built in New Zealand, all by small groups; no types were ever put into serial production. After 1943, little such equipment was produced in the country, and RNZN warships were then provided with British outfits to replace the earlier New Zealand sets.

One final radar activity should be mentioned. In mid-1942, an Operational Research Section was formed in the RDL. A major part of its activity was in the analysis of radar operations. Dr. Elizabeth Alexander led this activity.

Elizabeth Somersville Alexander (1908-1959) was born in England but raised in India. She returned to England to earn undergraduate and Ph.D. degrees in geology from Newnham College, Cambridge University, the doctorate in 1934. She married a physicist from New Zealand, and they took up residency in Singapore. There she worked for British Naval Intelligence with the rank of Captain.

Elizabeth Alexander Analyzing Data

In early 1942, as the Japanese took over Singapore and interned her husband, she escaped to New Zealand. There she joined the RDL and initiated studies relating radar performance and meteorological data. At the start of 1945, radar operators reported a great increase in signal noise just at sunrise. From this, she developed an extensive program that within a year proved that the Sun was a major source of electromagnetic noise. Because of this, she is often called "the first female radio astronomer."

After the war, Elizabeth Alexander and her husband were reunited and eventually settled in Singapore where she served as the State Geologist. Ernest Marsden became Scientific Liaison Officer in London, and was Knighted in 1958.

Charles Watson-Munro turned to nuclear power. Following periods in Canada and Great Britain, he returned to New Zealand, first as an official at the DSIR, then as Professor of Physics at Victoria University College (here he was also awarded a D.Sc. degree). Later he became the Chief Scientist at the Australian Atomic Energy Commission.

CANADA

Dr. John T. Henderson, head of the Radio Section of the National Research Council of Canada (NRCC), attended the briefings in Great Britain in early 1939 on their RDF developments. This ultimately led to Canada's making significant contributions to radar technology.

John Henderson

John Tasker Henderson (1905-1983) was born in Montréal, Canada. He received the B.S. and M.S. degrees in engineering physics from McGill University, and then went to England, where he studied under Professor E. V. Appleton. He received the Ph.D. degree in physics from the University of London in 1932, and followed this with post-doctoral studies at the Sorbonne in Paris and the Technische Hochschule at Munich. In 1933, he joined the staff of the NRCC in Ottawa. His work there included the building of a radio direction-finding set that used a CRT for its display; this was installed in Nova Scotia in 1938.

Upon Henderson's return from the briefings in England, he was authorized to set up an RDF development laboratory in the NRCC Radio Section. His initial report included the following statement:

> RDF represents such an advance in the science of war that it is difficult, at this point, to determine or approximate its strategical or tactical effects in the future.

This activity had the strong support of Major General Andrew McNaughton, then the President of NRCC and an electrical engineer by training, but essentially no development funds were provided by the Canadian Department of Defense.

Early Radar Development in Other Countries

Canada declared war on Germany in September 1939, and McNaughton took leave to go on active military service. Dr. C. J. Mackenzie, Dean of Engineering at Saskatchewan University, was named as the Acting President of the NRCC. The fall of France convinced a group of businessmen of the seriousness of the war, and, shocked by the inaction of the government, they set up a fund of several hundred thousand dollars for war support; a portion went to the RDF efforts at the NRCC. Treasury funding soon followed.

In 1938, the NRCC had acquired 250 acres of rural land in the vicinity of Ottawa for a Field Station. As the RDF development started, this Station was made secure and improved for the testing of antennas and other equipment.

Using commercial components and with essentially no further assistance from Great Britain, Henderson led a team (shortly reaching 22 persons) in developing RDF systems. In March 1940, the Royal Canadian Navy (RCN) requested surface-warning equipment. The first version, called Night Watchman (NW), was a fixed apparatus designed to protect the entrance to the Halifax Harbor. This simple set was successfully tested at Herring Cove near Halifax in July. It operated at about 200 MHz (1.5 m) and had a 1-kW output power with a pulse length of 0.5 μs.

In August 1940, a Crown company, Research Enterprises, Ltd. (REL), was established to manufacture RDF and optical equipment. An Ontario industrialist, Lt. Col. W. E. Phillips, was hired as the General Manager, and a 50,000-square-foot facility was built at the town of Leaside near Toronto to accommodate up to 500 employees. A site for field testing was set up at Scarborough on the shore of Lake Ontario. There were no longer funding problems – only shortages of personnel and materials. The facilities had the highest levels of security.

The REL built the next RDF version, a ship-borne set designated Surface Warning 1st Canadian (SW1C). The basic electronics of this set were essentially the same as the NW, but it had a more sensitive receiver and initially used a Yagi antenna that was turned using an automobile steering wheel. It was also better packaged than the NW and designed to stand the environment at sea. It was first tested aboard the corvette HMCS *Chambly* in mid-May 1941, at which time it detected a Dutch submarine at a range of 2.7 miles (4.3 km) in heavy fog. The project engineer from the

SW1C RDF Controls

NRCC was H. Ross Smith, who remained in charge of projects for the RCN throughout the war.

The REL had considerable difficulties in getting the SW1C into production, and by the end of 1941 only a few corvettes and merchant ships had been outfitted. In early 1942, the frequency was changed to 215 MHz (1.4 m) to make it compatible with the recently adopted IFF sets. An electric drive was added to rotate the antenna. With other minor changes, it was known as the SW2C and produced by the REL for corvettes and mine sweepers. A lighter version, designated SW3C, followed for small vessels such as motor torpedo boats. A CRT plan-position indicator (PPI) display was added in 1943. Several hundred SW sets were eventually produced by the REL.

The Canadian Army had the responsibility of defending the Nation's coast. As the surface-warning RDF sets were being developed for the RCN, the Army requested that coastal-defense sets be built for their firing battalions. Henderson's group started this work in September 1939; four men from the Royal Canadian Corps of Signals were attached to the team to give the Army's perspective.

Much of the first year was spent in examining an existing 500-MHz (60-cm) radio altimeter from Western Electric that might be modified for this application. Because of the low radiated power, modification of this set was eventually abandoned, and a 200-MHz (1.5-m) set was developed. Designated the CD, this had electronics similar to the Navy's surface-warning sets, but required a much larger antenna. By October 1940, a suitable antenna had been erected at the Field Station.

CD Antenna & Tower

After a satisfactory demonstration was given to Army officials, a 70-foot wooded tower was built for a permanent installation of the CD on the coast at Duncan Cove, Nova Scotia. The firing battalion was located some distance away, and the system included a mechanical "displacement corrector" to automatically compensate for the difference in positions.

The installation and checkout activity continued through January 1942, at which time the accuracy of the CD system was about 25 yards for range and 4.6 minutes of arc for azimuth. A second installation was later made at Osborne Head. R. D. Harrison served as the CD project engineer.

In the summer of 1940, A. V. Hill, a member of the Tizard Committee that had oversight of RDF development in Great Britain, had visited

Canada to review its war preparedness. Hill was very impressed with what Canada was doing in RDF; in his report he stated:

> Everything I saw and heard convinced me that there had been a grave lack of imagination and foresight on our part in failing to make full use of the excellent facilities and personnel available in Canada.

As a result of Hill's report, when the Tizard Mission came to the United States in August 1940, they also visited Canada. In fact, Tizard personally stopped in Canada for a few days before going on to Washington. Henderson was invited to participate in the Washington briefings where the high-power magnetron was disclosed. Subsequently two other Mission members, Dr. Edward Bowan and Brigadier F. C. Wallace, visited Canada in October; Wallace was later assigned to remain in Canada and assist in the developments.

From the meetings in Washington, it was decided that Canada would build a microwave gun-laying system for the Canadian Army, as well as for possible use elsewhere. There was a similar development in Great Britain with a system designated GL Mk III, and then the Radiation Laboratory (Rad Lab) at MIT started a third system. To ensure that the different projects were properly identified, they came to be known as GL IIIA, GL IIIB, and GL IIIC; the A, B, and C representing America, Britain, and Canada.

A Microwave Group led by Dr. F. H. Sanders was added to the Radio Section, and in December 1940, six men from the group were sent to the Rad Lab to learn and participate in the rapidly emerging microwave radar developments. A local source of magnetrons was vital, and in February 1941, the National Electric Company (NEC) in Montreal began manufacturing these devices.

By early 1941, both the REL and the Radio Section had staffs of engineers engaged in the design – the REL handling the mechanical equipment, and the Radio Section the electronics. J. W. Bell, who had accompanied members of the Tizard Mission on tours of facilities in the United States, was the Project Engineer. In March, experimental sets operating at 10 cm (3 GHz) obtained echoes from planes at 17,000 yards.

The GL IIIC was housed in two trailers, one with a rotating cabin and one fixed. The rotating one was called the Accurate Position Finder (APF) and held the primary equipment and separate antennas with parabolic reflectors antennas for transmitting and receiving. The antenna aiming required rotating the cabin because means for coupling the microwave signals had not yet been developed. The other trailer carried the Zone Position Indicator (ZPI, a 150-MHz (2-m) radar that found the position of all aircraft within the system's coverage of 60,000 yards. A

GL IIIC APF Trailer

PPI for the ZPI radar was in the APF cabin. Power generators, maintenance equipment, and spare parts were carried in separate trailers

In mid-1941, the REL received orders for 660 GL IIIC systems, mainly to be sent to Great Britain under the Lend-Lease Program from the United States. In July, a very satisfactory demonstration of the prototype system was held in the Field Station. By December, the first six systems had been built at the Radio Section.

During 1942 and into 1943, there were many technical and administrative problems concerning the GL IIIC, and only a few sets were eventually built by the REL. In September 1943, the decision was made that the Canadian Army overseas would use the British GL IIIB in liberating Europe. Consequently, the large order for the REL was never filled.

Success in March 1941 at the Radio Section in getting a 10-cm experimental set for the Army led the RCN to request a ship-borne, early-warning microwave set. Development of a 10-cm (3-GHz) set designated RX/C was initiated in September while the 200-MHz SW was still being finalized. All microwave radars have much in common; thus, the development of the Army's GL IIIC also benefited the RX/C.

The RCN, however, continuously changed the requirements, and by the end of 1942 the REL did not have the RX/C in production. At that time, the Radio Branch (upgraded from a Section) was asked to build eight sets, and these were completed by July 1943. The fielded set incorporated many of the characteristics of the SW sets, but had a PPI display and a parabolic-reflector antenna that could be hand- or motor-directed. Further sets were produced by the REL and used throughout the war.

As the 10-cm RX/C was being developed, the Microwave Section (upgraded from a Group) looked into using even shorter wavelengths. Further, the Admiralty in Great Britain asked about Canada's interest and capability in manufacturing 3-cm magnetrons. This led to the development of a 3-cm device by the NEC and a full 3-cm (10-GHz) radar for small crafts. In May 1942, the Admiralty gave a formal purchase order for these developments. The set was designated Type 268 (not to be confused with the SCR-268 from the U.S. Signal Corps), and was particularly designed to detect a *schnorkle* (a tube for providing air when the submarine submerged.)

Early Radar Development in Other Countries

The development centered on minimizing the size, weight, and power requirements. Since it would also be used in the tropics, resistance to high humidity and temperature conditions were imposed. Since repairs at sea would be almost impossible, high reliability was required. Dr. K. C. Mann served as the Project Engineer. (Mann, like many of the "engineers" at the NRC, was actually a physicist by education.)

A preliminary model was tried in October 1942 with excellent performance. The Radio Branch then built eight sets, and one was sent by plane to England for testing. Many changes were requested, and it was over a year before another version was ready. This was sea-tested on Lake Ontario in July 1944, and full-scale production began at the REL in December with Mann as the Production Manager, About 1,600 Type 268 sets were manufactured before the end of the war.

Type 268 Radar

Type 268 Antenna

As the work of the Radio Branch enlarged, Brigadier F. C. Wallace was made Administrative Manager, but Henderson, as the Chief Scientific Officer, remained in control of the technical activities. By mid-1942, the overall staff of the Radio Branch had grown to near 200 persons, of which some 60 were engineers or physicists. There were electronic and mechanical workshops managed by D. L. West with over 100 technicians, fully capable of building prototypes and few-of-a-kind sets. In addition, at any given time more than 100 military personnel were participating at the main facility and the Field Station.

While the Army was basically satisfied with the 200-MHz CD systems that went into service in early 1942, they did ask about an improvement to 10-cm operation, primarily to give a higher degree of discrimination between targets. Since the Radio Branch was then well experienced in these microwave systems, they easily provided a design. Before even a prototype was built, the Army gave an order to the REL for a number of sets designated CDX.

CDX Radar

The CDX incorporated many elements of the GL IIIC, and used the displacement converter of the CD. A preliminary "lash-up" model was tested in comparison with the CD at Osborne Head, Nova Scotia, in October 1942; the CDX was far better than the CD in range and accuracy, showing the superiority of 10-cm systems. The prototype was completed in February 1943. The progress of the war, however, was going well and the production order was reduced to 14 sets with 5 additional sets built for the Russians.

Long-range detection systems require high power, and to obtain this level the early systems were limited to meter-wave operation. Also, very large antenna arrays were required to concentrate the energy in a search beam. By mid-1942, the firm NEC developed a magnetron capable of producing 300 kW at 10.7 cm (2.8 GHz); the Microwave Section then started an internal project using this magnetron.

In the spring of 1943, German submarines started operating just outside the St. Laurence Seaway – the primary ship route from Canada to Great Britain. To counter this, the RCAF asked that 12 sets of a long-range microwave system be built by the Radio Division. To make the system more flexible, they asked for a height-finder capability. There was a similar project at the Rad Lab, and close coordination between the two projects was maintained.

The first system was designated MEW/AS (Microwave Early Warning Anti Submarine) and did not have the height finder. Using a slotted antenna 32 by 8 feet in size designed by Dr. William H. Watson at McGill University, this radiated a narrow horizontal beam to sweep the sea surface. The transmitting and receiving equipment was located behind the antenna, and the assembly could be rotated at up to 6 RPM. The controls and PPI display was in a nearby fixed building. This could detect targets at up to 120-miles (196-km) range.

MEW/AS Radar

Early Radar Development in Other Countries

The second version, designed for detecting high-flying aircraft, was designated MEW/HF (Height Finding). In this, the power could be switched to a smaller, rotating antenna that gave a vertical beam 20 by 1.5 degrees and tilted upward by 10 degrees. The initial sets were tested at the Field Station, then the sets went into operation at several sites in Newfoundland, Quebec, and Ontario. No single person was responsible for the MEW project, but the final report was prepared by H. A. Ferris and Dr. D. W. R. McKinley.

One of Canada's major contributions to WWII radar was the training of personnel in electronics. In June 1940, Canada agreed to provide RDF (radar) technicians for worldwide assignments, as well as to give training for Commonwealth and other selected nations. The Royal Canadian Coastal Services offered the first RDF course in November 1940 at Halifax, Nova Scotia.

An RDF school was established at Clinton, Ontario, in May 1941, by the Air Council of Canada. In addition to students from Canada and other Commonwealth nations, initial classes included U.S. Army, Navy, and Marine students. Training at this highly secret #31 RAF school had eight weeks on ground radars and six on airborne systems. Canadian students were required to have 10 to 12 weeks in difficult preparatory courses provided by colleges and technical schools across Canada.

During the war, about 6,500 Commonwealth students and 2,300 Americans went through courses at Clinton. At its high point in early 1945, the #31 School had a staff of 478. The airborne radar training program set up by the U.S. Navy at Annapolis in late 1941 was patterned on this school.

In 1942, the Canadian Army set up RDF maintenance schools at Debert, Nova Scotia, and at Camp Barriefield, near Kingston, Ontario. The Barriefield school was enlarged in 1944 to become the Royal Canadian Electrical and Mechanical Engineers (RCEME) training center.

The Royal Canadian Navy's Signal School at St. Hyacinthe, Quebec, was the central location for training Canada's shipboard and shore-based communications personnel. Moved from an original location in Halifax in the summer of 1941, the St. Hyacinthe school had, at a given time, about 3,200 officers, enlisted men, and Wrens (Women's Royal Canadian Naval Service), involved in all phases of training in communications. The radar course for enlisted men was 42 weeks in length. Officers taking a similar, but somewhat shorter, course were all university graduates in electrical engineering, physics, or similar fields.

In addition to the radar sets previously described, many others were designed at the NRCC's Radar Branch during the war years – a total of 30 of all types. Of these, 12 types were turned over to the REL where they were built in quantities varying from a few to hundreds; altogether, some 3,000 were produced before the REL was closed in September 1946.

Dr. John Henderson remained with the NRCC until he retired. In 1949, he set up new absolute electrical standards for Canada, and shortly thereafter became the Principal Research Officer of the NRCC. In 1957, he was elected the President of the IRE, and in 1960, he was awarded the inaugural McNaughton Medal (given for Canadian accomplishments in engineering).

SOUTH AFRICA

The lead in RDF development in South Africa was taken by the Bernard Price Institute (BPI) for Geophysical Research, a unit of the University of the Witwatersrand (Wits) in Johannesburg. Dr. Basil Schonland, a world-recognized authority on lightning detection and analysis, headed the effort.

As a scientifically advanced member of the British Commonwealth, South Africa would have been invited to send a representative to Great Britain in early 1939 for briefings on their RDF developments. The records do not show that such an invitation was received, but, in any case, no representative from South Africa was sent. However, General Jan Smuts, the new Prime Minister and the commander of the military force, asked Schonland to meet with Dr. Ernest Marsden, who had represented New Zealand at the briefings in Great Britain.

Marsden returned from England to New Zealand aboard the SS *Winchester Castle*. In mid-September as the boat went around South Africa, Schonland boarded it at Cape Town and traveled three days with Marsden along the coast to Durban. The two scientists were well acquainted and Marsden, after administering his version of the Official Secrets Act, gave Schonland verbal information on RDF developments. Also, while the ship was in port at Durban, Schonland was able to have some pages of Marsden's secret notes copied on glass slides at the local University of Natal.

Basil Ferdinand Jamieson Schonland (1896-1971) was born in Grahamstown, South Africa, where his father was Professor of Biology at Rhodes University College. Schonland entered Rhodes at age 14, and graduated at 17 with a B.A. degree in physics and first-place Honors in South Africa. During 1914-15, he studied at Caius College of Cambridge University in England, then, as WWI started, he volunteered for the Signal (Wireless) Services of the Royal Engineers. As a Second Lieutenant, he saw action and was wounded in France. By the end of the war, he had been promoted to Captain, had 30 officers and 900 enlisted men under him, and was responsible for 300 wireless sets.

Returning to Cambridge at the start of 1919, Schonland studied atomic physics under Professor Ernest Rutherford at the Cavendish Laboratory. Upon completing his doctorate in 1922, he went home to South Africa and joined the physics faculty of the University of Cape Town (UCT). Over the next 15 years, he conducted extensive research in atmospheric electricity, developing new types of instrumentation for detecting and recording lightning.

Basil Schonland

By 1937, Schonland had gained an international reputation in his field and was selected to become the founding Director of the Bernard Price Institute of Geophysics (BPI) at Witwatersrand University in Johannesburg. Among other activities, he and graduate assistant David B. Hodges built a radio direction-finding set similar to the one used by Robert Watson Watt at the RRS, Slough, in England (see Chapter 2) and established the capability for warning aircraft of electrical storms.

Immediately following his meeting with Marsden, Schonland received funding to establish an RDF project at the BPI. As with Watson Watt in England, the equipment used for monitoring lightning provided a start. An outstanding team was assembled for the project and was briefed on the Secrets Act. With nothing more than poor copies of some "rather vague documents" and notes taken by Schonland during his brief sea trip with Marsden, the team set about developing an RDF system. Their primary references were the *ARRL Radio Amateur's Handbook* and Terman's *Radio Engineering*, both American publications.

Dr. Christopher P. Gane of the BPI developed the transmitter. The receiver was built by Dr. Guerino R. Bozzoli, also of the BPI, and Professor W. Eric Phillips from the University of Natal. Dr. Noel Roberts from the UCT designed the timing and synchronization circuits. The antennas were personally developed by Schonland. Research assistants Hodges and Frank J. Hewitt (who both later completed their doctorates) were also major participants. Jock Keiller, the leader in BPI's shop, made "invaluable contributions."

The transmitter operated at 90 MHz (3.3 m) and had a power of about 500 W. The pulse was 20-μs in width and the PRF was 50 Hz, tied to the power-line. The receiver was super-regenerative, using type 955 and 956 Acorn tubes in the front end and a 9-MHz IF amplifier. Separate,

Initial JB

rotatable antennas with stacked pairs of full-wave dipoles were used for transmitting and receiving. The beams were about 30 degrees wide, but the azimuth of the reflected signal was determined more precisely by using a goniometer. Pulses were displayed on the CRT of a commercial oscilloscope.

By the end of November 1939, just two months after the start, the various elements of the system were completed, all by using locally available components. Before the end of the year, a full system had been assembled and detected a water tank at a distance of about 10 km.

BPI with JB Receiving Antenna

With the initial success, the Secretary for Defense declared the BPI a Special Defense Section and placed Wits under military control. All of the RDF participants were commissioned as Army Officers. Improvements were made on the receiver, and the transmitter pulse-power was increased to 5 kW. Hewitt designed the very important monitor. Designated JB (for Johannesburg), the prototype system was taken to near Durban on the coast for operational testing. There it detected ships on the Indian Ocean, as well as aircraft at ranges to 80 km.

There was a pressing need to protect supply shipping coming down the east coast of Africa. In early March 1940, the first JB was deployed to Mombrui on the coast of Kenya, joining an anti-aircraft Brigade for intercepting attacking Italian bombers.

Paintings by Geoffrey Long of Improved JB Radar System in Operation

Improved JB systems were built at the BPI, the most important change being the use of a transmit-receive device allowing a common antenna. In mid-1941, six systems were deployed to East Africa and Egypt; at the latter, they were used in protecting RAF installations near the Suez Canal. Captain Frank Hewitt was responsible for these installations. JB systems were also placed at the four main South African ports, guarding the highly important shipping lanes around the Cape of Good Hope. Eventually, British and American radars were also installed, but the JBs remained in operation throughout the war.

Shortly after the JB went into service, the University of Natal initiated special training courses in radar operation and maintenance. This effort was led by Professor Hugh Clark, head of the Electrical Engineering Department.

In late 1941, Schonland, as a Colonel in the South Africa Defense Force, went to Great Britain to serve as Superintendent of the Army Operational Research Group and later the scientific advisor to Field Marshall Bernard Montgomery. Schonland's accomplishments during this period are covered in Chapter 4.

Returning to South Africa after the war, Schonland founded the Council for Scientific and Industrial Research (CSIR) and served as its first President. Dr. Frank Hewitt was later the CSIR Deputy President. In opposition to the apartheid policies of the then South African government, Schonland went back to England in 1954. There he headed the Atomic Energy Research Establishment at Harwell and was Knighted by the Queen in 1960 for his services to British science.

In 1999, 27 years after his death, Sir Basil Schonland was elected South Africa's "Scientist of the Century."

Sir Basil Schonland

HUNGARY

During World War I, the Austrian-Hungary Empire sided with Germany. At the close of this war, the Empire was divided into a number of states, including an independent Hungary that was much reduced in size. During the Great Depression, Hungary returned to closer ties with Germany, including the adoption of laws that particularly affected the status of Jews.

Historically, Hungary had excellent educational institutes, and, although its industry was not comparable to that in Western European nations, it was the most advanced of the Balkan states. However, as political conditions declined, large numbers of the scientific community fled the country, and both industries and universities deteriorated.

When World War II started, Germany persuaded Hungary to join the Axis, promising the return of land lost at the end of WWI. In opposition to this, many Axis-leaning heads of state were removed and negotiations were started with the Allies. Therefore, for most of the war years, Hungary was caught between trying to please both the Axis and the Allies.

Two of the more technically advanced firms in pre-war Hungary were ETVERT (in English, United Incandescent Lamps and Electrical Company), and HIKI (Research Institute of the Telecommunications Industry). In 1936, Dr. Zoltan Bay was made Director of Research at ETVRT. For pursuing research and development for the Hungarian Defense Ministry, the two firms jointly formed an operation called IZZO; Bay served as the Chief Engineer.

Zoltán Bay

Zoltán Lajos Bay (1900-1992) was born in the small town of Gyulavári, Hungary, the son of a Lutheran clergyman. He attended a Presbyterian boarding school, where he developed a deep interest in science. Studies at the Royal Hungarian Péter Pázmány University (later named Technical University of Budapest) resulted in a diploma in education and, in 1926, the doctorate in physics. This was followed by four years of study and research, primarily in chemical physics, at several institutes in Berlin.

Returning to Hungary in 1930, he was appointed Professor of Physics at Szeged University. Later, he moved his academic affiliation to Technical University of Budapest, serving as Director of the Nuclear Physics Department. During all of this time, he was also employed in industry.

Work at IZZO in radio-location (*rádiólokáció*, radar) was started in late 1942. Bay's own words describe this origin:

> The Minister of Defense summoned me and set forth that I have to develop the basic radar principles and techniques for the Hungarian Army. I accepted the task with the idea in mind that if I have to [do] military work in this unjust war, I prefer working on defense [rather] than offensive weapons.

Bay's development group included 12 engineers and physicists, mainly from Budapest University, and 30 technicians. They were given no assistance or even information from the Germans. The HIKI already had transmitting and receiving equipment suitable for modifications. ETVRT was a producer of radio tubes, and took the task of improving those for this application.

They were aware of developments in klystrons and magnetrons, and knew of the potential benefits of operating at microwave frequencies, but decided that this would require too long for the development and stayed at VHF wavelengths. The group turned to journal papers on ionospheric measurements for information (such as using pulsed transmission for distance determination.

A system called *Sas* (Eagle) was developed and installed atop Mount János, located in the center of the defended area. No detailed information is now available, but the equipment operated at 120 MHz (2.5 m) and was in a cabin with separate transmitting and receiving dipole arrays attached; the assembly was all on a rotatable platform. According to published records, the system had a range of "better than 500 km." A second *Sas* was installed at another location. There is no indication that the systems were ever in regular service.

Hungarian Sas Radio Locator

In 1944, Bay began the planning of a first-of-a-kind experiment: reflecting a radar signal off of the Moon. For this, he devised a highly novel detection technique to be used with the *Sas*. Knowing that the returned pulse would be too weak for display on a CRT, and that the frequency stability was not sufficient for electronic integration of pulses, he developed a novel electro-chemical signal accumulator.

The output from the receiver was fed into a series of delay circuits, with the output of each in turn feeding into one of ten coulometer cells. With time, even the extremely small currents from received pulses could cause measurable electrolysis of a KOH solution in a cell. "Active" signals would be accumulated over a 30-minute period, followed by a "blind" 30-minute period during which the antenna was turned away from the Moon.

Near the end of the war, the lunar experiment was ready to be performed. The conquering Soviets, however, took away the equipment and it had to be rebuilt. On February 6, 1946, the experiment was finally performed. Any pulse returned from the Moon would be received about 2.5 seconds after transmission. In the cell where the 2.5-s delayed signal was accumulated, the gas from the electrolysis was about 4.4 percent greater in the active period than in the blind period, indicating that the Moon had been "detected" by the radar.

Because of the delay caused by the Soviets taking his initial set, Bay's first test unfortunately occurred two weeks behind a successful experiment on January 20 by the U.S. Army Signal Corps Engineering Laboratories in New Jersey. The American experiment used a modified SCR-271, the radar of Pearl Harbor.

As Communism took hold in Hungary, Dr. Bay left in 1948, accepting a position with George Washington University in America. He later became head of the Department of Nuclear Physics at the U.S. Bureau of Standards. There he devised a new means of measuring the velocity of light, leading in 1983 to a new world standard for the meter.

Bibliography and References for Chapter 8

Austin, Bryan; *Schonland: Scientist and Soldier*, Inst. of Physics Pub., 2002

Avery, Donald H.; *The Science of War: Canadian Scientists and Military Technology During the Second World War*, U. of Toronto Press, 1999

Blanchard, Yves; "Microwave Radar in Europe: A Historical Perspective," *Microwave Journal*, Oct., p. 94, 2008

Brain, Peter, Sheilagh Lloyd, and Frank Hewitt; *South African Radar in World War II*, SSS Radar Book Group, 1993

Bremer, Frans O. J., and Piet van Genderen (editors); *Radar Development in The Netherlands*; Thales Nederland, 2004

Brown, Lewis; *A Radar History of World War II – Technical and Military Imperatives*, Inst. of Physics Pub., 1999

Calamia, M., and R. Palandri; "The History of the Italian Radio Detector Telemetro," in *Radar Development to 1945*, ed. by Russell Burns, Peter Peregrinus, 1988

David, Pierre; *Le Radar (The Radar)*, Presses Universitaries de France, 1949 (in French)

Giboin, E.; "L'Evolution de la Detection Electromagnetique dans la Marine National, (Evolution of Electromagnetic Detection in the National Marine)" *Onde*, vol. 31, p. 53 (in French), 1951

Girardeau, Emile; "The Radar, French Invention," *France-America Magazine*, p. 95, 1953

Guierre, Maurice; "The History and the Creation and Progress of the Radar [in France],"in *Waves and Men, History of the Radio*, Julliard, 1951

Gutton, H., M. Elie, M. Hugon, and M. Ponte; "Detection d'obstacles, (Detection of Obstacles)" *Bulletin de la Soc. Francais des Electriciens*, vol. 9, April, 1939 (in French)

Hewitt, F. J.; "South Africa's Role in the Development and Use of Radar in World War II," *Military History Journal*, vol. 3, June, 1975

Hewitt, Frank J.; "50 Years of Radar in South Africa," *The Radioscientist*, vol. 1, p. 71, 1990;

Kovács L.; "Zoltán Bay and the First Moon-Radar Experiment in Europe," Science and Education, vol. 7, May, p. 313, 1998

Le Pair, C. (Kees); "Radar in the Dutch Knowledge Network," Telecommunication and Radar Conference, EUMW98, Amsterdam, 1998; http://www.clepair.net/radar-web.htm

Lloyd, Sheilah, and Phyllis Lloyd; "Fifty Years of SA Radar," South African Military History Society, 1990;
http://samilitaryhistory.org/diaries/radarfif.html

Mason, Geoffrey B.; "New Zealand Radar Development in World War 2"; http://www.naval-history.net/xGM-Tech-NZRadar.htm

Molyneux-Berry, R. B.; "Henri Gutton, French radar pioneer," in *Radar Development to 1945*, ed. by Russell Burns, Peter Peregrinus, 1988

Moorcroft, Don; "Origins of Radar-based Research in Canada," Univ. Western Ontario, 2002;
http://www.physics.uwo.ca/~drm/history/radar/radar_history.html

Posthumus, K; "Oscillations in a Split-Anode Magnetron, Mechanism of Generation, " *Wireless Engineer*, vol. 12, p. 126, 1935

Ponte, Maurice; "Sur des Apports Francais a la Technique de la Technique de la Detection Electromagnetique (On French Contributions to the Technique of Electromagnetic Detection)," *Revue Technique Thomson-CSF*, vol. 1, p. 433, 1955 (in French)

Renner, Peter; "The Role of the Hungarian Engineers in the Development of Radar Systems," *Periodica Polytechnica Ser. Soc. Man. Sci*, vol. 12, p. 277, 2004; http://www.pp.bmc.hu/so/2004_2/pdf/so2004_2_12.pdf

Romano, Salvatore; "History of the Development of Radar in Italy," Regia Marina Italiana [Royal Italian Navy], 2004; http://www.regiamarina.net/others/radar/radar_one_us.htm

Staal, M., and J. L. C. Weiller; "Radar Development in the Netherlands before the war," in *Radar Development to 1945*, ed. by Russell Burns, Peter Peregrinus, 1988

Swords, Sean S.; *Technical History of the Beginnings of Radar*, Peter Peregrinus, Ltd., 1986

Tiberio, U.; "Some historical data concerning the first Italian naval radar," *IEEE Trans. AES*, vol. 15, Sept., p. 733, 1979

Unwin, R. S.; "Development of Radar in New Zealand in World War II," *IEEE Antennas and Propagation Magazine*, vol. 34, June, p. 31, 1992

Waters, S. D.; "Development of Radar," Ch. 28 of *The Royal New Zealand Navy in the Second World War*, Historical Publications Branch, 1956

INDEX – NAMES

Adams, Renie 221
Alder, Leonard S. B. 73
Alekseyev, N. F. 289, 300
Alexander, Elizabeth Somersville 360, 361
Alexanderson, Ernst F. W. 10, 12, 13, 25
Alfred, R. V. 77
Allen, Alexander J. 175
Alvarez, Luis W. 163, 176, 214-216
Amato, Ivan 115
Anderson, S. Herbert 120
Appleton, Edward N. 40, 41, 47, 51, 66, 70, 71, 354, 376, 362
Armstrong, Edwin H. 12, 16, 133
Arnold. Harold D. 12
Arnold, Henry H. 134
Atkinson, J. R. 146
Baba, Shoji 332
Bacher, Robert F. 175
Bain, Alexander 24
Bainbridge, Kenneth T. 175
Bainbridge-Bell, L. H. (Labouchere Hillyer) 40, 41, 47-49, 106, 184
Baird, John Logie 24
Baldwin, Stanley 42
Banneitz, Friedrich Wilhelm 250
Banwell, Congreve J. 96, 97, 99, 100, 357
Bareford, C. F. 74, 77
Barker, G. H. 53
Barfield, Robert H. 38, 39
Barkhausen, Heinrich G. 20, 141, 232, 293, 313, 315
Barrow, Wilmer L. 22
Batt, Reginald G. 147
Bay, Zoltán Lajos 374-376
Bedford, Lesley H. 93, 100
Belin, Edouard 24
Bender, Louis E. 121
Bennett, Edwin L. 113
Bentley, Jetson O. 31
Berg, Aksel Ivanovich 306
Beria, Lavrentiy P. 275

Black, D. H. 90
Blackett, Patrick M. S. 43, 93, 156
Blair, William R. 33, 119, 120, 123, 130-133
Blumlein, Alan D. 87, 97, 173, 189
Bohr, Niels 293
Bonch-Bruyevich, Mikhail A. 279, 280, 282, 286-289, 291
Bondi, Hermann 187, 188
Booker, Henry G. 194
Boot, Henry (Harry) Albert 143-145, 316, 352
Bose, Jagdish Chandra 5
Bowen, B. Vivian 102, 193
Bowen, Edward G. 47-53, 55, 56, 59, 62-68, 70, 73, 74, 78, 83-88, 100, 142, 155-161, 163, 173, 174, 176, 194, 199, 224, 355
Bowen, Harold G. 117, 125, 156
Bowles, Edward L. 22, 158, 160, 161
Bozzoli, Guerino R. 371
Brandimarte, Alfeo 343, 345
Brandt, Leo 241
Branley Edouard 7
Braude, Semion Y. 294, 301, 308
Braun, Karl Ferdinand 6, 11, 232
Breit, Gregory 23, 31, 41, 123, 131, 205
Breuning, Ernst 261, 162
Bridge, Maurice 349
Brown, Jim 57
Burcham, W. E. 146
Burgess, W. B. 114
Burnett, Miles A. F. 40
Bush, Vannevar 153-157, 160, 161, 164, 203
Butement, William Alan S. 31, 32, 68, 70, 89-91, 93, 197, 205, 224
Calpine, H. C. 76
Campbell-Swinton, Alan A. 26
Carlow, Lord 79, 80
Carnegie, Andrew 23
Carrara, Mello 342, 343, 345
Castellani, Arturo 345
Chamberlain, A. Neville 59, 80, 230
Chaplygin, S. A. 280
Chernyshev, A. A. 280

Christ, Fritz	245	Du Mont, Alan B.	28
Churchill, Winston L. S.	80, 81, 99, 148, 149, 151, 155, 173, 183, 184, 190, 193, 300	Eastman, Melville	180
		Eastwood, W. G.	53
		Eddy, William C.	118
Clark, Hugh	373	Edison, Thomas A.	5, 11, 22, 112, 251
Clarke, William J.	8		
Cleeton, C. E.	201, 202	Edwards, George C.	172
Coales, John F.	76, 77, 184, 186	Eisenhower, Dwight D.	148, 150
Cockburn, Robert	104, 105, 107, 192	Elias, G. J.	339
Cockcroft, John D.	91, 94, 155-161, 168, 169, 177, 195, 198, 224	Elie, Maurice	350
		Elizabeth, Queen	85, 99
Cockrance, C. A.	166	Elji, Maurice	350
Cole, Ralph I.	132	Elliot, George	134
Cole, R. C,	197	Erbsloh, Paul-Gunther	234
Colton, Roger B.	131-133, 178, 180	Escobar, Frank A.	202
Cook-Yarbrough, E. H.	192	Espley, Dennis C.	170
Cooke, Arthur H.	165, 172, 175, 176	Faraday, Michael	3
Compton, Karl T.	22, 154, 157, 317	Farnsworth, Philo Taylor	27, 28
Cox, C. W. H.	107	Faulkner, H. W.	155
Croney, J. R.	166	Fawcett, Eric W,	85
Cunningham, John	99	Fellers, Eugene H.	221
Curran, Joan Strothers	192	Ferris, H. A.	369
Daniels, Josephus	112	Fessenden, Reginald A.	8, 12
Davenport, Lee L.	177, 178	Fink, Donald G.	181
David, Pierre	345-351	Fisher, John	15
Davis, N. E.	187	Fiske, Bradley A.	5
Davis, T. M.	114	Fleming, John A.	11, 313
Dee, Philip I.	97, 145-147, 164, 166, 168, 172, 189, 190, 191	Fridman, Alexander	299
		Fry, Donald W.	172
		Fuchs, Klaus	309
De Forest, Lee	8, 11, 12, 232	Furth, F. R.	128
Devyatkov, N. D.	289, 305	Gane, Christopher P.	371
Dewhurst, R. H.	79	Garrard, Derek J.	106
Diehl, Hermann	238	Gebhard, Louis A.	16, 23, 31, 114-116, 123
Dippy, Robert J.	102, 103, 181		
Dirac, Paul	293	Genishta, E. N.	299
Dixon, E. J. C.	53	George VI, King	85, 99, 184
Dobbs, John M.	56	Getting, Ivan A.	177, 178
Dolbear, Amos E.	5	Gibson, Harold	170, 190
Dominik, Hans	2	Gibson, Reginald O.	85
Doolittle, James H.	321	Girardeau, Emil	346, 347, 352
Dowding, Hugh	46, 54, 176	Goebbels, Joseph	191
Draper, Charles S.	22	Gold, Thomas	187, 188
Drury, W. F.	77	Goldschmidt, Rudolph	14
DuBridge, Lee Alvin	162-164	Göring, Hermann	80, 230, 237, 255, 256, 270
Duckworth, John C.	83		
DuCretet, Eugène	278		
Duke, S. M.	145	Gratama, S. G.	339

Index Names

Griffiths, Frank C. 96
Guthrie, Robert C. 124, 125, 223
Gutton, Camille 345, 347
Gutton, Henri 145, <u>347</u>-350, 352
Hammond, John Hays, Jr. 22
Hanbury Brown, Robert 65, 73, 85, 86, 98, 194, 202
Harding, Lawrence M. 181
Harper, S. D. 359
Harris, N. L. 97
Hartree, Douglas R. 197
Hart, Raymund George 55, 58, 61, 62
Hashimoto, Chuji 326
Hay, D. Stanley 197
Hayes, Harvey C. 113, 117, 124, 125
Heaviside, Oliver W. 9
Heil, Oskar E. and Agnesa A. 141
Henderson, John Tasker 61, 156, 157, <u>362</u>-367, 370
Hensby, Geoffrey S. 173, 189
Herd, James F. 38, 40, 47, 100, 101
Heron, K. M. 58
Hershberger, W. Delmar 131, 132
Hertz. Gustav 270
Hertz, Heinrich Rudolf 3, 6, 7, 29, 144, <u>231</u>, 278
Hewitt, Frank J. 371-373
Hill, Archibald Vivian 43, 364
Himmler, Heinrich 230
Hitler, Adolph 59, 78, 81, 120, 147, 149, 151, 192, 229. 230, 237, 241, 268, 275, 291
Hodges, David B. 371
Hodgkin, Alan L. 171, 172
Hoffmann-Heyden, Adolf-Echard 259
Hogben, H. E. 184, 187
Hollingsworth, J. 38, 39
Hollmann, Hans Erich 234, 240, 248, 249, <u>251</u>, 25, 259, 269
Horton, Cecil E. 74, 166
Hoyle, Frederick 187, 188
Hulburt, Edward O. 113, 116
Hull, Albert W. 19, 142, 153, 234, 314
Hülsmeyer, Christian 29, <u>231</u>, 232,
337, 346
Hyland, Lawrence A. 31, 121-123, 240
Ioffe, Abram Fedorovich <u>280</u>, 283-285, 287-290, 293, 300, 302, 309
Ito, Tsuneo 314, 316
Ito, Yoji 314, <u>315</u>, 316, 318, 325, 326, 328, 330, 335
Jackson, Henry 37, 38
Jenkins, Charles Francis 25
Jenkins, John W. 57
Johnson, Alfred W. 125
Johnson, Lawrence H. 215
Jones, Frank E. 103
Jones, R. V. (Reginald Victor) 104-107, 192, 197, 270
Kaiser Wilhelm II 229
Kammhuber, Josef 107, 256
Kanner, Morton H. 216
Keiller, Jock 371
Kelly, Mervin J. 208
Kempton, A. E. 170, 196
Kennelly, Arthur R. B. 9
Kiepenheuer, Karl-Otto 198, 255
Kimura, Shunkiti 312
Kinjiro, Okabe 234, 313-315, 324, <u>325</u>, 348
Kitsee, Isidor 5
Kiyoshi, Morita 314
Kobayashi, Masatsugu 319-322
Korn, Arthur 24
Korolyov, Sergei Pavlovich 308
Korovin, Yu. K. 283, 289, 292
Kramer, Ernst 246
Krug, K. A. 280
Kühnhold, Rudolph 231, <u>233</u>-236, 239, 242, 254, 269
Kurchatov, Igor Vasilievich 309
Kurz, Karl 20, 141, 232
Labanov, M. M. 291, 288
Landale, S. E. D. 166, 184, 185
Landau, Lev Davidovich 293, 295, 296
Langmuir, Irvin 12
Larnder, Harold 58, 67, 93
Latmiral, Gaetano 345
Latour, Maurice 14

Lawrence, Ernest O.	150, 160, 161, 163	Mater, Hans	258
Lazarev, P. P.	280	Mauborgne, Joseph C.	156
Leahy, William D.	126	Mayer, Emil	239
Lee, Roland J.	97, 99	Marx, Karl	273
Leibniz, Gottfried	279	Massie, Walter W.	8
Leigh, Humphrey de Verde	88	Maxwell, James Clerk	3
Lenin, Vladimir Illich Ulyanov	274, 278, 279, 286, 288	McKinley, D. W. R.	369
Leshchinsky, V. M.	280	McMillan, Edwin M.	175, 176
Lewis, Frank D.	154, 161, 163	McNaughton, Andrew	362,363
Lewis, Wilfred B.	71, 72, 82, 86, 101, 189	Megaw, Eric C. S.	144, 145, 159, 362
Lelyakov, Paul P.	294, 302	Merchant, P. A.	58
Linder, Ernest C.	32, 131	Mesny, René	346
Lindermann, Frederick A.	51, 173	Miller, J. M.	114
Lobanov, M. M.	281, 288	Mirick, Carlos B.	114, 117, 121
Lockard, Joseph	135	Montgomery, Bernard L.	147, 149, 197, 198, 373
Lodge, Oliver Joseph	5	Morison, Samuel Eliot	212
Long, Geoffrey	372	Morita, Kiyoshi	314, 31
Loomis, Alfred Lee	22, 154, 157, 158, 160-163, 175, 178, 180	Morita, Masanori	322
Loomis, F. Wheeler	163	Morse, Samuel F. B.	2
Loomis, Mahlon	5	Mueller, George E.	216
Lorenz, Carl	243	Muir, Alex J.	49
Lovell, A. C. Bernard	72, 86, 145-147, 166, 168,171, 173, 188-190	Müller, Gottfried	243
Lowell, Clarence A.	179	Mussolini, Benito	80, 147, 230, 341
Löwy, Heinrich	30	Muth, Hans	242
Ludlow, John H.	56	Naismith, Robert	41
Luebke, Emmeth A.	214	Nakajima, Shigeru	314, 315, 325
Lugg, Sidney	85	Newhouse, Russell C.	31
Mackenzie, C. J.	156, 363	Nipkow, Paul J. G.	24
Madsen, John P. V.	353	Nissen, Jack	53
Malyarov, D. E.	289, 300	Noyes, Robert H.	132
Mann, K. C.	367	Oatley, Charles W.	197
Marchetti, John W.	136	Oberlin, Edgar G.	113, 117, 122
Marconi, Guglielmo	4, 5- 8, 10, 30, 232, 278, 341, 342	Obreimov, , Ivan V.	280, 293
Marks, William J.	132	Ohnesorge, Wilhelm	250
Marsden, Ernest	61, 355-358, 362, 370, 371	Okabe, Kinjiro	234, 313-315, 324, 325, 348
Marshall, Lauritsen C.	204	O'Kane, Bernard J.	173, 189
Martini, Wolfgang	237, 238, 241, 244, 245, 253, 257, 258,. 261, 262, 266, 269, 270	Oliphant, Marcus L. E.	142-145, 164-166, 168, 186
Martyn, David F.	61, 352, 353, 355	Oshchepkov, Pavel Kondratyevich	283, 284, 285, 287, 288-290, 307
		Owen, C. S.	166
		Oxford, Alan	197
		Page, Robert Morris	124-126, 128, 199, 201, 223
		Paris, E. Talbot	68, 90

Index Names

Parkinson, David B. 179
Pawsey, Joseph L. 353, 355
Pearce, F. L. 155
Penley, William H. 83, 97, 100, 193
Perskyi, Constantin D. 24
Peter I ("Peter the Great") 273
Petlyakov, Vladimr 305
Phillips, W. Eric 371
Phillips, W. C. 363
Philpott, LeVern R. 124
Pickard, Greenleaf W. 11
Piddington, John H. 353-355
Pierce, John A. 181
Pierret, Emile 347
Plendl, Johannes (Hans) E. 198, 246, 254, 257, 268, 270
Pollard, Ernest C. 178, 212
Pollard, P. E. 32, 68, 69, 89, 90, 92, 168, 186, 198
Ponte, Maurice 345, 347, 348, 352
Popov, Alexandr Stepinevich 5, 277, 281, 292
Posthumus, Klass 142, 234, 338
Poulsen, Valdemar 10, 13
Preece, William H. 4
Preist, Donald H. 69, 107
Pringle, John W. C. 85
Pulley, Oliver Owen 360
Rabi, I. I. (Isidor Isaac) 153, 175, 207, 223
Raeder, Erich 235
Ramsey, Norman F. 215
Randall, John Turton 143-145, 316, 352
Ratcliffe, John A. 62, 91, 92, 97
Reeves, Alec H. 103
Reppart, H. E. 125
Rice, Chester W. 33, 34
Ridenour, Louis N. 177
Righi, Augusto 4
Ritson, F. J. U. 97
Roberts, Noel 371
Robinson, Denis M. 168, 194
Robson, J. M. 196
Röhrl, Anton 235
Rose, H. E. 196
Rosing, Boris Lvovich 25, 26

Roosevelt, Franklin D. 61, 81, 93, 148, 149, 151, 154-156, 160, 164
Round, Henry J. 12, 16
Rowe, A. P. (Albert Percival) 43, 45, 46, 59, 60, 61, 67, 78, 82, 83, 87, 96, 07, 101, 147, 164, 168, 175, 189, 224, 324
Rozhdestvensky, D. S. 280
Rumyantsev, Dmitry A. 282, 283, 289, 292, 293
Runge, Carl 239
Runge, Wilhelm Tolmé 234, 239, 240, 242, 243
Ryle, Martin 192
Samuel, Arthur L 19
Sanders, F. H. 365
Satake, Kinji 318-320
Sayers, James 164, 186
Scheiler, Otto 246
Scherl, Richard 29
Schonland, Basil F. J. 61, 107, 197, 198, 370, 371, 373
Schottky, Walter 19
Schultes, Jakob Theodor J. 234, 24
Sedyakin, A. I. 288
Shearing, George 73
Shembel, B. K. 283, 285-287
Shire, Edward S. 90, 94, 169
Shone, David H. 204
Shestako, A. I. 290
Shteinberg, D. S. 292
Skinner, Herbert W. B. 97, 145, 146, 166, 172
Slaby, Adolph K. H. 5
Slack, Frederic W. 179
Sliozberg, Mikhail L. 300, 301
Slutskin, Abram A. 292, 293, 294, 303, 308
Somerfield, James 91
Squier, George O. 118, 131
Staal, C. H. J. A. 338
Staal, Max 340
Stalin, Josif Vissarionovich Dzhugashvili 149, 151, 274-276, 279, 288, 293, 296, 297, 307, 308
Stagov, D. E. 289
Stone, John S. 8

Stratton, Julius A.	181
Stubblefield, Nathan B.	5
Takayanagi, Kenjiro	26, 326, 332
Taylor, Albert Hoyt	16, 30, 31, 113-117, 121-125, 223
Taylor, Dennis	82
Terman, Frederick E.	203, 371
Tesla, Nikola	1, 5, 6, 43, 334, 339
Thiessen, Peter	270
Thompson, Amberst Felix H.	90
Thompson, J. J.	40
Tiberio, Ugo	342, 343, 344
Tikhomirov, Victor V.	299, 305
Tizard, Henry Thomas	42, 60, 65, 71, 88, 94, 155-158, 365
Tomlin, Donald H.	169, 195
Touch, A. Gerald	63-65, 86
Trinkle, Fritz	245
Truten, Ivan D.	303, 304, 308
Tsimbalin, V. V.	287
Tucker, S. M.	128
Tucker, W. S.	89
Turner, Louis A.	175
Tuve, Merle A.	31, 41, 123, 131, 205
Uda, Shintaro	313, 314
Usikov, Aleksandr Y.	294, 301, 303, 304, 308
Vecchiacchi, Francesco	345
Vail, Alfred M.	118
Vallauri, Giancarlo	342
Valley, George E.	215
Van Soest, J. L.	339
Varela, Arthur A.	124, 126
Varian, Russell H. and Sigurd F.	141, 153
Vilbig, Fredrich	250, 263
Volmer, Max	270
Von Ardenne, Manfred	23, 25, 39, 235, 253, 270
Von Arco, Georg Graf	13, 14
Von Braun, Wernher	255
Von Handel, Paul	245, 262
Von Weiler, J. L. W. C	339
Von Willisen, Hans-Karl	234
Wallace, F. C.	155, 365, 367
Ward, A. G.	146
Warneck, M. Robert	145, 348, 350, 352
Watson, Paul E.	132
Watson, William H.	368
Watson-Munro, Charles Norman	357, 360, 362
Watson Watt (also Watson-Watt), Robert Alexander	37-39, 40, 44-47, 49-53, 55, 59-63, 65, 68, 71-74, 78, 82, 91, 92, 98, 100, 147, 155, 184, 221, 224, 269, 351, 353, 371
Wehnelt, Arthur R. B.	11
Weisner, Jerome B.	20
West, D. L.	357
Westcok, Carl H.	97
Wheatstone, Charles	24
Whipple, Fred L.	204
White, Brian	65
White, Eric L. C.	87, 97
White, Frederick W. G.	355, 356, 359, 360
Whitehead, J. Rennie	61, 193
Wilkins, Arnold Frederic	44-53, 55, 59, 61, 62, 224, 351
Willett, B. R.	166, 184
Williams, Frederick C.	83, 97, 102, 103, 193, 201
Willis, George	48, 49
Wilson, Carroll	157, 158, 161
Wimperis, H. E. (Henry Egerton)	43-45, 60
Wolff, Irving W.	32, 131
Wood, Albert B.	68, 90
Wood, Keith A.	63, 65, 85
Woodring, Henry A.	132
Woodward Nutt, Arthur E.	155
Worlledge, Peter	6
Wright, Charleds S.	73, 142
Yagi, Hidetsugu	313-315, 317, 324
Yakovlev, Alexander S.	296
Yamamoto, Tadaoki	26
Young, Leo Clifford	16, 23, 30, 31, 114, 115, 116, 121-126, 223, 240
Zahl, Harold A.	120, 134, 136
Zhukovsky, N. E.	280
Zworykin, Vladimir Kosma	26-28

INDEX - SUBJECTS

A. C. Cossor, Ltd 37, 54, 56, 57, 66, 69, 71, 93, 100
Academy of Sciences, National (U.S.) 118
Academy of Sciences, Russian 279, 283, 293
Academy of Sciences, St. Petersburg 276
Academy of Sciences, USSR 208
Acorn Tube 19, 64, 131, 141, 153, 235, 243, 294, 340, 344, 371
Admiralty Compass Observatory 38
Admiralty Signal Establishment (ASE) 183-187, 191
ADRDE (Air Defense Research and Development Establishment) 88, 94, 99, 169, 170, 183, 188, 194-198
AEG (General Electricity Company) 239, 251
Air Armament Establishment 55, 62
Aircraft (Related to Radars):

American
Lockheed Hudson Bomber 84, 85
Lockheed P-38 Lightning 177, 210, 215
Martin B-10 Bomber 133
Martin PBM-5 Mariner Patrol 216
Douglas A-20 Havoc 159, 200
Boeing B-17 (Army), PB-1 (Navy), Flying-Fortress Bomber 209
Douglas B-18 Bomber 176
Grumman TBF Avenger 200, 209
Northrop P-61 Black Widow 210
Boeing 247D Transport 176
Consolidated PBY Catalina Patrol 159, 177, 199, 200, 210
Consolidated PB2Y Coronado Bomber 213
Consolidated B-24 Liberator Bomber 213
North American B-25 Mitchell Bomber 321

British
Avro Anson 64, 65, 67
Avro Lancaster 191
Bristol Beaufighter 87, 88, 96, 98, 99, 172
Bristol Blenheim 67, 79, 84, 86, 87, 96, 164, 171-174
De Havlland Mosquitoe 96, 210
Consolidated Vultee Liberator 191
Fairey Battle 67
Hadley-Page Harrow 67
Hadley-Page Halifax 174, 189 180
Hadlay-Page Hayford 46, 63-65
Short Stirling 191, 259
Vickers Wellington 172

German
Dornier Do-19 Heavy Bomber 87
Graf Zeppelin II, LZ-130 70
Messerschmitt Me-110 Fighte 243
Messerschmitt Me-262 (First Jet Fighter) 250
Junkers Ju-88 Fighter 243, 258
Focke-Wulf FW-200 Bomber 246

Japanese
Sei ran, Float-Plane Bomber 329
Kawanishi Flying Boat 330
Kyuysu Q1W1 Tokai 330
Nakajima J1N1-S Gekko 331

Russian
Petlyakov Pe-2 Dive Bomber 305
Petlyakov Pe-3 Fighter 304
Yakovlev Yak-3, 9 Fighter 306
Lavochkin La-5 Fighter 306
Ilyushin Il-2 Ground-Assault 306
Aircraft Radio Laboratory, U.S. Naval 16, 30, 113, 114, 116
Aircraft Radio Laboratory, U.S. Signal Corps 118, 119, 199, 223
Aircraft Radio Company 21

Radar Origins Worldwide

Air Defense Experimental
   Establishment (ADEE)   68, 71,
   79, 88-94, 96, 99, 160, 168, 177
Air Defense Research (ADR)
   Committee   43
Air Defense Research and
   Development Establishment
   (see ADRDE)
Air Defense System, Germany   256
Air Materiel Command   223
Air Ministry Research
   Establishment (AMRE)   82-84,
   92, 106
Aldershot Wireless Station   37-39
Alexanderson Alternator   13
All-Union Electro-Technical
   Institute   260, 281
Allen West & Company   167
Alma-Ata, USSR   296. 301, 302
Amateur Radio Operator ("Ham")
   16, 18, 115, 116, 175, 234, 250, 268
American Physical Society   116
American Radio Relay League
   (ARRL)   115, 371
American Telephone & Telegraph
   (AT&T)   12, 17, 21
Anacostia (D.C.) Naval Air Station
   16, 30, 114, 119
Antennas:
   Bedspring   126, 128, 235
   Cheese   166, 167, 185, 186
   Lobe-Switching   70, 71, 77, 93,
   132, 134, 200,208, 235, 236,
   238, 242, 287, 322, 323, 358
   Mattress   235
   Orange-Slice   217
   Phased-Array   106, 216, 238,
   241, 265
   Scheiller   246
   Yagi (Yage-Uda)   70, 71, 77, 86,
   126, 129, 200, 223, 234, 290, 291,
   299, 305, 311, 313, 317, 325, 327,
   328, 330, 331, 358, 359, 361, 363
Antiaircraft Artillery Board   178
Applied Physics Laboratory   206
Arc Transmitter   10, 13, 14, 16

Army (British) Cell (SEE)   58, 68,
   69, 71, 70, 89, 90
Army (British) Operational
   Research Organization   198
Army (U.S) Air Corps (Became
   Army Air Forces in June 1941)
   134, 159, 176
Army (U.S.) Air Forces   135, 148,
   177, 200, 208, 214-216, 220
Army (U.S.) Signal Corps Reserves
   119
Army (U.S.) Signal Corps   11, 32,
   113, 118, 119, 130, 160, 219, 223, 326
ARRL Radio Amateur's Handbook 371
Atomic Bomb   1, 3, 152, 155, 164,
   222, 270, 300, 308, 333
Atomic Energy Research
   Establishment, Harwell   224, 373
Audion (Triode)   11
Australian Radio Research Board
   352
Aviapribor Plant, Moscow   299
Bagneux, France   345
Balban and Katz Theater   218
Barkhausen-Kurz Tube   20, 33,
   141-143, 232-234, 236, 239, 258,
   283, 293, 315, 341, 347
Battle of Britain   3, 61, 80
Battle of the Atlantic   79, 149, 190
Bawdsey Research Station/Manor
   House   52-56, 58-63, 67-69, 71,
   72, 74, 78, 79, 81, 82, 89, 93, 99,
   100, 102, 142, 143, 164
Bawdsey Village   52
BBC   18, 25, 28, 40, 41, 46, 74, 105
Bell System Technical Journal   21
Bell Telephone Laboratories (BTL)
   19, 21, 32, 111, 123, 126, 127,
   157-162, 174, 178-180, 199, 208,
   210, 212-217, 223, 236
Bentley Priory RAF Station   58
Berlin, Germany   23, 25, 39, 148,
   149, 151, 191, 230, 233, 248, 250,
   252, 256, 261, 266, 269, 318, 324
Biggin Hill RAF Airfield   59, 89, 90
Blackett's Circus   93

Index Subjects

Blaupunkt, GmbH 25
Bletchley Park 106
Blitz 81, 168, 183
Blitzkrieg 275
Boar War 8
Boca Raton Airport (U. S. Signal Corps) 220
Boffins 55
Braun Tube 40, 235, 236, 244
Braun-Siemens-Halske 10
Brest, France 351
British Expeditionary Force 69, 80
British Post Office 4, 7, 359
British Royal Navy 7, 10, 318
British Thompson-Houston (BTH) 20, 37, 169, 169, 196
Bruneval Raid (Operation Biting) 107, 183, 197, 267
Bukhara, USSR 297, 301-304
Bureau of Engineering, U.S. Navy 30, 31, 117, 121-123, 125, 127
Camp Barriefield, Ontario 222
Camp Charles Wood 219
Camp Crowder 219
Camp Davis (U. California Davis Campus) 219
Camp Edison 218
Camp Kohler 219
Camp Murphy 220
Camp Springs (later Andrews AFB) 201
Camp Vail 16, 118
Canadian National Research Council (CNRC) 169
Cape of Good Hope, South Africa 373
Cape Town, South Africa 370
Carnegie Institution of Washington (CIW) 23, 31, 153, 205
Cathode-Ray Tube (CRT) 23, 25-28, 38-40, 57, 58, 63, 66-68, 71, 83, 100, 101, 125, 138, 162, 163, 171-173, 180, 181, 187, 198, 216, 232, 235, 242, 244, 249, 252, 257, 266,303, 323, 332, 344, 350, 352, 357-360, 364, 371, 373

Cat's Whisker 11, 146
Cavindish Laboratory (Cambridge) 159
Central Institute for Solar Research 255
Central NZ Post Office 356-360
Christchurch Airfield 89, 96, 172
Christchurch, Dorset 71, 79, 82, 88
Christchurch, Australia 356
Chrysler Motors 179
Clarendon Laboratory (Oxford) 63, 165
Clinton, Ontario 221, 222, 354, 369
Coast Artillery Board, US Army 23, 122, 133
Coastal Command, RAF 64, 65, 85, 101, 190
Coastal Defense Apparatus 32, 89
Co-axial Cable 84, 85, 248, 259, 286
Cockoo 69
Coherer Detector 4-7, 11, 29, 231, 278
College de France 278
Collins Radio Company 2
Combined Research Group (CRG), NRL 193, 201-203
Committee for the Coordination of Valve Development 142
Commonwealth, British 220, 221, 270, 352, 354, 369, 370
Compton Report 317
Copenhagen Telephone Company 10
Corregidor, Philippines 320, 321
Crown and Castle Hotel 49
CSIR (Council for Scientific and Industrial Research), Australia 352, 353, 355
CSIR, South Africa 373
Dachau Concentration Camp 255
Darwin, Australia 354
Daventry Experiment 46, 240
Death Ray 1, 4, 43, 334, 339, 341
Deben Estuary 52
Debert, Nova Scotia 222
Defford RAF Airfield 189

Department of Terrestrial
  Magnetism (DTM)   23, 31, 33, 205
Detection Devices:
  Acoustic/Sound   42, 52, 89, 90,
    119, 231, 239, 282, 283, 289, 316
  Ignition/Spark   41, 130, 282, 346
  Infrared/Optical   22, 41, 120,
    130, 131, 133, 134, 184, 195, 244
    281, 282-285, 287, 299, 319, 363
Differential Analyzer   197
Diode, Crystal   145, 146, 165,
    175, 232, 252, 257, 265, 332
Diode, Vacuum Tube   11, 146
Direction Finder, Type 10   313
Ditton Park   38, 41
Doorbnob Tube   19, 63, 65, 77, 141,
    153, 236
Doppler (see Interference Beat)
Dover Heights, Australia   354
DSIR (Department of Scientific and
  Industrial Research), Great
  Britain   20, 37-39, 43
DSIR, New Zealand   355-360
DuCretet-Popov   278
DuCretet-Roger   10
DuMont Laboratories   28
Dundee, Scotland   71, 78, 79
Dunkirk Evacuation   69, 80, 318
Duplexer   126, 134, 165, 200, 291
  (Also see Transmit-Receive/TR
  Switch)
Durban, South Africa   370, 372
Durnford House   97
Dusseldorf, Germany   29
DVL (German Laboratory for
  Aviation)   249, 262
E1188, E1189, E1198 Magnetrons
    145
Eddy Test   218-220, 228
Edison Effect   11
Eindhoven, Netherlands   237
E. K. Cole Company   85, 86, 87
Electric Telescope   24
Electrical Experimenter   26
Electronics Training Program (U.S.
  Navy)   218

EMI (formerly Gramaphone)
    28, 37, 53, 62, 63, 66, 83, 87, 97,
    101, 169, 173, 189, 190, 196, 353
Empire State Building   32
Enigma Machine   106
ETVRT (Hungarian Firm)   374
Experimental Department, HMSS
    15, 20, 37, 54, 73, 142, 143, 166,
    170, 173, 183, 184, 187
Evans Signal Laboratory, SCL
    135-138, 199, 215
Fabbrica Italiana Valvove Radio
  Elettriche (FIVRE), Italy   343
Factories (USSR)
  No. 92, 112, Gorky   296
  No. 115, Moscow, Kubyshev   296
  No. 183, Kharkov   296
  No. 339, Moscow   299, 301, 305
  Svelana, Leningrad   284, 285,
    290, 292
Fat Boy (Atomic Bomb)   222
Felixstowe Township   53
Felixstowe Air Station (RAF)   50
Ferranti, Ltd.   37, 61, 83, 102, 169
    172, 196
FFO (German Air Radio Research
  Institute)   249
Fighter Control Centers,
  Germany   256, 266
Filter Center/Room   57-59
Five-Year Plans   275
Forres School, Swanage   97, 220
Fort Hancock, New Jersey   132, 134
    135
Fort Monmouth, New Jersey   32
    118, 119, 130, 133, 178, 188,
    217, 219, 223
Fort Monroe, Virginia   133, 178
Fort Schafter, Hawaii   135
Fort Vail, New Jersey   118
Fruit (Slot) Machine   58
Fuji Electrical   326
Fukuoka, Japan   320
Funkmessgerät (Radar - Radio
  Measuring Device)   229, 231, 233,
    234, 258, 260

Index Subjects

Fuse, Acoustical and Doppler 90, 91
Galena Crystal 11
GAU (Main Artillery Adm., Army, USSR) 280-286, 300, 302
GEE (see Navigation Systems)
GEMA 233-238, 240, 241, 244, 248, 253, 256-258, 262, 268, 319, 349
General Electric (GE) 10, 12, 17-20, 25, 31, 33, 111, 121, 128, 172, 153, 157, 173, 179, 201, 214, 314, 315
General Electric Company (GEC) 21, 37, 83, 84, 86, 97, 101, 143-146, 165, 166, 170, 171, 172, 186, 352
General Electric Review 33
General Radio Company 21, 180, 20
Gilfillan Brothers 215
Giraffe Houses, ADRDE 195
Glaxo (Automatic Follower) 196
Goniometer 54. 55, 57, 58, 93, 301, 372
Gramaphone (EMI) 37
Great Depression 111, 117, 120, 229, 312, 373
Great Lakes Naval Radio Station 115
Great Malvern (see Malvern)
Great Patriotic War 276
Great Purge 275, 288, 295, 306
Greenwich Time Signal (Pips) 18
Gulag (Penal Labor Camp) 288, 305, 307
Halifax, Nova Scotia 156, 221, 222, 363, 369
Hallicrafter Radio Company 21, 108, 204
Hammarlund Radio Company 21
Hallicrafter S-27 Receiver 108
Hammond Radio Research Lab. 22
Harvard Pre-Radar School 220
Hareltine Research Corp. 202
Hazemeijer Fabriek van Signaalapparaten (Firm) 340
Heer (German Army) 229, 230, 241, 244, 248, 275
Heinrich Hertz Institute 234, 252
Heterodyne Principle 8, 12
HIKI (Hungarian Institute) 374

Hiroshima, Japan 152, 222, 335
His Majesty's Signal Office (HMSO) 40
His Majesty's Signal School (HMSS) – See Experimental Department
HMV Gramophone 354
Hollmann Tube 33
Honshu, Japan 312, 333
Hurn RAF Airfield 96, 174, 189
Hytron Radio Company 206
ICI Company 84, 85
Iconoscope 27, 28
Identification Friend or Foe:
  American/British
  IFF 61, 83, 85, 102, 129, 159, 178, 184, 187, 193, 201,210, 220, 238, 256, 299, 324, 364
  Mk I 61, 159
  Mk II 61, 83, 199, 201
  Mk III 102, 193, 202
  Mk IIIG 202
  Mk X 203
  UNBIFF 203
  ABA, SCR-515 202
  ABD, ABE, SCR-535 130, 201
  ABF, SCR-69 202
  XAE Recognition System 129
  German
  Erstling 238, 242, 249, 257, 258
  Zwilling 258
  Neuling 258
  Wespe 258
  Japanese
  Taki-15, 17 (Army) 324
  M-13 (Navy) 324
  USSR
  SCH-3 299
  IG Farben 259
  Image Dissector 27, 28
  Imperial German Navy 29
  Infrared Detection 22, 41, 51, 120, 130, 133, 134, 184, 195, 244, 281, 284, 287, 329, 339
  Institute of Radio-Physics and Electronics (IRE, USSR) 308

Institute of Higher Military
  Transmissions, Italy   342
Instrumentation Lab., MIT   22
Interference (Doppler) Beat   30, 31,
  33, 122, 123, 130, 131, 235,
  278, 285, 319, 320, 346, 348
Inventions Book, Royal Engineer 32
Ionosphere   9, 23, 31, 40, 45, 48, 49,
  55, 116, 124, 131, 162, 182, 235
  252, 254, 287, 315, 353, 355
Isle of Wight   67, 73, 89, 95, 98, 143
Italian Naval Academy   342
IZZO (Hungarian Research
  Institute)   374
Japan Victor   326
Johannesburg, South Africa   370
JRC (Japan Radio Company)   314-
  316, 324, 325, 328, 334
Julius Pintisch (Firm)   233-235
Kamikaze Attacks   152, 209
Kammhuber Line   107
Kammhuber Cinemas   256
Katakana (Japanese Alphabet)   312,
  324, 334
Katsu, Japan   326
Kazan, USSR   296, 299
Kharkov, USSR   273, 280, 284, 286,
  292, 293, 295, 296, 301, 302
Kiev, Germany   230, 235
Kiev, USSR   292
King Edward's School   183, 184
Kinescope   26
Klystron   76, 141, 143-146, 153,
  157, 163, 165, 166, 168, 172,
  174, 175, 189, 190, 249, 258,
  264, 267, 289, 305, 360, 375
KKK (Communications Research
  Institute, Germany)   265
Kriegsmarine (German Navy)   79,
  165, 230, 235, 236, 243, 244
Kubyshev, USSR   296, 297
Ku-go (Death Ray)   334
Laboratory for Aviation (German)
  236, 245
Laboratori National de Radio-
  electricite (LNR)   345-347, 351
La Havre, France   350, 351

Lawrence Livermore Lab.   141
Le Bourget Airport, Paris   356
Leeson House/TRE Leeson   97, 98,
  102, 164, 166, 170
Leigh Light   88, 265
Le Matériel Téléphonique (LMT –
  French Firm)   351, 352
LEMO (Laboratory of
  Electromagnetic Oscillations)
  293-296, 301-304, 308
Lend-Lease Program   61, 301, 307,
  360, 366
LEPI (Leningrad Electro-Physics
  Institute)   280, 283, 285, 286
Lighthouse Tube   19
Little Boy (Atomic Bomb)   222
Lloyd's of London   10
LMS-10 (German 10-cm
  Magnetron)   259
Logan Airport, Boston   176, 199
Loomis Laboratories   22, 154, 158
Loomis Radio Navigation   180
LORAN (Acronym)   131
Lorenz, AG   104, 241, 243-247, 249,
  258, 265, 260
Lorenz Guidance System   104,
  105, 246, 254
Leningrad, USSR   147, 274-276, 280,
  282-284, 286, 288, 289, 291-293,
  296, 298, 300, 305
Leningrad Physical-Technical
  Institute (LPTI)   280
Leningrad Electro-Physics Institute
  (LEPI)   280, 283, 285, 286
Leningrad Physics and Technology
  Institute (LPTI)   260, 284, 287,
  290-293, 298, 300, 303, 306, 307
LFO (Laboratory for Physical
  Development, Holland)   339
Liverno, Italy   342
Luftwaffe (German Air Force)   3, 78,
  80-83, 99, 105, 147, 182, 192, 230,
  233, 236, 237, 240, 241, 244-247,
  249, 253-259, 261-269, 275, 281,
  297, 298, 352
Luftwaffe Radar Laboratories   249,
  297, 298, 352

Index Subjects

Lupolen H (Material) 259
Lythe Hill House, Hyslemere 184
Maginot Line (France 351
Magnetrons:
  Early 19, 32, 33, 74, 131, 142, 153, 258
  British/American 94, 143-147, 153, 155-160, 163-166, 168, 171, 174, 175, 185, 186, 189, 191, 200, 189, 191, 200, 207, 208, 214, 236, 257, 260, 264
  Dutch 142, 234, 235, 338
  French 145, 148, 150, 152
  German 249, 252, 259, 260, 264, 266
  Japanese 250, 311-316, 318, 324, 325, 328, 331, 334
  USSR 273, 284, 286, 289, 293, 294, 300, 302-305, 308
Malvern, Worceshire 183
Malvern College 184
Manhattan Project 3, 164, 308, 334
Marconi Hotel/Evans Lab. 135
Marconi Officine 345
Marconi Wireless Telegraph Company of America (Marconi America) 7, 10, 15-17, 114, 135
Marconi's Wireless Telegraph Company (Marconi Company) 5, 12-15, 21, 28, 37, 56, 232, 347
Marsdiep, Netherlands 338
Martlesham Heath Airfield 50, 53, 63, 78, 78
McNaughton Medal (Canada) 370
Mechanical Transmitter 14
Meetgebouw (Measurements Building), Netherlands 339
Melbourne, Australia 354
Meteorological Office 37-39
Metox Société 251
Metropolitan-Vickers (Met-Vick 20, 37, 54, 56, 57, 64, 67, 69, 71, 93, 168, 169, 196
Micropup Valve 83, 86, 170, 357
Microwave Committee, NDRC 154, 157, 158, 160, 162, 168, 175

Ministry of Aircraft Production Research Establishment (MAPRE) 94
Ministry of Supply, War Office 89, 197
Morse Code 2, 4-6, 10, 15, 168, 278, 312, 257, 359
Moscow, USSR 147, 273-276, 279, 280, 285, 290, 291, 296-302, 304, 305
Most Secret War 270
Mount János, Hungary 375
Murmansk, USSR 300, 304
Mystery Ray 2
Nagasaki, Japan 152, 223, 335
Nash & Thompson 171, 172, 190
National (German) Air Ministry 258
National (U.S.) Bureau of Standards (NBS) 15, 17, 113, 258, 376
National Carbon Company 206
National Defense Research Committee (NDRC) 20, 154-162, 164, 205
National Guard Hangar, Boston 176, 199
National Physics Laboratory, DSIR (NPL) 20
National Research Council of Canada (NRCC) 156, 362
National Electric Company (NEC, Montreal) 365, 366
National Radio Company 21
Natlab (Natuurkundig Laboratorium) 337, 338
Nature 26, 55, 198
Naval Communications Reserve 16
Naval Consulting Board (NCB) 112, 113
Navy Pier (U.S. Navy Radar School) 219
Navy Radio Laboratory 15, 113, 11
Naval Research Laboratory (NRL) 20, 23, 31, 111, 112-135, 157, 159, 193, 199, 200-204, 211-215, 218, 223, 235
NRL Heat and Light Division 113, 116

391

NRL Radio Division 113-115, 117, 199
NRL Sound Division 113, 117, 124, 125
Navigation Systems:
 British
  GEE (AMES 7000) 103, 104, 161, 173, 180, 181, 263, 264
  Oboe (AMES 9000) 103, 104, 173
 American
  Loomis Radio Navigation 180
  LORAN (Long-Range Navigation) 181, 182
 German
  Leitstrahl (LEF, Guiding Beam) 245
  X-Leitstrahlbake (Directional Beacon, "Bent Leg") 247
  X-Gerät, Y- Gerät (Device) 247
  Sonne (Sun, "Console") 247
NEC (National Electric Company, Canada) 365, 366, 368
NEC (Nippon Electric Company, Japan) 28, 319, 321-323, 326
Nederlandse Seintoestellen Fabfrek, Netherlands 338
NEDALO, Netherlands 340
NEDISCO, Netherlands 340
Negative Feedback 338
New South Wales, Australia 355
NHK (Japan Broadcasting Corp.) 26, 326, 332
NII-9 (Scientific Res. Institute No. 9, USSR) 286, 288, 289, 291, 292, 298, 300, 305
NII-20 (R&D Institute of Radio Industry No. 20, USSR) 290, 291, 299-301, 305
NIIIS-KA (Scientific Res. Institute of Signals, Red Army 280, 283, 288, 289, 294, 301, 302, 304
NIIS-KS (Scientific Res. Institute of Signals, (Red Fleet) 306
Nike Missile 223
NKVD (Internal Affairs Police) 275, 288, 296

Nipkow Disc 24-26
Nizhny-Novgorod Radio Laboratory 280, 282, 286
Nobel Prize/Laureate 6, 21. 40, 41, 43, 71, 72, 91, 93, 160, 163, 171, 192, 222, 224, 232, 270, 293, 334, 355
NSW Railways Engrg. Group 354
NT46 Valve (Tube) 49, 50, 73
NT57 Valve (Tube) 54
NTRI (Naval Technical Research Institute, Japanese) 315-317, 321, 325, 326, 328, 330, 331, 333-335
NVA (Experimental Institute of Communication Systems, German) 230, 231, 233-236, 247
Office of Scientific Research and Development (OSRD) 164, 203
Official Secrets Act 47, 370
OKW (High Command of the Armed Forces, Germany) 230
Omak, USSR 297
ONATD (Oppama Naval Air Technical Depot), Japan 330
Operation Barbarossa 275
Operational Amplifier 193
Operational Research (British) 58, 93, 107, 197, 198, 355, 361, 373
Operations Research (American) 58, 93, 129
Orford, Suffolk 49
Orfordness Peninsula/Island 46-52, 54, 55, 60, 63, 65, 69, 73, 74
Osaka, Japan 320
Osborne, Nova Scotia 368
Oscillite 27
Oscillograph/Scope 39, 40, 45, 122, 124, 152, 290, 294, 343, 348, 372
Ottawa, Canada 155, 156, 362, 363
Pale Manor Farm, Malvern 188
Parrot (IFF) 61
Paris, France 7, 16, 24, 28, 119, 150, 229, 247, 271, 265, 278, 346, 351
Pathfinder Force 191
Pearl Harbor 1, 3, 128, 135, 148, 200, 217, 219, 320, 354, 376
Pearl Harbor Radar Warning 135
Penang, Malaysia 319

Index Subjects

Pennemünde 248, 255
Perspex 167, 171, 174, 190
Petrograd (USSR) 274
Philco Corporation 21, 27, 190, 200, 201, 213, 215, 216
Philips Company 142, 234, 235, 338
Physical Review 116
Physical-Technical Institute for Realm 259
Plan Position Filtering 58
Plan-Position Indicator (PPI) 34, 76, 99-101, 125, 128, 134, 178, 185, 186, 193, 209, 212-215, 249, 252, 253, 256, 257, 266, 332, 339, 364, 366, 368
Poldhu, Cornwall 8
Polyethylene 85, 259
Polyisobutylene 259
Polyrods 216
Popular Mechanics 14
Post-Design Services, TRE 97
Post Office (British/General) 4, 7, 10, 16, 37, 44, 46, 53, 54, 58, 359
Post Office (New Zealand) 356-358, 360
Post Office (German/Realm) 250
Potomac River 30, 113, 124
Predictor-Corrector, M-9 179, 180, 211, 307
Prestwick, Scotland, Radio School 220
Principles of Radar 220
Proc. of the IRE 234, 314, 348
Proc. of the Physical Society 4
Project 88 120
Project Cadillac, I & II 209, 210
Project Diana 229
Project Paperclip 269, 270
Proximity Fuse, Variable Time (VT) 205-207
Prozhzvuk (Sound Detector) 281
Puazo (Optical Range Finder) 282, 285, 298, 300
Purbeck Peninsula (Isle of Purbeck) 94
Pye (W. G.) Company 21, 37, 66, 67, 85-87, 357, 359

PVO (Air Defense Forces of the Red Army) 283-288, 300
RADAR (Acronym) 2, 111, 128
Radar Countermeasures:
  British & American
    Chaff 192, 204, 267
    Rope 193
    Window 192
    D-Day Invasion 192
    Jammers 192
      AN/APT-2, 3; SPT-2, 3 204
      SLQ, SPT (Navy) 204
      Tuba 203, 204
    Spoofers
      Mandrel 192
      Moonshine 192
    Warning/Intercept Receivers
      AN/SPR-2, AN/SQR-2 204
      ARC-1 204
      ARC-7, SCR-587 204
      Bagfull, Boozer, Monica 192, 193
  German
    Jammers
      Barrage Jamming 262
      Breslan 262
      Olga I, II 262
      Caruso 263
      Karl I, II (Communications) 263
      Heinrich I, II (GEE) 263
      Feuerstein, Feuerhilfe (GEE) 263
      Jamming Villages 264
      Kobold 264
      Roderich (10-cm) 264
      Roland (10-cm) 264
      Feuerhall (10-cm) 264
    Warning/Intercept Receivers
      Matox R203, R600 265, 266
      Samoz RS 1/5 265, 266
      Fanö 265
      Naxos (10-cm) 265, 266
      Naxburg (10-cm) 265
      Korfu 912 (10-cm) 266
      Tuniz, Altos (3-cm) 266
      Frequency Diversity 267

Other German Countermeasure
  KHP, Small Heidelberg
    Parasite 257
  Frequency Diversity 267
  Düppel (Chaff) 267
  Nürzburg 267
  K-Laus Signal Processor 267
    Japanese
E-17 Shipboard Warning
  Receiver 331
FT-B, FT-C Aircraft Warning
  Receivers 332
Naxos-Type Microwave Warning
  Receiver, Shipboard 332
Taki-11 Army Airborne Jammer 324
Radar Production Shop (RPS), TRE 190
Radar Range Equation 70
Radar Research and Development
  Establishment (RRDE) 195, 196, 224
Radar School, MIT 220
Radar Systems, American:
  Designations 207-208
  AI-10 (Experimental Microwave) 174
  AN/APQ-7 (Imaging, "Eagle") 214
  AN/APQ-13 (3-cm Imaging, "Mickey") 216, 223
  AN/APS-2 (ASG, "George") 213
  AN/APS-3 (ASD, "Dog") 215
  AN/APS-6 (AIA-1) 215
  AN/APS-6 (ASH) 215
  AN/APS-13 (Tail Warning) 223
  AN/APS-15 (3-cm Imaging, "Mickey") 216, 260
  AN/APS-20 (Airborne Surveillance) 209
  AN/CPS-1 (Early Warning) 216
  AN/CPS-4 (Height Finder) 216
  AN/CPS-5 (Ground-Control Intercept) 216
  AN/GPM-2, 6, 18 (Mobil Ground Search) 214

AN/TPS-3 (Portable Early Warning) 136
AN/TPS-10 (Height Finder, "Li'l Abner") 216
ASA (Airborne SA) 129
ASB (Airborne Surveillance) 200, 213
ASC (Airborne Surveillance) 210, 213
ASD (Airborne Search) 215
ASE (Air-to-Surface) 200
ASG (Airborne Surveillance) 213
ASH (Lightweight ASG) 215
AYA/SCR-518 (Radar Altimeter) 129, 135
CXAM/SK (Shipboard Search) 128
CXAS/FA (Shipboard Fire Control) 128
CXBL/ SM, SP (Shipboard SCR-584) 211
CXZ (Shipboard, Experimental) 127
Cadillac, Airborne Early Warning 209
Combat Information Center (CIO) 209, 210
FA/Mark 1(Fire Control) 208
FC/Mark 2 (Fire Control) 208
FD/Mark 3 (Fire Control) 208
FH/Mark 8 (3-cm Fire Cont.) 216
FPS-16 (Post-War, Tracking) 201
GCA (Ground Control Approach) 213
Monopulse Technique 201
SCR-268 (Antiaircraft Director) 133, 134, 136, 178, 237, 320, 322, 323, 327, 354, 366
SCR-270 (Mobile Early Warning) 134, 135, 214, 320, 323
SCR-271 (Fixed-Site Early Warning) 134, 214
SCR-512 (Air Force Version of ASB) 200
SCR-516 (Low-Altitude Early Warning) 134

Index Subjects

SCR-517, SCR-717 (Airborne Surveillance) 210, 213
SCR-518 (Microwave Altimeter) 129, 135
SCR-520 (Airborne Interception) 177, 210
SCR-527 (Ground Control Intercept) 201
SCR-540 (Army Version of AI Mk IV) 201
SCR-545 (Gun Laying, Mobile, BTL) 180
SCR-582, SCR-682 (Harbor Defense) 214
SCR-584 (Gun Laying, Mobile) 166, 170, 178, 179, 196, 210, 211
SCR-720 (Improved SCR-520 AI) 177
SCR-784 (Lightweight SCR-584) 211
SD (Submarine, Search) 129
SG (Shipboard, 10-cm Search) 212
SE, SF, SH, SL, SO (Specialized Descendants of SG) 212
SJ, SV (Submarine, 10-cm Search) 213
SK (Fielded CXAM, Shipboard Search) 128
SA, SC (Smaller Versions of SK) 128
SM, SP (Navy Versions of SCR-584) 210
SS (Submarine, 3-cm Search) 217
XAF (Experimental, Shipboard) 126
XAT (Experimental, Airborne) 200
XT-1 (Prototype SCR-584) 178
Radar Systems, British:
Air Interception (AI) 65
AI Mk I 66, 67, 68
AI Mk II 67, 78, 83, 84, 87
AI Mk III 86
AI Mk IV 87, 88, 96
AI Mk V 97, 98
AI Mk VII 176
AI Mk X (U.S. SCR-720) 177, 210

AIS (Microwave) 172, 173
Air Ministry Experimental Station (AMES):
Type 1 (CH) 62
Type 2 (CHL) 82
Type 9 Mk 1 (TRU) 69
Type 9 Mk 2 (TRU) 70
Type 9(T) (MRU) 69
Type 11 (mobile GCI/CHL) 193
Type 13 (NHF) 185, 1193, 194
Type 14 (HPS) 170
Type 15 (GCI) 201
Type 16 (FDS) 193
Type 20 (DHF) 193
Type 24 (LRHF) 193
Type 52 (LAD) 196
Type 54 (CHEL) 170
Type 70 (GCC) 194
Air-to-Surface Vessel (ASV)
ASV Mk I 65, 67, 70, 76, 84-86, 100, 156, 173
ASV Mk II 86, 96, 159, 190, 199-201, 245, 264, 265
ASV Mk III 190, 191
ASVS 172, 190
DMS-1000 191
Chain Home (CH) 52-58, 59, 61, 62, 66, 69, 71, 72, 79, 81-83, 87, 91-93, 95, 98, 100, 143, 170, 196, 220, 257, 261, 262, 358
Initial CH Locations 56
Chain Home Low (CHL) 61, 62, 82, 83, 92, 95-101, 157, 193, 220, 262
Chain Home Extra Low (CHEL) 170
Chain Home U-Boat (CHU) 91, 92
Coastal Defense (CD)
CD Mk I, II, III 69-71, 82, 89, 90-93
CD Mk IV, V, VI 170
Convoy/Group Control Center (GCC) 194
Deci. Height Finder (DHF) 193
Fighter Direction Station (FDS/GCI) 193

Radar Origins Worldwide

Ground Controlled Interception
  (GCI)   95, 98-101, 193, 201, 202, 214, 359
GMY4 (Missile Guidance) 185,186
Gun Laying (GL)
  GL Mk I, II (Metric)   69, 70, 75, 87-95, 198, 300, 301, 304, 318, 320, 322
  GL3, 3B (Microwave)   167-169, 177, 180, 196, 365
H2S (10-cm Imaging)   174, 187, 189-191, 216, 259, 264, 265, 266
H2X (3-cm Imaging)   191, 260
High-Power Surveillance (HPS) 196
High-Power Transmitter (HPT) 196, 197
Long-Range Height-Finder
  (LRHF)   193
Low-Altitude Detection (LAD) 196
Mobile Radio [Radar] Unit
  (MRU)   69, 80, 318, 322
Nodding Height-Finder (NHF/
  GCI)   185,193
Radar Convoy   194
Searchlight Control (SLC)   93-95, 320, 321
Transportable Radio [RDF] Unit
  (TRU)   69
Variable Elevation Beam (VEB/
  DHF)   100
Shipboard, Decimeter
  Type 79 (Air Warning)   74, 75
  Type 267 (Air Warning)   76
  Type 279 (Fire Control/Air Warning)   75
  Type 280 (Air Warning)   75
  Type 281 (Air Warning)   75, 76
  Type 282, 283, 284, 285 (Fire Control)   77
  Type 86/286 (Air Warning)   76
  Type 291 (Air Warning)   76
Shipboard, Microwave
  Type 261 (3-cm Fire Cont.)   192
  Type 262 (10-cm Fire Cont.)   186
  Type 268 (Target Indicator)   186
  Type 271 (Surface Warning) 166,167, 170, 184, 185, 211, 307
  Type 274, 275, 276 (Gun Director)   186
  Type 277 (Nodding Height Finder)   185, 193, 194
Radar Systems, German:
  Berlin (Airborne 10-cm)   260
  Berlin-S (Shipboard 10-cm)   260
  Darmstadt (Ground-Based Air Warning)   240
  DeTe-I (Ship/Shore Surface Search)   236, 237
  DeTe-II (Shore-Based Air Warning)   236, 237
  EFA (Experimental Detection Device)   243
  Elefant-Rüssel (Long-Range Tracking)   251
  Flakleit-G (DeTe-I with Height Finder)   237
  Freya (Luftwaffe DeTe-II)   106, 192, 204, 237, 238, 243-245, 256-258, 267, 343
  Hohentwiel (Ait-to-Surface)   245, 246
  Hohentwiel-U (Submarine Search)   246
  Jagdschloss (Hunting Lodge Ground Surveillance)   248, 249
  Jagdschloss-Z (10-cm Experimental)   260
  Jagdwagen (Mobile Jagdschloss)   249
  Klumbach (10-cm Experimental)   260
  Kurpfulz (Mobile Gun Durector)   245
  Kurmark (Transportable Gun Director)   245
  Lichtenstein (Night-Fighter Set)   242, 243
  Mannheim (Ground-Based Air Warning)   241, 242
  Mannheim-Riese (Giant Mannheim)   242

Index Subjects

Manmut (Array of 16 Freyas) 106, 192, 238, 239
Marbach (10-cm Flak Aiming) 259
Meddo (3-cm Experimental) 260
Neptune (Jet Fighter Set) 250
Neptune V-2 (Pecision Tracking) 248
Panorama (Ground Surveillance) 248, 248, 253, 257
Rotterdam-X (3-cm Experimental) 260
Seetakt (Shipboard, Search) 106, 236-238, 258, 267
Wassermann (Array of 8 Freyas) 192, 238, 239
Würzburg (Ground-Based Air Warning) 106, 192, 240-242, 244, 245, 255-258, 260, 265, 267, 318, 319, 324, 328, 329
Würzburg-Riese (Giant Würzburg) 241, 242, 256, 257

Radar Systems, Japanese:
Designations 321, 326
BDID, Bi-Directional Doppler Interference Detector 320, 321
FD-3 (25-cm Aircraft) 330
FH-1 Radar Altimeter 333
Tachi-4 (Based on SCR-268) 323
Tachi-6 (Based on SCR-270) 323
Tachi-7 (Transportable Tachi-6) 323
Tachi-18 (Portable Tachi-6) 323
Tachi-20 (Height Finder) 324
Tachi-24 (Würzburg based) 324

Tachi-35 (Height Finder) 324
Tase-1, 2 (Army, Shipboard) 324
Tase-1 (Army, Aircraft) 324
Type 11 (Land Based) 326
Type 12 (Light Type 11) 327
Type 13 (Land/Ship Based) 327
Type 14 (Large Ships) 327
Type 21 (Lighter Type 12) 327
Type 22 (10 cm Ship/Sub.) 328
Type 23 (Würzburg copy) 328
Type 24 (10-cm Small Ships) 328
Type 31 (10-cm Fire Control) 329

Type 32 (10 cm Fire Control) 329
Type 33 (10 cm Fire Control) 329
Type 41 (Based on SCR-268) 327
Type 42 (Revised Type 41) 328
Type 64 (Nightfighter) 330

Radar Systems, USSR:
Bistro (Rapid, Experimental) 285-287, 289
Burya (Storm, Experimental, 18-cm) 286, 287
Gneis (Origin, 16-cm Experimental) 305
Genis 2 (Airborne) 305
Gyuis-1 (Shipboard) 307
Mars-1, 2 (Shipboard) 307
Reconnaissance Electromagnetic Station 283
Redan-1, 2 (Shipboard) 307
Redut (Redoubt, Experimental) 290
Redut-K (Shipboard RUS-2) 306
Reven (Rhubarb, Experimental) 289
Rubin (Developmental) 295
RUS-1 (CW, Bi-static) 289, 290
RUS-2 (Pulsed, Mobile) 290, 291, 300, 302, 303, 306
RUS-2s (Fixed-Position, also called Pegmatit) 298, 299
Son (Sleep, 15-cm CW) 300
Son-2A (Copy of GL Mk II) 301
Zenit (Gun Directing) 294-296, 301-303

Radar Systems, Other Nations:
Netherlands
Electrical Listening Device 339
Italy
EC-1, 2 (Experimental) 343
EC-3; Folaga and Gufo, (Shipboard) 344
Lince (Ground Based) 345
Radioecometro (Marconi) 341
France
Barrage Electromagnétique (Electromagnetic Curtain) 346

Radar Origins Worldwide

Early Warning Detection
  System   351
Système de Réperages
  d'Obstacles (System for
  Location of Obstacles)   348, 350

Australia

AW Mark I (Air Warning)   354
LW/AW Mark II   354
LW/AWH   355
LW/AWH Mark II (10 cm)   355
LW/GCI   355
LW/LFC   355
MAWD (Modified Air Warning
  Device, SCR-268)   355
ShD (Shore Defense)   353

New Zealand

CW (Coast Watching)   357
CD (Coast Defense)   358
LWAW (Long-Range Air
  Warning)   361
ME (Mobile, Microwave)   360
SWG (Ship Warning, Gunnery)   359
SW (Ship Warning)   359

Canada

CD (Coastal Defense)   364
CDX (10-cm CD)   367, 368
GL IIIC (Mobile, Microwave)   365
API (Accurate Position
  Finder)   365
ZPI (Zone Position
  Indicator)   365
MEW/AS (Microwave Early
  Warning Anti Submarine)   368
MEW/HF (Height Finding)   368
NW (Night Watchman)   363
RX/C (10-cm Shipboard)   366
SW1C (Surface Warning 1st
  Canadian)   363
  SW2C (Rotating Antenna)   364
  SW3C (Lightweight)   364
Type 268 (3-cm Sea Search)   366

South Africa

JB (for Johannesburg)   372

Hungary

Sas (Eagle)   375
Radial Time-Base Display   101, 125
Radiation Laboratory, Berkeley   153
Radiation Laboratory (Rad Lab),   20
  160-164, 168, 172, 176-182, 190,
  191, 194, 198, 199, 203, 207, 209-
  212, 214, 216, 217, 223, 224, 360
Formation   160-164
Project 1 (Airborne Radar)
  175-177
Project 2 (Mobile Gun-Laying
  Radar)   177-180
Project 3 (Long-Range
  Navigation)   180-182
Radio Artificer   222
Radio Astronomy   198, 335, 355
Radio Communications Co.   37
Radio Corp. of America (RCA)   17,
  19, 21, 24, 27, 28, 32, 64, 111, 121,
  122, 126-131, 135, 153, 157, 162,
  200 235, 294, 314, 344, 357, 368
Radio-Detection
  (Radioobnaruzehenie)   285
Radio-Detection Telemetry (Radar
  in Italy) see RDT
Radio Development Laboratory
  (RDL, New Zealand)   360, 261
Radio Direction Finding (Radar in
  Great Britain), SEE RDF
Radio Engineering, Terman   371
Radio Frequency (RF) Spectrum   9
Radio-Location (Rádiólokáció,
  Radar in Hungarian)   374
Radio-Location (Radiolokatsiya,
  Radar in Russian)   273, 284, 306 +
Radio Marelli (Italian Firm)   344
Radio Materiel School, U.S. Navy
  Radar School   113, 217
Radio Physics Board, N.Z.   359
Radio Position Finding (RPF)   130
  RPF Section, SCL   132
Radio Range-Finder
  (Radiodal'nomer)   291
Radio Range Finder (RRF, Radar in
  Japan)   311

Index Subjects

Radio Research Board (RRB)   37, 38, 41
Radio Research Station (RRS), Slough   20, 38, 39, 41, 45-50, 57, 224, 371
Radio Screen   123, 285, 320
Radio Test Shop, Washington   10, 16, 113, 114
Range Cutting   54
Raytheon Corporation   21, 201, 206, 212
RDF (Radar in Great Britain)   2, 37, 50, plus many more.
RDF-1, RDF-1.5, RDF-2   63
RDT (Radar in Italy, Telemetro Radiofonico del Rivelatore)   342
Research Lab. for Electron Physics   23, 253
Regenerative Oscillator   12
Regenerative Receiver   12, 234, 235, 285
Research Enterprises, Ltd. (REL, Canada)   363-367, 369, 370
Ring Oscillator   126, 132, 322
RKKA (Soviet Army)   274, 280
RKKF (Soviet Navy)   274
Rochefort   10
Rohde & Schwartz   251
Rome, Italy   149, 230, 265, 341, 342, 345
Rotary Spark Transmitter   13
Rotterdam-Gerät   259, 264
Royal Aircraft Establishment (RAE)   15, 20, 37, 39, 67, 81, 103, 104, 106, 171
Royal Air Force (RAF)   41, 50, 52-59, 61, 63, 65. 79-81, 84, 86, 89, 92, 95, 96, 98, 100, 103-197, 148-149, 161, 170-174, 176, 181, 189, 190, 194-196, 211, 220-222, 236, 245, 263, 354, 369, 373
RAF #31 School   221, 222, 354-369
Royal Canadian
  Air Force (RCAF)   156, 176, 368
  Coastal Service   221, 369
  Corps of Signals   364
  Elect. and Mech. Engrs.   222, 369

Navy (RCN)   363
Navy's Signal School   222
Women's Royal Naval Service (Wrens)   222, 369
Royal Dutch Navy (Koninklijke Marine)   338
Royal Institute for Electro-technics and Communications (RITC), Italy   342, 343
Royal Italian Navy (Regia Marina Italiana)   76, 147, 316, 343, 344
Royal New Zealand Air Force (RNZAF)   3, 358
Royal New Zealand Navy (RNZN)   359-361
Royal Radio-technical Institute, Italian Air Force   345
Royal Society Mond Laboratory   91
RRF (Radio Range Finder)   311, 318, 320, 321
SAFAR (Firm)   345
San Burno Company   200
Sandy Hook, New Jersey   132
Sanitas GmbH   259
Savage & Parsons   88
Scapa Flow, Scotland   63
Scapa Flying Boat   50
SCB (Special Construction Bureau, USSR)   283, 288
Schnorkle   366
Scientific American   4, 25
Scone Airfield   78, 84
Sevastopol, USSR   290, 292, 306
Ships and Submarines:
  American
  California, USS   128
  Chicago, USS   128
  Hornet, USS   321
  Leary, USS   126
  Lexington, USS   211
  Los Angeles, USS   122
  New York, USS   127
  Nordic, MS   136
  Northhampton, USS   128
  Pensacola, USS   128
  Seattle, USS   115
  Semmes, USS   212

Radar Origins Worldwide

Texas, USS	127
Yorktown, USS	128

British

Bulwark, HMS	186
Courageous, HMS	65
Didi, HMS	75
Dorchester, SS	30
Duchess of Richmond, SS	156
King George V, HMS	77
Lusitania, HMS	16
Marigold, HMS	167
Nelson, HMS	77
Newton, HMS	77
Ocean, HMS	186
Orchis, HMS	167
Prince of Wales, HMS	75
Rodney, HMS	75
Sardonyx, HMS	77
Sea Lion, HMS	172
Sheffield, HMS	75
Southdown, HMS	77
Southhampton, HMS	65
Titantic, RMS	14
Titlark, HMS	166
Valiant, HMS	76
Wessex, HMS	340
Winchester Castle, SS	370

French

Normandie, Ocean Liner	348, 349

German

Admiral Graf Spee, Pocket Battleship	73, 79, 100, 184, 237
Togo, Aircraft Warning Ship	256
Welle, Research Ship	235, 236

Italian

Littorio (Battleship)	344

Japanese

I-30 (B-Class Submarine)	318, 319
Hyuga (Battleship)	329
Ise (Battleship)	329
Kazegumo (Destroyer)	328
Makigumo (Destroyer)	328

New Zealand

Archilles (Cruiser)	359
Leander (Cruiser)	359, 360
Monowai (Armed Merchant)	359

USSR

Molotov (Light Cruiser)	306
Sidney, Australia	354
Siemens & Halske	8, 10, 13, 233, 239, 248, 249, 250, 251, 253, 260, 264, 268, 278
Signal Corps Laboratories (SCL)	20, 33, 111, 119, 120, 122, 129-137, 178, 199. 211, 214, 217, 223, 252
SCL RDF Section	132
Signal Corps Radio Lab.	15, 118
Signals Experimental Establishment (SEE)	15, 20, 31, 37, 68, 70, 89, 197
Signals Intelligence Services, German	262, 264
Signals Research and Development Establishment (SRDE)	197
Singapore	319, 321, 332, 359, 361
Skip Distance	116
Slaby-Arco (Firm)	10
Slough, Berkeshire	38, 39, 44, 46, 47, 49, 50, 53-55, 60, 100, 224, 371
Société Française Radioélectrique (SFR, French Radioelectric Company)	346, 347, 351
Solar Radio Astronomy	198
Sopley, Hampshire	95, 99
Southern Signal Corps School	220
Solving the Naval Radar Crisis	119
Somerford Grange	89
Soviet Air Forces (VVS)	274
Soviet Army (RKKA)	274
Soviet Navy (RKKF)	274
Spark Gap	4, 5, 14, 29, 97, 231, 277
Squier Hall, Fort Monmouth	131
St. Aldhelm's Head	146
St. Athan Air Base, South Wales	84-86, 88, 96
St. Hyacinth, Quebec`	222, 369
St. John's, Newfoundland	9
St. Petersburg, USSR	147, 273, 274, 277-279
Steamer Point, Christchurch	89, 94, 168
Standard Telephones Ltd	169, 196

Index Subjects

Stalingrad, USSR   150, 276, 306
Strahlenzieler (Ray Pointer)   29
Sunday Soviets   60, 96, 165, 173, 183
Sunitomo Communications   321
Super-Heterodyne Receiver   19, 64, 124, 143, 266, 300, 303, 328, 340, 35
Surface-Ducting Effect   304
Sutton Klystron   142, 146, 172
Swanage, Dorset   83, 88, 94, 96-98, 101, 106, 145, 160, 166, 173, 182, 183, 188, 189, 211, 220
Tachibana (Mandarin, Magnetron Type)   316
Tamagana, Japan   321
Tashkent, USSR   297
Telecommunications Flying Unit   96, 189
Telecommunications Research Establishment (see TRE)
Telefunken   10, 13, 14, 28, 233, 234, 239-243, 244, 245, 248-252, 254-260, 263, 265, 267, 340, 345, 349
Telemetro Radiofonico del Rivelatore (Radio–Detector Telemetry, RDT)   342
Telefunken Italiana (Firm)   345
Telemobilskop   29, 231, 337, 346
Television   2, 19, 23-28, 32, 62, 63, 66, 74, 87, 105.125, 2-9, 218, 250, 286, 319, 326, 332, 353, 356, 357, 359
Television Laboratories   27
Teplopelengator (Aircraft Thermal Tracker, USSR)   282
Tetrode   19
Thames Estuary   51, 52
The Hague, Netherlands   340
The New York Times   1, 43, 334, 339
The New York Times Magazine   112
The Electrical Magazine   29, 231
Thermionic Valve   11
Tizard Committee (Scientific Survey for Air Defense)   41-46, 50, 51, 58, 60, 66, 93, 364, 365
Tizard Mission (Technology Exchange)   94,155-160, 165, 169, 182, 199, 205, 212, 365

Toksovo, USSR   291, 298
Tokyo, Japan   297, 311, 312, 315, 320, 321, 425, 326, 334
Tokyo Denki Company   28
Tonographie (Firm)   234
Toronto, Ontario   363
Toshiba (Tokyo Shibaura Denki)   28, 315, 321, 322, 323
Toulon, France   351
Transmit-Receive (T-R) Switch   97, 100, 126, 134, 159, 165, 259, 291, 303, 339, 373
Treasure Island (U.S. Navy Radar School)   218
Trchkent, USSR   302
TRE (Telecom. Research Establishment)   94
TRE Swanage/Worth Matravers   94-99, 101-107, 145, 146, 160, 161, 164-170, 172, 173, 211, 220, 357
TRE Malvern   183, 188-195, 201, 202, 224
TsRL (Central Radio Laboratory)   282, 284, 286
TTRI (Tama Technological Research Institute, Tokyo)   318, 320, 321, 324
Tuned Radio Frequency (TRF) Receiver   12
Tuxedo Park   22, 154, 158, 160
TVA (Torpedo Res. Establishment, German)   230, 231, 237
Ukrainian Institute of Physics and Technology (UIPT)   283, 284, 286, 93, 295, 296, 301, 302, 308
Union of Soviet Socialist Republic (USSR or Soviet Union)   274
Universities and Other Institutions Involved in Radar Development:
Australia
Melbourne Technical College   354
University of Sidney   354
Canada
McGill University   368
France
Nancy University   345, 347

Radar Origins Worldwide

Great Britain
Birmingham Univ. 143, 145, 316
Bristol University 143, 183
Cambridge University 71, 91, 106
Oxford University 142, 165
University College, Dundee 78
Hungary
Technical Univ. of Budapest 374
Italy
Italian Naval Academy 342
Nancy University 345, 347
University of Naples 345
Japan
Osaka University 314, 324
Tohoku Unversity 313, 314, 316
Tokyo Inst. of Tech. 314
Tokyo University 312
Netherlands
Delft University 340
Leiden University 340
New Zealand
Auckland University 356
Canterbury University 356
South Africa
Bernard Price Institute (BPI),
  Witwatersrand Univ. 370-373
University of Natal 370, 373
United States
Bowdoin College 220
Harvard University 204, 220
Johns Hopkins Univ. 207
Mass. Inst. of Tech. 155, 161, 204
USSR
Kharkov University 283, 290

Unterseeboots (U-Boats) 149
U.S. Air Force 224
U.S. Weather Bureau 10, 19
V-1, German Flying Bomb 150, 170, 179

V-2, German Long-Range Rocket
  150, 155, 170, 250, 308
Velocity Modulation 19, 141, 23
V-E Day 1, 151, 198
V-J Day 1, 153
Vichy France 352
VT-158 Tube 136
VVS (Soviet Air Forces) 274, 275, 305, 306
Wireless Station No. 10 197
Woodbridge Village 53
Ward Island (U.S. Navy Airborne
  Radar School) 219
War Office, United Kingdom
  20, 32, 41, 73
Wijk aan Zee, Netherlands 338
World War I 15-16
World War II 78-81, 147-153
Western Electric 12, 17-21, 37, 63,
  65, 68, 77, 127, 128, 132, 133,
  142, 176, 177, 201, 208, 210,
  214, 215, 217, 322, 364
Westinghouse 5, 17, 21, 26, 27,
  132, 134, 162, 179, 204
Westinghouse, British 37
Windsor Great Park 41
Wireless Engineer 380
Woolwich Commons, London 15, 31, 68, 89
World War II 1, 78-81, 147-153, 230
Worth Matravers, Dorset 94, 95, 97
Wright Air Development Ctr. 224
Wright Field, Ohio 113, 119, 159, 118, 177, 199, 200, 210, 223
Yatesbury Radio School 220
Yokosuka, Japan 330
Zenith Corporation 136

The Author

Raymond C. (Ray) Watson, Jr., was born in 1926 at Anniston, Alabama, and raised on a small dairy farm near Fort McClellan. He gained an early interest in radio from "hanging around" the communications station at this U.S. Army installation, and built his first receiver before he entered his teens. After completing a special curriculum through the Government's long-forgotten Engineering, Science, and Management War Training (ESMWT) program, he obtained a commercial radio license in 1942 and began work as a radio engineer.

With World War II expanding, Watson volunteered into the Navy in 1944 and entered the Electronics Training Program (described in Chapter 4). After a year of intensive study in radar and other advanced electronics, he was assigned to the instructional staff at the highly secret Advanced Airborne School at Ward Island (near Corpus Christi, Texas), and remained there until being discharged.

After the Navy, Watson returned to work as a radio engineer and soon formed his own firm, Dixie Service Company. Although still lacking a bachelor's degree, he was recognized as a consulting engineer by the Federal Communications Commission. For the next 10 years, he provided services to commercial radio and television stations. He participated in establishing television in Atlanta, and in starting the Alabama Educational Television Network, the first in the Nation.

In parallel with post-Navy engineering work, he also returned to educational pursuits. He studied at a number of institutions and eventually earned five degrees, including a Ph.D. in engineering, all as he was working and/or teaching full time.

In 1953, Watson began a career in academia while continuing with engineering work. His positions during the next two decades included professorships in engineering and physics; department and division chairmanships; director of a research institute; and researcher in radio astronomy as a National Science Foundation Faculty Fellow.

On the industrial side, in 1960 he joined Brown Engineering Company (later Teledyne Brown Engineering) in Huntsville, Alabama, eventually becoming Vice President for the Science and Engineering Group, an organization of near 1,000 persons. In later years, he served as Directing Manager for the American L.L.C. of an Israeli electronics firm, and as the President/CEO of Vision Technologies Kinetics, a subsidiary of a giant Singaporian operation.

In personal technical work, he was an early participant in U.S. ballistic missile defense programs and had extensive experience in the intelligence field, both activities drawing on his Navy-initiated radar knowledge. He led in developing the laser Doppler velocimeter, forerunner of laser radar. In 1969, he received the Public Service Award from NASA for his contributions to the Apollo lunar exploration, America's greatest technological achievement.

In 1976, Watson was a primary founder and served as President and Professor of Southeastern Institute of Technology, a graduate-level professional school primarily for employees of government agencies and space and defense industries. Here he personally mentored over 100 students in master's and doctoral degrees and is still involved in this activity. He is also presently Chair of the Engineering and Technology Advisory Board at Alabama A&M University.

Throughout his careers, Watson has authored over 430 papers, reports, and major presentations, including a number on historical subjects.

In 1991, he established R. C. Watson and Associates as a vehicle for his independent engineering and management consulting, and he continues today full time in this endeavor.

Watson's home has been in Huntsville, Alabama, since 1960, but his work has involved activities across America and in many other countries. He is married and has four children, eight grandchildren, and a growing number of great-grandchildren.

Dr. Watson may be contacted via e-mail at RCW-Assoc@comcast.net